T0366199

INTRODUCTION TO QUANTUM EFFECTS IN GRAVITY

This is the first introductory textbook on quantum field theory in gravitational backgrounds intended for undergraduate and beginning graduate students in the fields of theoretical astrophysics, cosmology, particle physics, and string theory. The book covers the basic (but essential) material of quantization of fields in expanding universe and quantum fluctuations in inflationary spacetime. It also contains a detailed explanation of the Casimir, Unruh, and Hawking effects, and introduces the method of effective action used for calculating the backreaction of quantum systems on a classical external gravitational field.

The broad scope of the material covered will provide the reader with a thorough perspective of the subject. Complicated calculations are avoided in favor of simpler ones, which still contain the relevant physical concepts. Every major result is derived from first principles and thoroughly explained. The book is self-contained and assumes only a basic knowledge of general relativity. Exercises with detailed solutions are provided throughout the book.

VIATCHESLAV F. MUKHANOV is Professor of Physics at Ludwig-Maximilians University, Munich. His main result is the theory of inflationary cosmological perturbations. Professor Mukhanov is author of *Physical Foundations of Cosmology* (Cambridge University Press, 2005). He also serves on the editorial boards of leading research journals and is Scientific Director of the *Journal of Cosmology and Astroparticle Physics* (*JCAP*).

SERGEI WINITZKI is Research Associate in the Department of Physics at Ludwig-Maximilians University, Munich. His main areas of research include quantum cosmology, the theory of dark energy, the global structure of spacetime, and quantum gravity.

INTRODUCTION TO QUANTUM EFFECTS IN GRAVITY

VIATCHESLAV MUKHANOV AND SERGEI WINITZKI

Ludwig-Maximilians University, Munich

CAMBRIDGE UNIVERSITY PRESS
Cambridge, New York, Melbourne, Madrid, Cape Town,
Singapore, São Paulo, Delhi, Mexico City

Cambridge University Press
The Edinburgh Building, Cambridge CB2 8RU, UK

Published in the United States of America by Cambridge University Press, New York

www.cambridge.org
Information on this title: www.cambridge.org/9780521868341

First published 2007
Third printing 2010

A catalogue record for this publication is available from the British Library

ISBN 978-0-521-86834-1 Hardback

Contents

Preface

This book is an expanded and reorganized version of the lecture notes for a course taught at the Ludwig-Maximilians University, Munich, in the spring semester of 2003. The course is an elementary introduction to the basic concepts of quantum field theory in classical backgrounds. A certain level of familiarity with general relativity and quantum mechanics is required, although many of the necessary concepts are introduced in the text.

The audience consisted of advanced undergraduates and beginning graduate students. There were 11 three-hour lectures. Each lecture was accompanied by exercises that were an integral part of the exposition and encapsulated longer but straightforward calculations or illustrative numerical results. Detailed solutions were given for all the exercises. Exercises marked by an asterisk (*) are more difficult or cumbersome.

The book covers limited but essential material: quantization of free scalar fields; driven and time-dependent harmonic oscillators; mode expansions and Bogolyubov transformations; particle creation by classical backgrounds; quantum scalar fields in de Sitter spacetime and the growth of fluctuations; the Unruh effect; Hawking radiation; the Casimir effect; quantization by path integrals; the energy-momentum tensor for fields; effective action and backreaction; regularization of functional determinants using zeta functions and heat kernels. Topics such as quantization of higher-spin fields or interacting fields in curved spacetime, direct renormalization of the energy-momentum tensor, and the theory of cosmological perturbations are left out.

The emphasis of this course is primarily on concepts rather than on computational results. Most of the required calculations have been simplified to the barest possible minimum that still contains all relevant physics. For instance, only free scalar fields are considered for quantization; background spacetimes are always chosen to be conformally flat; the Casimir effect, the Unruh effect, and the Hawking radiation are computed for massless scalar fields in suitable

1 + 1-dimensional spacetimes. Thus a fairly modest computational effort suffices to explain important conceptual issues such as the nature of vacuum and particles in curved spacetimes, thermal effects of gravitation, and backreaction. This should prepare students for more advanced and technically demanding treatments suggested below.

The authors are grateful to Josef Gaßner and Matthew Parry for discussions and valuable comments on the manuscript. Special thanks are due to Alex Vikman who worked through the text and prompted important revisions, and to Andrei Barvinsky for his assistance in improving the presentation in the last chapter.

The entire book was typeset with the excellent LyX and TEX document preparation system on computers running Debian GNU/Linux. We wish to express our gratitude to the creators and maintainers of this outstanding free software.

Suggested literature

The following books offer a more extensive coverage of the subject and can be studied as a continuation of this introductory course.

N. D. BIRRELL and P. C. W. DAVIES, *Quantum Fields in Curved Space* (Cambridge University Press, 1982).

S. A. FULLING, *Aspects of Quantum Field Theory in Curved Space-Time* (Cambridge University Press, 1989).

A. A. GRIB, S. G. MAMAEV, and V. M. MOSTEPANENKO, *Vacuum Quantum Effects in Strong Fields* (Friedmann Laboratory Publishing, St. Petersburg, 1994).

Part I

Canonical quantization and particle production

1

Overview: a taste of quantum fields

Summary Quantum fields as a set of harmonic oscillators. Vacuum state. Particle interpretation of field theory. Examples of particle production by external fields.

We begin with a few elementary observations concerning the vacuum in quantum field theory.

1.1 Classical field

A classical field is described by a function $\phi(\mathbf{x}, t)$, where $\mathbf{x}$ is a three-dimensional coordinate in space and t is the time. At every point the function $\phi(\mathbf{x}, t)$ takes values in some finite-dimensional "configuration space" and can be a scalar, vector, or tensor.

The simplest example is a real scalar field $\phi(\mathbf{x}, t)$ whose strength is characterized by real numbers. A free massive scalar field satisfies the Klein–Gordon equation

$$\frac{\partial^2 \phi}{\partial t^2} - \sum_{j=1}^{3} \frac{\partial^2 \phi}{\partial x_j^2} + m^2 \phi \equiv \ddot{\phi} - \Delta \phi + m^2 \phi = 0, \tag{1.1}$$

which has a unique solution $\phi(\mathbf{x}, t)$ for $t > t_0$ provided that the initial conditions $\phi(\mathbf{x}, t_0)$ and $\dot{\phi}(\mathbf{x}, t_0)$ are specified.

Formally one can describe a free scalar field as a set of decoupled "harmonic oscillators." To explain why this is so it is convenient to begin by considering a field $\phi(\mathbf{x}, t)$ not in infinite space but in a box of finite volume V, with some boundary conditions imposed on the field ϕ. The volume V should be large enough to avoid artifacts induced by the finite size of the box or by physically irrelevant boundary conditions. For example, one might choose the box as a cube with sides of length L and volume $V = L^3$, and impose the *periodic* boundary conditions,

$$\phi(x=0, y, z, t) = \phi(x=L, y, z, t)$$

and similarly for y and z. The Fourier decomposition is then

$$\phi(\mathbf{x}, t) = \frac{1}{\sqrt{V}} \sum_{\mathbf{k}} \phi_{\mathbf{k}}(t) e^{i\mathbf{k}\cdot\mathbf{x}}, \tag{1.2}$$

where the sum goes over three-dimensional wavenumbers $\mathbf{k}$ with components

$$k_x = \frac{2\pi n_x}{L}, \quad n_x = 0, \pm 1, \pm 2, \ldots$$

and similarly for k_y and k_z. The normalization factor $\sqrt{V}$ in equation (1.2) is chosen to simplify formulae (in principle, one could rescale the modes $\phi_{\mathbf{k}}$ by any constant). Substituting (1.2) into equation (1.1), we find that this equation is replaced by an infinite set of decoupled ordinary differential equations:

$$\ddot{\phi}_{\mathbf{k}} + \left(k^2 + m^2\right) \phi_{\mathbf{k}} = 0,$$

with one equation for each $\mathbf{k}$. In other words, each complex function $\phi_{\mathbf{k}}(t)$ satisfies the harmonic oscillator equation with the frequency

$$\omega_k \equiv \sqrt{k^2 + m^2},$$

where $k \equiv |\mathbf{k}|$. The "oscillators" with complex coordinates $\phi_{\mathbf{k}}$ "move" not in real three-dimensional space but in the *configuration space* and characterize the strength of the field ϕ. The total energy of the field ϕ in the box is simply equal to the sum of energies of all oscillators $\phi_{\mathbf{k}}$,

$$E = \sum_{\mathbf{k}} \left[\frac{1}{2} \left| \dot{\phi}_{\mathbf{k}} \right|^2 + \frac{1}{2} \omega_k^2 \left| \phi_{\mathbf{k}} \right|^2 \right].$$

In the limit of infinite space when $V \to \infty$ the sum in (1.2) is replaced by the integral over all wavenumbers $\mathbf{k}$,

$$\phi(\mathbf{x}, t) = \int \frac{d^3\mathbf{k}}{(2\pi)^{3/2}} e^{i\mathbf{k}\cdot\mathbf{x}} \phi_{\mathbf{k}}(t). \tag{1.3}$$

1.2 Quantum field and its vacuum state

The quantization of a free scalar field is mathematically equivalent to quantizing an infinite set of decoupled harmonic oscillators.

Harmonic oscillator A classical harmonic oscillator is described by a coordinate $q(t)$ satisfying

$$\ddot{q} + \omega^2 q = 0. \tag{1.4}$$

The solution of this equation is unique if we specify initial conditions $q(t_0)$ and $\dot{q}(t_0)$. We may identify the "ground state" of an oscillator as the state without motion, i.e. $q(t) \equiv 0$. This lowest-energy state is the solution of the classical equation (1.4) with the initial conditions $q(0) = \dot{q}(0) = 0$.

When the oscillator is quantized, the classical coordinate q and the momentum $p = \dot{q}$ (for simplicity, we assume that the oscillator has a unit mass) are replaced by operators $\hat{q}(t)$ and $\hat{p}(t)$ satisfying the Heisenberg commutation relation

$$[\hat{q}(t), \hat{p}(t)] = [\hat{q}(t), \hat{\dot{q}}(t)] = i\hbar. \tag{1.5}$$

The solution $\hat{q}(t) \equiv 0$ does not satisfy the commutation relation. In fact, the oscillator's coordinate always fluctuates. The ground state with the lowest energy is described by the normalized wave function

$$\psi(q) = \left[\frac{\omega}{\pi\hbar}\right]^{\frac{1}{4}} \exp\left(-\frac{\omega q^2}{2\hbar}\right).$$

The energy of this minimal excitation state, called the *zero-point energy*, is $E_0 = \frac{1}{2}\hbar\omega$. The typical amplitude of fluctuations in the ground state is $\delta q \sim \sqrt{\hbar/\omega}$ and the measured trajectories $q(t)$ resemble a random walk around $q = 0$.

Field quantization In the case of a field, each mode $\phi_{\mathbf{k}}(t)$ is quantized as a separate harmonic oscillator. The classical "coordinates" $\phi_{\mathbf{k}}$ and the corresponding conjugated momenta $\pi_{\mathbf{k}} \equiv \dot{\phi}^*_{\mathbf{k}}$ are replaced by operators $\hat{\phi}_{\mathbf{k}}$, $\hat{\pi}_{\mathbf{k}}$. In a finite box they satisfy the following equal-time commutation relations:

$$\left[\hat{\phi}_{\mathbf{k}}(t), \hat{\pi}_{\mathbf{k}'}(t)\right] = i\delta_{\mathbf{k},-\mathbf{k}'},$$

where $\delta_{\mathbf{k},-\mathbf{k}'}$ is the Kronecker symbol equal to unity when $\mathbf{k} = -\mathbf{k}'$ and zero otherwise. In the limit of infinite volume the commutation relations become

$$\left[\hat{\phi}_{\mathbf{k}}(t), \hat{\pi}_{\mathbf{k}'}(t)\right] = i\delta\left(\mathbf{k} + \mathbf{k}'\right), \tag{1.6}$$

where $\delta(\mathbf{k} + \mathbf{k}')$ is the Dirac δ function. To simplify the formulae, we shall almost always use the units in which $\hbar = c = 1$.

Vacuum state The *vacuum* is a state corresponding to the intuitive notions of "the absence of anything" or "an empty space." Generally, the vacuum is defined as the state with the lowest possible energy. In the case of a classical field the vacuum is a state where the field is absent, that is, $\phi(\mathbf{x}, t) = 0$. This is a solution of the classical equations of motion. When the field is quantized it becomes impossible to satisfy simultaneously the equations of motion for the operator $\hat{\phi}$ and the commutation relations by $\hat{\phi}(\mathbf{x}, t) = 0$. Therefore, the field always fluctuates and has a nonvanishing value even in a state with the minimal possible energy.

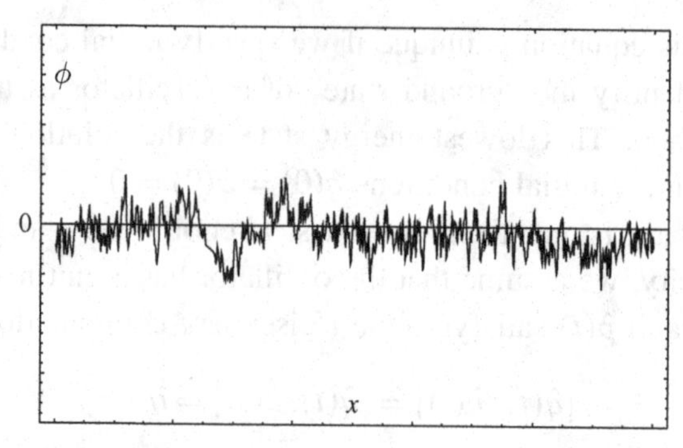

Fig. 1.1 A field configuration $\phi(x)$ that could be measured in the vacuum state.

Since all modes $\phi_{\mathbf{k}}$ are decoupled, the ground state of the field can be characterized by a *wave functional* which is the product of an infinite number of wave functions, each describing the ground state of a harmonic oscillator with the corresponding wavenumber **k**:

$$\Psi[\phi] \propto \prod_{\mathbf{k}} \exp\left(-\frac{\omega_k |\phi_{\mathbf{k}}|^2}{2}\right) = \exp\left[-\frac{1}{2}\sum_{\mathbf{k}} \omega_k |\phi_{\mathbf{k}}|^2\right]. \tag{1.7}$$

The ground state of the field has the minimum energy and is called the vacuum state. Strictly speaking, equation (1.7) is valid only for a field quantized in a box. Note that if we had normalized the Fourier components $\phi_{\mathbf{k}}$ in equation (1.2) differently, then there would be a volume factor in front of ω_k.

The square of the wave function (1.7) gives us the probability density for measuring a certain field configuration $\phi(\mathbf{x})$. This probability is independent of time t. The field fluctuates in the vacuum state and the field configurations can be visualized as small random deviations from zero (see Fig. 1.1).

When the volume of the box becomes very large, we have to replace sums by integrals,

$$\sum_{\mathbf{k}} \to \frac{V}{(2\pi)^3}\int d^3\mathbf{k}, \quad \phi_{\mathbf{k}} \to \sqrt{\frac{(2\pi)^3}{V}}\phi_{\mathbf{k}}, \tag{1.8}$$

and the wave functional (1.7) becomes

$$\Psi[\phi] \propto \exp\left[-\frac{1}{2}\int d^3\mathbf{k}\,|\phi_{\mathbf{k}}|^2\,\omega_k\right]. \tag{1.9}$$

Exercise 1.1

The vacuum wave functional (1.9) contains the integral

$$I \equiv \int d^3\mathbf{k}\,|\phi_{\mathbf{k}}|^2\sqrt{k^2+m^2}, \tag{1.10}$$

where $\phi_{\mathbf{k}}$ are defined in equation (1.3). This integral can be expressed directly in terms of the function $\phi(\mathbf{x})$,

$$I = \int d^3\mathbf{x}\, d^3\mathbf{y}\, \phi(\mathbf{x})\, K(\mathbf{x}, \mathbf{y})\, \phi(\mathbf{y}).$$

Determine the required kernel $K(\mathbf{x}, \mathbf{y})$.

1.3 The vacuum energy

Let us compute the energy of a free scalar field in the vacuum state. Each oscillator $\hat{\phi}_{\mathbf{k}}$ is in its ground state and has energy $\frac{1}{2}\omega_k$, so that the total zero-point energy of the field in a box of finite volume V is

$$E_0 = \sum_{\mathbf{k}} \frac{1}{2}\omega_k.$$

Taking the limit $V \to \infty$ and replacing the sum by an integral according to (1.8), we obtain the following expression for the vacuum energy density,

$$\frac{E_0}{V} = \int \frac{d^3\mathbf{k}}{(2\pi)^3} \frac{1}{2}\omega_k. \tag{1.11}$$

The integral diverges at the upper bound as k^4. Taken at face value, this would indicate an infinite vacuum energy *density*. If we impose an ultraviolet cutoff, for example, at the Planckian scale, where one expects quantum gravity to induce new physics, then the vacuum energy density is of order unity in the Planck units. This corresponds to a mass density of about $10^{94}\,\mathrm{g/cm^3}$. We recall that the mass of the entire observable Universe is only $\sim 10^{55}\mathrm{g}$! Therefore, if the vacuum energy contributes to the gravitational field, such a huge energy density is in obvious contradiction with observations.

The standard way to resolve this problem is to *postulate* that the vacuum energy density given in (1.11) does not contribute to the gravity. Another way to avoid this problem is to consider a supersymmetric theory where every field has a supersymmetric partner that contributes an equal amount to the vacuum energy with an opposite sign. However, experiments show that supersymmetry must be broken at some energy scale that is larger than the energy currently accessible to particle accelerators. This leads to a mismatch of the superpartner contributions to the vacuum energy of order the supersymmetry breaking scale, which is still too large when compared with observational limits. Therefore the supersymmetric solution of the vacuum energy problem is not immediately successful.

1.4 Quantum vacuum fluctuations

Amplitude of fluctuations As we found above, the typical amplitude of quantum fluctuations for the mode $\mathbf{k}$ is

$$\delta\phi_{\mathbf{k}} \equiv \sqrt{\left\langle |\phi_{\mathbf{k}}|^2 \right\rangle} \sim \omega_k^{-1/2}. \tag{1.12}$$

Field values cannot be measured at a point; in a realistic experiment, only their values, averaged over a finite region of space, are measured. Let us consider the average value of a field $\phi(\mathbf{x})$ in a cube-shaped region of volume L^3,

$$\phi_L \equiv \frac{1}{L^3} \int_{-L/2}^{L/2} dx \int_{-L/2}^{L/2} dy \int_{-L/2}^{L/2} dz\, \phi(\mathbf{x}).$$

Exercise 1.2

Justify the following order-of-magnitude estimate of the typical amplitude of fluctuations $\delta\phi_L$,

$$\delta\phi_L \sim \left[(\delta\phi_{\mathbf{k}})^2 k^3 \right]^{1/2}, \quad k = L^{-1},$$

where $k \equiv |\mathbf{k}|$ and $\delta\phi_{\mathbf{k}}$ is the typical amplitude of fluctuations in the mode $\phi_{\mathbf{k}}$.

Hint: The "typical amplitude" δx of a variable x fluctuating around 0 is $\delta x = \sqrt{\langle x^2 \rangle}$.

Taking into account that for vacuum fluctuations, $\delta\phi_{\mathbf{k}}$ is given in (1.12), we find that the typical amplitude of $\delta\phi_L$ is

$$\delta\phi_L \sim \sqrt{\frac{k_L^3}{\omega_{k_L}}}, \quad k_L \equiv L^{-1}. \tag{1.13}$$

We conclude that $\delta\phi_L$ diverges as L^{-1} for small $L \ll m^{-1}$ and decays as $L^{-3/2}$ for large $L \gg m^{-1}$.

Observable effects of vacuum fluctuations Quantum vacuum fluctuations have observable consequences that cannot be explained by any other known physics. The three well-known effects are the spontaneous emission of radiation by atoms, the Lamb shift, and the Casimir effect. All of them have been measured experimentally.

The spontaneous emission of a photon by a hydrogen atom in vacuum occurs as a result of the transition between the states $2p \to 1s$. This effect can only be explained if we consider the interaction of electrons with the vacuum fluctuations of the electromagnetic field. Without these fluctuations, the hydrogen atom would have remained forever in the $2p$ state.

The Lamb shift is a small difference between the energies of the $2p$ and $2s$ states of the hydrogen atom. This shift occurs because the electron "clouds" have different geometrical shapes for the $2p$ and $2s$ states and hence interact differently with vacuum fluctuations of the electromagnetic field. The measured energy difference, corresponding to the frequency $\approx 1057\text{MHz}$, is in a good agreement with the theoretical prediction.

The Casimir effect is manifested as an attractive force between two parallel *uncharged* conducting plates. The force decays with the distance L between the plates as $F \sim L^{-4}$. This effect is explained by considering the shift of the energy of zero-point fluctuations of the electromagnetic field due to the presence of the conductors.

1.5 Particle interpretation of quantum fields

The classical concept of particles involves point-like objects moving along specific trajectories. Experiments show that this concept does not actually apply on subatomic scales. For an adequate description of photons and electrons and other elementary particles, one needs to use a relativistic quantum field theory (QFT) in which the basic objects are not particles but quantum fields. For instance, the quantum theory of photons and electrons (quantum electrodynamics) describes the interaction of the electromagnetic field with the electron field. Quantum states of the fields are interpreted in terms of corresponding particles. Experiments are then described by computing probabilities for specific field configurations.

The energy levels of a "quantum oscillator $\phi_{\mathbf{k}}$" are $E_{n,\mathbf{k}} = \left(\frac{1}{2}+n\right)\omega_k$ where $n = 0, 1, \ldots$ At level n the energy $E_{n,\mathbf{k}}$ is greater than the ground state energy by $\Delta E = n\omega_k = n\sqrt{k^2+m^2}$, which is equal to the energy of n relativistic particles of mass m with momentum $\mathbf{k}$. Therefore the excited state with energy $E_{n,\mathbf{k}}$ is interpreted as describing n particles of momentum $\mathbf{k}$. We refer to such states as having the *occupation number* n.

A classical field corresponds to states with large occupation numbers, $n \gg 1$. In this case, quantum fluctuations can be very small compared to expectation values of the field.

A free, noninteracting field with given occupation numbers will remain in the corresponding state forever. On the other hand, in an interacting field occupation numbers can change with time. An increase in the occupation number for a mode $\mathbf{k}$ is interpreted as production of particles with momentum $\mathbf{k}$.

1.6 Quantum field theory in classical backgrounds

"Traditional" QFT deals with problems of finding cross-sections for transitions between different particle states, such as scattering of one particle on another. For instance, typical problems of quantum electrodynamics are:

(i) Given the initial state (at time $t \to -\infty$) of an electron with momentum $\mathbf{k}_1$ and a photon with momentum $\mathbf{k}_2$, find the cross-section for the scattering into the final state (at $t \to +\infty$) where the electron has momentum $\mathbf{k}_3$ and the photon has momentum $\mathbf{k}_4$.

This problem is formulated in terms of quantum fields in the following manner. Suppose that ψ is the field representing electrons. The initial configuration is translated into a state of the mode $\psi_{\mathbf{k}_1}$ with the occupation number 1 and all other modes of the field ψ having zero occupation numbers. The initial configuration of "oscillators" of the electromagnetic field is analogous – only the mode with momentum $\mathbf{k}_2$ is occupied. The final configuration is similarly translated into the language of field modes.

(ii) Initially there is an electron and a positron with momenta $\mathbf{k}_1$ and $\mathbf{k}_2$. Find the cross-section for their annihilation with the emission of two photons with momenta $\mathbf{k}_3$ and $\mathbf{k}_4$.

These problems are solved by applying perturbation theory to a system of infinitely many weakly interacting quantum oscillators. The required calculations are usually rather tedious because of the vacuum polarization effects which are due to the couplings of the excited "oscillators" with infinitely many "oscillators" in the ground state.

In this book we study quantum fields interacting only with a strong external field called the *background*. It is assumed that the background field is adequately described by a classical theory and does not need to be quantized. In other words, our subject is *quantum fields in classical backgrounds*. A significant simplification comes from considering quantum fields that interact *only* with classical backgrounds but not with other quantum fields. Such quantum fields are also called *free* fields, even though they are coupled to the background.

Typical problems of interest to us are:

(i) Computation of probabilities for transitions between various configurations of quantum field under the influence of a classical background field, which describe the process of particle production by the external field.

(ii) Determination of the energy level shifts for the quantum fluctuations due to the presence of the background. Since the vacuum contribution to gravity is assumed to have been subtracted already, it is likely that these energy shifts contribute to gravity.

(iii) Calculation of the backreaction of a quantum field on the classical background. For example, the external gravitational field influences the vacuum fluctuations shifting their zero-point energy levels. As a result, the vacuum fluctuations begin to contribute to a gravitational field. Their contribution can be described by an effective energy-momentum tensor, which is determined by the strength of the external gravitational field.

1.7 Examples of particle creation

A quantum oscillator in an external classical field A nonstationary gravitational background influences quantum fields in such a way that the frequencies

ω_k become time-dependent, $\omega_k(t)$. We shall examine this situation in detail in Chapter 6. For now, we simplify our task and consider the behavior of a single harmonic oscillator with a time-dependent frequency $\omega(t)$. The energy of such an oscillator is not conserved and the oscillator exhibits transitions between different energy levels.

Let us assume that an oscillator satisfies the following equation of motion:

$$\ddot{q}(t)+\omega_0^2 q(t)=0, \text{ for } t<0 \text{ and } t>T; \tag{1.14}$$

$$\ddot{q}(t)-\Omega_0^2 q(t)=0, \text{ for } 0<t<T,$$

where ω_0 and Ω_0 are real constants.

Exercise 1.3

Given the solution of equation (1.14), $q(t)=q_1\sin\omega_0 t$ for $t<0$, and assuming that $\Omega_0 T\gg 1$ verify that for $t>T$

$$q(t)=q_2\sin(\omega_0 t+\alpha),$$

where α is a constant and

$$q_2\approx\frac{1}{2}q_1\sqrt{1+\frac{\omega_0^2}{\Omega_0^2}}\exp(\Omega_0 T). \tag{1.15}$$

It follows from (1.15) that the oscillator has a large amplitude, $q_2\gg q_1$, for $t>T$. Thus the final state has much larger energy than the initial state and it can be then interpreted as a state with many particles produced within the time interval $T>t>0$.

Exercise 1.4

Estimate the number of particles produced, assuming that the oscillator is initially in the ground state.

The Schwinger effect A static electric field can create electron–positron (e^+e^-) pairs. This effect, called the *Schwinger effect,* is currently on the verge of being experimentally verified.

To understand the Schwinger effect qualitatively, we may imagine a virtual e^+e^- pair in a constant electric field of strength E. If the particles move apart from each other a distance l, they receive energy leE from the electric field. In the case when this energy exceeds the rest mass of the two particles, $leE\geq 2m_e$, the pair becomes real and the particles continue to move apart. The typical separation of the virtual pair is of order the Compton wavelength $2\pi/m_e$. More precisely, the probability of separation by a distance l turns out to be $P\sim\exp(-\pi m_e l)$. Therefore the probability of creating an e^+e^- pair is

$$P\sim\exp\left(-\frac{m_e^2}{eE}\right). \tag{1.16}$$

The exact formula for P can be obtained from a full consideration in quantum electrodynamics.

Exercise 1.5

Consider the strongest electric fields available in a laboratory today and estimate the corresponding probability for producing an e^+e^- pair.

Hint: Rewrite equation (1.16) in SI units.

Particle production by the gravitational field Generally, a static gravitational field does not produce particles. Both particles in a virtual pair have the same "gravitational charge," i.e. mass, so they fall together and are never separated far enough to become real particles. However, a time-dependent gravitational field generally produces the particles.

One would expect that a nonrotating black hole could not produce any particles because its gravitational field is static. Therefore it came as a surprise when, in 1973, Hawking discovered that static black holes emit particles with a blackbody thermal spectrum of temperature

$$T = \frac{\hbar c^3}{8\pi GM},$$

where M is the mass of the black hole and G is Newton's constant.

One can qualitatively understand Hawking radiation by appealing again to the picture of the virtual particle–antiparticle pairs. It may happen that one of the virtual particles is just outside the black hole horizon while the other particle is inside. The particle inside the horizon inevitably falls to the center of the black hole, while the other particle can escape and may be detected by an observer far from the black hole. The existence of the horizon is crucial in this case.

The Unruh effect This effect concerns an accelerated particle detector in empty space. Although all fields are in their vacuum states, the accelerated detector will nevertheless find particles with a thermal spectrum (a heat bath). The temperature of this heat bath is called the *Unruh temperature* and is equal to $T = a/(2\pi)$, where a is the acceleration of the detector (both the temperature and the acceleration are expressed in Planck units).

In principle, the Unruh effect can be used to heat water in an accelerated container. The energy for heating the water comes from the agent that accelerates the container.

Exercise 1.6

A glass of water is moving with a constant acceleration. Determine the smallest acceleration that makes the water boil due to the Unruh effect.

2

Reminder: classical and quantum theory

Summary Action principle. Functional derivatives. Lagrangian and Hamiltonian formalisms. Canonical quantization. Operators and vectors in Hilbert space. Schrödinger equation.

2.1 Lagrangian formalism

Quantum theory is built by applying a quantization procedure to a classical theory. The starting point of a classical theory is the action principle.

The action principle The state of a classical system is described by a (set of) generalized coordinate(s) $q(t)$. The variable q is a "symbolic combined notation," which includes all degrees of freedom of the system. For a system with N degrees of freedom we have $q \equiv \{q_1, q_2, \ldots q_N\}$. A field theory describes systems with an infinite number of degrees of freedom. For example, the configuration of a classical real scalar field at a given moment of time is characterized by a function $\phi(\mathbf{x})$. In this case $q(t) \equiv \{\phi_{\mathbf{x}}(t)\}$, where the spatial coordinate $\mathbf{x}$ can be considered as an "index" enumerating the infinite number of degrees of freedom of the scalar field.

The classical trajectory $q(t)$ connecting the states at two moments of time t_1 and t_2 is an extremum of an action functional[1]

$$S[q(t)] = \int_{t_1}^{t_2} L(t, q(t), \dot{q}(t), \ldots)\, dt. \tag{2.1}$$

The function $L(t, q, \dot{q}, \ldots)$ is called the *Lagrangian.* In the case of a field theory, the Lagrangian is a functional too. Different Lagrangians describe different systems. For a harmonic oscillator with unit mass and frequency ω we have

$$L(q, \dot{q}) = \frac{1}{2}\left(\dot{q}^2 - \omega^2 q^2\right). \tag{2.2}$$

[1] See Appendix A1.1 for more details concerning functionals.

If ω is time independent this Lagrangian has no explicit time dependence.

Equations of motion The requirement that the function $q(t)$ extremizes the action leads to a differential equation for $q(t)$. Let us derive this equation for the action

$$S[q] = \int_{t_1}^{t_2} L(t, q, \dot{q})\, dt. \tag{2.3}$$

If the function $q(t)$ is an extremum of the action functional (2.3), then a small variation $\delta q(t)$ of the trajectory $q(t)$ changes the value of $S[q]$ only by terms which are quadratic in $\delta q(t)$. In other words, the variation

$$\delta S[q, \delta q] \equiv S[q + \delta q] - S[q]$$

does not contain any first-order terms in δq. To obtain the differential equation for $q(t)$, let us compute δS:

$$\begin{aligned} \delta S[q; \delta q] &= S[q(t) + \delta q(t)] - S[q(t)] \\ &= \int_{t_1}^{t_2} \left[\frac{\partial L(t, q, \dot{q})}{\partial q} \delta q(t) + \frac{\partial L(t, q, \dot{q})}{\partial \dot{q}} \delta \dot{q}(t) \right] dt + O\left(\delta q^2\right) \\ &= \delta q(t) \frac{\partial L}{\partial \dot{q}} \bigg|_{t_1}^{t_2} + \int_{t_1}^{t_2} \left[\frac{\partial L}{\partial q} - \frac{d}{dt} \frac{\partial L}{\partial \dot{q}} \right] \delta q(t) dt + O\left(\delta q^2\right). \end{aligned} \tag{2.4}$$

If we impose the boundary conditions $q(t_1) = q_1$ and $q(t_2) = q_2$ and require that the perturbed trajectory also satisfy them, then $\delta q(t_1) = \delta q(t_2) = 0$. Hence the boundary terms in equation (2.4) vanish and we obtain

$$\delta S = \int_{t_1}^{t_2} \left[\frac{\partial L(t, q, \dot{q})}{\partial q} - \frac{d}{dt} \frac{\partial L(t, q, \dot{q})}{\partial \dot{q}} \right] \delta q(t) dt + O\left(\delta q^2\right). \tag{2.5}$$

The requirement that δS is second order in δq means that the first-order terms should vanish for any $\delta q(t)$. This is possible only if the expression inside the square brackets in equation (2.5) vanishes. Thus we obtain the *Euler–Lagrange equation*

$$\frac{\partial L(t, q, \dot{q})}{\partial q} - \frac{d}{dt} \frac{\partial L(t, q, \dot{q})}{\partial \dot{q}} = 0, \tag{2.6}$$

which is the classical equation of motion for a system with the Lagrangian $L(t, q, \dot{q})$. Note that in a field theory, where $L(t, q, \dot{q})$ is also a functional, the derivatives with respect to q and $\dot{q}$ in equation (2.6) are replaced by the functional derivatives (see below).

Example 2.1 For the harmonic oscillator with Lagrangian (2.2), the Euler–Lagrange equation reduces to

$$\ddot{q} + \omega^2 q = 0. \tag{2.7}$$

Generally the path $q(t)$ that extremizes the action and satisfies the boundary conditions is unique. However, there are cases when the extremum is not unique, and sometimes it does not even exist.

Exercise 2.1
Find the solution of equation (2.7) satisfying the boundary conditions $q(t_1) = q_1$ and $q(t_2) = q_2$. Determine when this solution is unique.

2.1.1 Functional derivatives

The variation of a functional can always be written in the form

$$\delta S = \int \frac{\delta S}{\delta q(t)} \delta q(t) dt + O\left(\delta q^2\right). \tag{2.8}$$

The expression denoted by $\delta S / \delta q(t)$ in (2.8) is called the *functional derivative* (or the *variational derivative*) of $S[q]$ with respect to $q(t)$.

Comparing the definition in (2.8) with (2.5), we find that the functional derivative of the functional $S[q]$ given in (2.3) is

$$\frac{\delta S}{\delta q(t)} = \frac{\partial L(t, q, \dot{q})}{\partial q} - \frac{d}{dt} \frac{\partial L(t, q, \dot{q})}{\partial \dot{q}}. \tag{2.9}$$

Example 2.2 For a harmonic oscillator with the Lagrangian (2.2) we have

$$\frac{\delta S}{\delta q(t)} = -\omega^2 q(t) - \ddot{q}(t). \tag{2.10}$$

It is important to keep track of the argument t in the functional derivative $\delta S / \delta q(t)$. A functional $S[q]$ generally depends on the values of q at all moments of time t. Discretizing the time interval $t_2 > t > t_1$ as $t_k = t_1 + \varepsilon k$, where k is integer, we may approximate the function $q(t)$ by its values at points t_i. Then the functional $S[q(t)]$ can be visualized as a function of many variables $q_k \equiv q(t_k)$,

$$S[q(t)] = \text{“}S(q_1, q_2, q_3, \ldots)\text{”}.$$

In the limit $\varepsilon \to 0$, the properly normalized partial derivative of this "function" with respect to one of its arguments, say $q_1 \equiv q(t_1)$, becomes the functional derivative $\delta S / \delta q(t_1)$. Clearly the derivative $\delta S / \delta q(t_1)$ is in general different from $\delta S / \delta q(t_2)$ and therefore $\delta S / \delta q(t)$ is a function of t.

For a functional $S[\phi]$ of a field $\phi(\mathbf{x}, t)$, the functional derivative with respect to $\phi(\mathbf{x}, t)$ retains both the arguments $\mathbf{x}$ and t and is written as $\delta S / \delta \phi(\mathbf{x}, t)$.

To calculate the functional derivatives, one has to convert the functionals to an integral form. When the original functional does not contain any integration, the Dirac δ function must be used. (See Appendix A1.1 to recall the definition and the properties of the δ function.) Below we demonstrate how the functional derivative is calculated in a few simple cases.

Example 2.3 For the functional

$$A[q] \equiv \int q^3 dt$$

the functional derivative is

$$\frac{\delta A[q]}{\delta q(t_1)} = 3q^2(t_1).$$

Example 2.4 The functional

$$B[q] \equiv 3\sqrt{q(1)} + \sin[q(2)] \\ = \int \left[3\delta(t-1)\sqrt{q(t)} + \delta(t-2)\sin q(t)\right] dt$$

has the functional derivative

$$\frac{\delta B[q]}{\delta q(t)} = \frac{3\delta(t-1)}{2\sqrt{q(1)}} + \delta(t-2)\cos[q(2)].$$

Example 2.5 For the functional

$$S[\phi] = \frac{1}{2}\int d^3\mathbf{x}\, dt (\nabla\phi)^2,$$

which depends on a field $\phi(\mathbf{x}, t)$, the functional derivative is found after an integration by parts:

$$\frac{\delta S[\phi]}{\delta\phi(\mathbf{x}, t)} = -\Delta\phi(\mathbf{x}, t).$$

The boundary terms have been omitted because the integration in $S[\phi]$ is performed over the entire spacetime and the field ϕ is assumed to decay sufficiently rapidly at infinity.

Remark: alternative definition The functional derivative may equivalently be defined with the help of the δ function:

$$\frac{\delta A[q]}{\delta q(t_1)} = \left.\frac{d}{ds}\right|_{s=0} A[q(t) + s\delta(t-t_1)].$$

One can prove that this definition is equivalent to (2.8).

Because the δ function is a distribution, the definition above can be understood in a more rigorous way as

$$\frac{\delta A[q]}{\delta q(t_1)} = \lim_{n\to\infty} \left.\frac{d}{ds}\right|_{s=0} A[q_n(t)],$$

where $q_n(t)$, $n = 1, 2, \ldots$ is a sequence of functions that converges to $q(t) + s\delta(t - t_1)$ in the sense of distributions.

Second functional derivative A derivative of a functional is a functional again. Therefore we can define the second functional derivative via

$$\frac{\delta^2 S}{\delta q(t_1)\, \delta q(t_2)} \equiv \frac{\delta}{\delta q(t_2)} \left\{ \frac{\delta S}{\delta q(t_1)} \right\}.$$

Exercise 2.2
Calculate the second functional derivative

$$\frac{\delta^2 S[q]}{\delta q(t_1)\, \delta q(t_2)}$$

for a harmonic oscillator with the action

$$S[q] = \int \frac{1}{2} \left(\dot{q}^2 - \omega^2 q^2 \right) dt.$$

2.2 Hamiltonian formalism

The starting point of a canonical quantum theory is a classical theory in the Hamiltonian formulation. The Hamiltonian formalism is based on the Legendre transform of the Lagrangian $L(t, q, \dot{q})$ with respect to the velocity $\dot{q}$.

Legendre transform Given a function $f(x)$, one can introduce a new variable p instead of x,

$$p \equiv \frac{df}{dx}, \tag{2.11}$$

and map the function $f(x)$ to a new function

$$g(p) \equiv px(p) - f(x(p)).$$

Here we imply that x has been expressed through p using (2.11); the function $f(x)$ must be such that p, which is the slope of $f(x)$, is uniquely related to x. The function $g(p)$ is called the *Legendre transform* of $f(x)$. A nice property of the Legendre transform is that when applied to the function $g(p)$, it restores the old variable $x = dg(p)/dp$ and the old function $f(x)$.

The Hamiltonian To define the Hamiltonian, one performs the Legendre transform of the Lagrangian $L(t, q, \dot{q})$ replacing $\dot{q}$ by a new variable

$$p = \frac{\partial L(t, q, \dot{q})}{\partial \dot{q}}, \tag{2.12}$$

called the *canonical momentum*. The variables t and q remain as parameters and the ubiquitously used notation $\partial/\partial\dot{q}$ means simply the partial derivative of $L(t, q, \dot{q})$ with respect to its third argument.

Remark If the coordinate q is a multi-dimensional vector, $q \equiv q_j$, the Legendre transform is performed with respect to each velocity $\dot{q}_j$ and the momentum vector p_j is introduced. In field theory there is a continuous set of "coordinates," so we need to use a functional derivative when defining the momenta.

Assuming that equation (2.12) can be solved for the velocity $\dot{q}$ as a function of t, q and p,

$$\dot{q} = v\,(p;\, q,\, t)\,, \tag{2.13}$$

one defines the *Hamiltonian* $H(p, q, t)$ by

$$H(p, q, t) \equiv \left[p\dot{q} - L\,(t, q, \dot{q})\right]_{\dot{q}=v(p;q,t)}\,. \tag{2.14}$$

In the above expression, $\dot{q}$ is replaced by the function $v\,(p;\, q,\, t)$.

Remark: the existence of the Legendre transform The possibility of performing the Legendre transform hinges on the invertibility of equation (2.12) which requires that the Lagrangian $L\,(t, q, \dot{q})$ should be a suitably nondegenerate function of the velocity $\dot{q}$. Many physically important theories with constraints, such as gauge theories or Einstein's general relativity, are described by Lagrangians that do not admit an immediate Legendre transform in the velocities. In those cases (not considered in this book) a more complicated formalism is needed to obtain an adequate Hamiltonian formulation of the theory.

The Hamilton equations of motion The Euler–Lagrange equations (2.6) are second-order differential equations for $q(t)$. In the Hamiltonian formalism they are replaced by first-order differential equations for the variables $q(t)$ and $p(t)$. Due to the definition (2.12), we can recast (2.6) as

$$\frac{dp}{dt} = \left.\frac{\partial L\,(t, q, \dot{q})}{\partial q}\right|_{\dot{q}=v(p;q,t)}, \tag{2.15}$$

where the substitution $\dot{q} = v$ must be carried out after the differentiation $\partial L/\partial q$. The other equation is (2.13),

$$\frac{dq}{dt} = v\,(p;\, q,\, t)\,. \tag{2.16}$$

Using the Hamiltonian $H(p, q, t)$, defined in (2.14), we can rewrite the above equations in a more symmetrical form. As a result of straightforward algebra, one has

$$\frac{\partial H}{\partial q} = \frac{\partial}{\partial q}(pv - L) = p\frac{\partial v}{\partial q} - \frac{\partial L}{\partial q} - \frac{\partial L}{\partial \dot{q}}\frac{\partial v}{\partial q} = -\frac{\partial L}{\partial q},$$

$$\frac{\partial H}{\partial p} = \frac{\partial}{\partial p}(pv - L) = v + p\frac{\partial v}{\partial p} - \frac{\partial L}{\partial \dot{q}}\frac{\partial v}{\partial p} = v,$$

and hence equations (2.15)–(2.16) become

$$\dot{q} = \frac{\partial H}{\partial p}, \quad \dot{p} = -\frac{\partial H}{\partial q}. \tag{2.17}$$

These are the well-known Hamilton equations of motion.

Example 2.6 For a harmonic oscillator described by the Lagrangian (2.2), the canonical momentum is equal to $p = \dot{q}$ and the corresponding Hamiltonian is

$$H(p,q) = p\dot{q} - L = \frac{1}{2}p^2 + \frac{1}{2}\omega^2 q^2. \tag{2.18}$$

The Hamilton equations are then

$$\dot{q} = p, \quad \dot{p} = -\omega^2 q.$$

The action principle The Hamilton equations follow from the action principle

$$S_H[q(t), p(t)] = \int [p\dot{q} - H(p,q,t)]\, dt, \tag{2.19}$$

where the *Hamiltonian action* S_H is a functional of two functions $q(t)$ and $p(t)$ which must be varied independently to extremize S_H.

Exercise 2.3

(a) Derive equations (2.17) by extremizing the action (2.19). Determine the boundary conditions which must be imposed on $p(t)$ and $q(t)$.

(b) Verify that the Hamilton equations imply $dH/dt = 0$ when $H(p,q)$ does not depend explicitly on the time t.

(c) Show that the expression $p\dot{q} - H$ evaluated for the classical trajectory (on-shell) is equal to the Lagrangian $L(q, \dot{q}, t)$.

2.3 Quantization of Hamiltonian systems

In quantum theory we replace the canonical variables $q(t)$, $p(t)$ by noncommuting operators $\hat{q}(t)$, $\hat{p}(t)$ for which one postulates the equal-time commutation relation

$$[\hat{q}(t), \hat{p}(t)] \equiv \hat{q}(t)\hat{p}(t) - \hat{p}(t)\hat{q}(t) = i\hbar\hat{1}. \tag{2.20}$$

(We shall frequently omit the identity operator $\hat{1}$ in formulae below.) Relation (2.20) reflects the impossibility of simultaneously measuring the coordinate and the momentum with unlimited accuracy (Heisenberg's uncertainty relation). Note that the commutation relations for unequal times, for instance $[\hat{q}(t_1), \hat{p}(t_2)]$, are not postulated but are *derived* using the equations of motion.

Exercise 2.4

Simplify the expression $\hat{q}\hat{p}^2\hat{q} - \hat{p}^2\hat{q}^2$ using the commutation relation (2.20).

The problem of operator ordering Consider the following classical Hamiltonian $H(p,q) = 2p^2 q$. Since $\hat{p}\hat{q} \neq \hat{q}\hat{p}$, it is not a priori clear whether the corresponding quantum Hamiltonian should be $\hat{p}^2\hat{q} + \hat{q}\hat{p}^2$, or perhaps $2\hat{p}\hat{q}\hat{p}$, or some other combination of the noncommuting operators $\hat{p}$ and $\hat{q}$. The difference between the possible Hamiltonians is of order $\hbar$ or higher, so the classical limit $\hbar \to 0$ is the same for any choice of the operator ordering. The ambiguity in the choice of the quantum Hamiltonian is called the *operator ordering problem.*

The operator ordering needs to be chosen by hand in each case when it is not unique. In principle, only a precise measurement of quantum effects could unambiguously determine the correct operator ordering in such cases.

Remark For frequently used Hamiltonians of the form

$$H(p,q) = \frac{1}{2m} p^2 + U(q)$$

there is no operator ordering problem.

2.4 Hilbert spaces and Dirac notation

The non commuting operators $\hat{q}$ and $\hat{p}$ can be represented as linear transformations ("matrices") in a suitable vector space (the space of *quantum states*). Since relation (2.20) cannot be satisfied by any finite-dimensional matrices,[2] the corresponding space of quantum states is necessarily infinite-dimensional and must be defined over the field $\mathbb{C}$ of complex numbers.

Infinite-dimensional vector spaces A vector in a *finite*-dimensional space can be visualized as a collection of components, e.g. $\vec{a} \equiv (a_1, a_2, a_3, a_4)$, where each a_k is a (complex) number. A vector in infinite-dimensional space has infinitely many components. An important example of an infinite-dimensional complex vector space is the space L^2 of square-integrable complex-valued functions $\psi(q)$ for which the integral

$$\int_{-\infty}^{+\infty} |\psi(q)|^2 \, dq$$

converges. One can check that a linear combination of two such functions, $\lambda_1\psi_1(q) + \lambda_2\psi_2(q)$, with constant coefficients λ_1 and $\lambda_2 \in \mathbb{C}$, is again an element of L^2. A function $\psi \in L^2$ can be thought of as a set of infinitely many "components" $\psi_q \equiv \psi(q)$, where the "index" q is continuous.

It turns out that the space of quantum states of a system with one degree of freedom is exactly the space of square-integrable functions $\psi(q)$, where q is a

[2] This is easy to prove by considering the trace of a commutator. If $\hat{A}$ and $\hat{B}$ are arbitrary finite-dimensional matrices, then $\mathrm{Tr}\,[\hat{A},\hat{B}] = \mathrm{Tr}\hat{A}\hat{B} - \mathrm{Tr}\hat{B}\hat{A} = 0$ which contradicts equation (2.20). In an infinite-dimensional space, this reasoning no longer holds because the trace is not well-defined for an arbitrary operator.

generalized coordinate of a system (for example, position of a particle moving in one dimensional space). In this case the function $\psi(q)$ is called the *wave function.* In the case of two degrees of freedom the wave function depends on both coordinates q_1 and q_2 characterizing the state of the corresponding classical system, $\psi = \psi(q_1, q_2)$. In quantum field theory, the "coordinates" are field configurations $\phi(\mathbf{x}) \equiv \phi_\mathbf{x}$ and the wave function depends on infinitely many "coordinates" $\phi_\mathbf{x}$; in other words, it is a functional, $\psi[\phi(\mathbf{x})]$.

The Dirac notation Linear algebra is used in many areas of physics, and the Dirac notation is a convenient shorthand for calculations in both finite- and infinite-dimensional vector spaces.

To denote vectors in abstract linear space, Dirac proposed to use symbols such as $|a\rangle$, $|x\rangle$, $|\lambda\rangle$ ..., which he called "ket"-vectors. Then the symbol $2|v\rangle - 3i|w\rangle$, for example, denotes a linear combination of the vectors $|v\rangle$ and $|w\rangle$.

Dual space In a vector space V one can define linear forms, which act on a vector to produce a (complex) number; $f: V \to \mathbb{C}$. A linear form is called *covector* or "bra"-vector and denoted by $\langle f|$. A complex number produced by linear form $\langle f|$ as a result of acting on a vector $|v\rangle$ is denoted by $\langle f|v\rangle$ (the mnemonic rule is: "bra"-vector acting on "ket"-vector makes a "bracket", which is a complex number). The fact that a form is linear means that

$$\langle f|(\alpha|v\rangle + \beta|w\rangle) = \alpha\langle f|v\rangle + \beta\langle f|w\rangle,$$

where α and β are arbitrary complex numbers. In the space of all linear forms one can define the multiplication by a complex number and a sum of two linear forms in a natural way as

$$(\alpha\langle f| + \beta\langle g|)|v\rangle \equiv \alpha\langle f|v\rangle + \beta\langle g|v\rangle,$$

valid for any vector $|v\rangle$. Then the space of all linear forms becomes a vector space called the *dual space*. At this stage one still has to distinguish the dual space from the original linear space because they have a different mathematical origin. For the reader familiar with differential geometry we point out that the dual space is an analog of the space of one-forms.

Hilbert space A linear space of quantum states must possess an extra structure, namely, a *Hermitian* scalar product. A scalar product maps any two vectors $|v\rangle$ and $|w\rangle$ into a complex number $(|v\rangle, |w\rangle)$. A complete, separable vector space[3] with a Hermitian scalar product is called a Hilbert space. The Hermitian

[3] A normed vector space is *complete* if all Cauchy sequences converge in it; then all norm-convergent infinite sums always have a unique vector as their limit. The space is *separable* if there exists a countable set of vectors $\{|e_n\rangle\}$ which is dense everywhere in the space. Separability ensures that all vectors can be approximated arbitrarily well by *finite* combinations of the basis vectors.

scalar product satisfies the usual axioms, while the hermiticity means that

$$(|v\rangle, |w\rangle) = (|w\rangle, |v\rangle)^*,$$

where the asterisk * denotes the complex conjugate.

The scalar product allows us to establish a one-to-one correspondence between "bra"- and "ket"-vectors. We say that the "bra"-vector $\langle v|$ corresponds to a "ket" vector $|v\rangle$ if

$$\langle v|w\rangle = (|v\rangle, |w\rangle)$$

for *any* vector $|w\rangle$. The scalar product of two "bra"-vectors $\langle w|$ and $\langle v|$ can then be defined as

$$(\langle w|, \langle v|) \equiv (|v\rangle, |w\rangle),$$

and the dual space also becomes a Hilbert space. Because the original Hilbert space and its dual space are isomorphic, we can "identify" them from now on and consider $|v\rangle$ and $\langle v|$ as simply different symbols designating the same quantum state. The scalar product $(|v\rangle, |w\rangle)$ can then be always written in somewhat more concise form as $\langle v|w\rangle$.

Exercise 2.5

Verify that the "bra"-vector $\alpha^* \langle v| + \beta^* \langle w|$ corresponds to a "ket"-vector $\alpha |v\rangle + \beta |w\rangle$.

2.5 Operators, eigenvalue problem and basis in a Hilbert space

Operators An operator $\hat{A}$ maps a vector space to itself; it transforms a vector $|v\rangle$ to the vector $|w\rangle \equiv \hat{A}|v\rangle$. For example, the identity operator $\hat{1}$ does not change any vectors: $\hat{1}|v\rangle = |v\rangle$. In quantum theory it is enough to consider only linear operators for which

$$\hat{A}(\alpha|v\rangle + \beta|w\rangle) = \alpha\hat{A}|v\rangle + \beta\hat{A}|w\rangle.$$

The product of two operators $\hat{A}$ and $\hat{B}$ is defined in a natural way as

$$\left(\hat{A}\cdot\hat{B}\right)|v\rangle \equiv \hat{A}\left(\hat{B}|v\rangle\right)$$

for any vector $|v\rangle$. Generically, operators do not commute with each other, that is,

$$[\hat{A}, \hat{B}] \equiv \hat{A}\hat{B} - \hat{B}\hat{A} \neq 0.$$

The notation $\langle v|\hat{A}|w\rangle$ denotes the scalar product of the vectors $|v\rangle$ and $\hat{A}|w\rangle$ and the quantity $\langle v|\hat{A}|w\rangle$ is also called the *matrix element* of the operator $\hat{A}$

with respect to the states $|v\rangle$ and $|w\rangle$. Acting on a "bra"-vector $\langle v|$ the operator $\hat{A}$ produces the "bra"-vector $\langle v|\,\hat{A}$, such that

$$\left(\langle v|\,\hat{A}\right)|w\rangle = \langle v|\,\hat{A}\,|w\rangle$$

for any $|w\rangle$.

Let us consider a "ket"-vector

$$|g\rangle = \hat{A}\,|v\rangle\,.$$

To the "ket"-vectors $|g\rangle$ and $|v\rangle$ there correspond the "bra"-vectors $\langle g|$ and $\langle v|$ respectively. Generically, $\langle g| \neq \langle v|\,\hat{A}$, and the vectors $\langle g|$ and $\langle v|$ are related by another operator $\hat{A}^\dagger$,

$$\langle g| = \langle v|\,\hat{A}^\dagger,$$

which is called the *Hermitian conjugate* of operator $\hat{A}$. Because $\langle g|w\rangle = \langle w|g\rangle^*$ we have

$$\langle v|\,\hat{A}^\dagger\,|w\rangle = \left(\langle w|\,\hat{A}\,|v\rangle\right)^*. \tag{2.21}$$

It is easy to see that the operation of Hermitian conjugation has the following properties:

$$(\hat{A}+\hat{B})^\dagger = \hat{A}^\dagger + \hat{B}^\dagger; \quad (\lambda\hat{A})^\dagger = \lambda^*\hat{A}^\dagger; \quad (\hat{A}\hat{B})^\dagger = \hat{B}^\dagger\hat{A}^\dagger.$$

The following subsets of all operators play a particularly important role in a quantum theory: *Hermitian* operators for which $\hat{A}^\dagger = \hat{A}$, *skew-Hermitian* operators satisfying $\hat{B}^\dagger = -\hat{B}$, and *unitary* operators: $\hat{U}^\dagger\hat{U} = \hat{U}\hat{U}^\dagger = \hat{1}$.

Eigenvalue problem Given an operator $\hat{A}$ one can consider the equation

$$\hat{A}\,|v\rangle = v\,|v\rangle\,. \tag{2.22}$$

The vector $|v\rangle \neq 0$ satisfying this equation is called an eigenvector of the operator $\hat{A}$ corresponding to the eigenvalue v.

According to quantum theory the result of any measurement of some quantity corresponding to an operator $\hat{A}$ is always an eigenvalue of this operator. Therefore, observables must be described by operators with real eigenvalues. Setting $|w\rangle = |v\rangle$ in equation (2.21) and assuming that $|v\rangle$ is an eigenvector of an operator $\hat{A}$ we find that if this operator is Hermitian, $\hat{A}^\dagger = \hat{A}$, then its eigenvalues are always real. This motivates an important assumption made in quantum theory: the operators corresponding to all observables are Hermitian.

Remark The operators of position $\hat{q}$ and momentum $\hat{p}$ must be Hermitian, $\hat{q}^\dagger = \hat{q}$ and $\hat{p}^\dagger = \hat{p}$. The commutator of two Hermitian operators $\hat{A}$, $\hat{B}$ is anti-Hermitian: $[\hat{A},\hat{B}]^\dagger = -[\hat{A},\hat{B}]$. Accordingly, the commutator of $\hat{q}$ and $\hat{p}$ contains the imaginary unit i. The operator $\hat{p}\hat{q}$ is neither Hermitian nor skew-Hermitian: $(\hat{p}\hat{q})^\dagger = \hat{q}\hat{p} = \hat{p}\hat{q} + i\hbar\hat{1} \neq \pm\hat{p}\hat{q}$.

Eigenvectors of an Hermitian operator corresponding to different eigenvalues are always orthogonal. This is easy to see: if $|v_1\rangle$ and $|v_2\rangle$ are eigenvectors of an Hermitian operator $\hat{A}$ with eigenvalues v_1 and v_2, then $\langle v_1|\hat{A} = v_1 \langle v_1|$ because v_1 is real. Consequently,

$$\langle v_1|\hat{A}|v_2\rangle = v_2 \langle v_1|v_2\rangle = v_1 \langle v_1|v_2\rangle ,$$

and $\langle v_1|v_2\rangle = 0$ if $v_1 \neq v_2$.

Basis in Hilbert space In an N-dimensional vector space, one can find a finite set of linearly independent vectors $|e_1\rangle, \ldots, |e_N\rangle$ and uniquely express any vector $|v\rangle$ as a linear combination of these vectors,

$$|v\rangle = \sum_{n=1}^{N} v_n |e_n\rangle .$$

The coefficients v_n are called the *components* of the vector $|v\rangle$ in the basis $\{|e_n\rangle\}$. In an orthonormal basis satisfying $\langle e_m|e_n\rangle = \delta_{mn}$, the scalar product of two vectors $|v\rangle$, $|w\rangle$ is expressed through their components v_n, w_n as

$$\langle v|w\rangle = \sum_{n=1}^{N} v_n^* w_n .$$

By definition, a vector space is infinite-dimensional if no finite set of vectors can serve as a basis. In that case, one might expect to have an infinite countable basis $|e_1\rangle, |e_2\rangle, \ldots,$ such that any vector $|v\rangle$ is uniquely expressible as

$$|v\rangle = \sum_{n=1}^{\infty} v_n |e_n\rangle . \tag{2.23}$$

However, the convergence of this infinite series is a nontrivial issue. For instance, if the basis vectors $|e_n\rangle$ are orthonormal, then the norm of the vector $|v\rangle$ is

$$\langle v|v\rangle = \left(\sum_{m=1}^{\infty} v_m^* \langle e_m|\right)\left(\sum_{n=1}^{\infty} v_n |e_n\rangle\right) = \sum_{n=1}^{\infty} |v_n|^2 . \tag{2.24}$$

This series must converge if the vector $|v\rangle$ has a finite norm. Therefore, for example, the sum $\sum_{n=1}^{\infty} n^2 |e_n\rangle$ does not correspond to a vector that belongs to a Hilbert space. The coefficients v_n must fall off sufficiently rapidly so that the series (2.24) is finite and only in this case is it plausible that the infinite linear combination (2.23) converges and uniquely specifies the vector $|v\rangle$. This does not hold in all infinite-dimensional spaces. However, as we have already mentioned, the required properties, known in functional analysis as completeness and separability, are fulfilled in a Hilbert space. When defining a quantum theory, one always chooses the space of quantum states as a separable Hilbert space. In some instances we have to "enclose" the system inside a large finite box and only then will a countable basis $\{|e_n\rangle\}$ exist. Once an orthonormal basis is chosen, any vector $|v\rangle$ is unambiguously represented by a collection of its components $(v_1, v_2, \ldots)$. Therefore a separable Hilbert space can be visualized as the space of infinite sets of complex numbers, $|v\rangle \equiv (v_1, v_2, \ldots)$, such that the sums $\sum_{n=1}^{\infty} |v_n|^2$ converge. The convergence guarantees that all scalar products $\langle v|w\rangle = \sum_{n=1}^{\infty} v_n^* w_n$ are finite.

Example 2.7 The space L^2 of square-integrable wave functions $\psi(q)$ defined on an interval $a < q < b$ is a separable Hilbert space, although it may appear to be "much larger" than the space of infinite rows of numbers. The scalar product of two wave functions $\psi_1(q)$ and $\psi_2(q)$ is defined by

$$\langle \psi_1 | \psi_2 \rangle = \int_a^b \psi_1^*(q) \psi_2(q) dq,$$

and the canonical operators $\hat{p}$, $\hat{q}$ can be represented as linear operators in the space L^2 that act on functions $\psi(q)$ as

$$\hat{p} : \psi(q) \to -i\hbar \frac{\partial \psi}{\partial q}, \quad \hat{q} : \psi(q) \to q\psi(q). \tag{2.25}$$

It is straightforward to verify that the commutation relation (2.20) holds.

Remark When one considers a field $\phi(\mathbf{x})$ in an infinite three-dimensional space, the corresponding space of quantum states is not separable. Therefore, to obtain a mathematically consistent theory, we need to enclose the field inside a finite box and impose appropriate boundary conditions.

Decomposition of unity Let $\{|e_n\rangle\}$ be a complete orthonormal basis in a separable Hilbert space. Then any vector $|v\rangle$ can be written as

$$|v\rangle = \sum_{n-1}^{\infty} v_n |e_n\rangle = \sum_{n=1}^{\infty} \langle e_n|v\rangle |e_n\rangle = \left(\sum_{n-1}^{\infty} |e_n\rangle \langle e_n| \right) |v\rangle .$$

Hence, the identity operator $\hat{1}$ can be decomposed as

$$\hat{1} = \sum_{n=1}^{\infty} |e_n\rangle \langle e_n| .$$

The symbol $|e_n\rangle \langle e_n|$ must be interpreted as the operator which acts on a vector $|v\rangle$ according to the rule

$$|v\rangle \to (|e_n\rangle \langle e_n|) |v\rangle \equiv \langle e_n|v\rangle |e_n\rangle .$$

It projects a vector $|v\rangle$ onto the one-dimensional subspace spanned by $|e_n\rangle$. Thus, the identity operator $\hat{1}$ can be written as a sum of projectors onto orthonormal basis vectors.

The choice of basis in Hilbert space corresponds to a choice of a certain representation in quantum theory (for example, coordinate or momentum representation). The decomposition of unity is very useful for establishing the relation between different representations ("coordinate systems").

Remark One must carefully distinguish the symbols $\langle w|v\rangle$ and $|w\rangle\langle v|$. The first one is a number, while the second is the operator which acts on the "ket"- and "bra"-vectors as $(|w\rangle\langle v|)|g\rangle = \langle v|g\rangle|w\rangle$ and $\langle g|(|w\rangle\langle v|) = \langle g|w\rangle\langle v|$ respectively.

2.6 Generalized eigenvectors and basic matrix elements

We can build a *basis* in a Hilbert space if we take all eigenvectors of a suitable Hermitian operator. This operator must have a discrete spectrum because its eigenvectors must form a countable set.

In calculations, however, it is often more convenient to use as a basis in Hilbert space the eigenbasis of the major operators $\hat{q}$ and $\hat{p}$. The position operator has a continuous spectrum and its eigenvalues are all possible generalized coordinates q. Therefore, it turns out that the operator $\hat{q}$ has no eigenvectors which belong to a separable Hilbert space. It is possible, however, to extend a Hilbert space to a larger vector space and introduce the "generalized vectors" $|q\rangle$ that are the eigenvectors of operator $\hat{q}$. Assuming the completeness of the basis $\{|q\rangle\}$ we can expand a vector $|\psi\rangle$ as

$$|\psi\rangle = \int dq\, \psi(q)\, |q\rangle\,. \tag{2.26}$$

Note that $|\psi\rangle$ belongs to the Hilbert space while the generalized vectors $|q\rangle$ do not. This is very similar to the situation in the theory of distributions (generalized functions), where for example the delta-function, $\delta(x-y)$, is well-defined only when applied to some function $f(x)$ from the space of base functions.

Since the operator $\hat{q}$ is Hermitian, its different eigenvectors $|q_1\rangle$ and $|q_2\rangle$ are orthogonal:

$$\langle q_1|q_2\rangle = 0 \text{ for } q_1 \neq q_2.$$

The generalization of the decomposition of unity for the case of continuous q must be

$$\hat{1} = \int dq\, |q\rangle\langle q|\,. \tag{2.27}$$

Substituting this decomposition into (2.26) we obtain

$$|\psi\rangle = \int dq\, \psi(q)\hat{1}\,|q\rangle = \int dq dq'\, \psi(q)\langle q'|q\rangle\,|q'\rangle = \int dq'\psi(q')\,|q'\rangle$$

and hence the identity

$$\psi(q') = \int dq \psi(q) \langle q' | q \rangle$$

must be satisfied for an arbitrary $\psi(q)$. This is possible only if

$$\langle q' | q \rangle = \delta(q' - q).$$

Thus, the basis $\{|q\rangle\}$ is normalized on the *delta-function* and in particular $\langle q|q\rangle = \delta(0)$ is undefined. Generally, the matrix elements such as $\langle q| \hat{A} |q'\rangle$ are also distributions.

The basis $\{|p\rangle\}$ of generalized eigenvectors of the momentum operator $\hat{p}$ is constructed in a similar way and has the same properties as the basis $\{|q\rangle\}$.

Below we will use the commutation relation (2.20) to calculate the basic matrix elements for the position and momentum operators.

The matrix elements $\langle q_1| \hat{p} |q_2\rangle$ and $\langle p_1| \hat{q} |p_2\rangle$ To determine $\langle q_1| \hat{p} |q_2\rangle$ let us consider the matrix element of the commutator $[\hat{q}, \hat{p}]$:

$$\langle q_1| [\hat{q}, \hat{p}] |q_2\rangle = \langle q_1| \hat{q}\hat{p} - \hat{p}\hat{q} |q_2\rangle = (q_1 - q_2) \langle q_1| \hat{p} |q_2\rangle .$$

Taking into account the commutation relation (2.20) we find that on the other hand

$$\langle q_1| [\hat{q}, \hat{p}] |q_2\rangle = \langle q_1| i\hbar |q_2\rangle = i\hbar\delta (q_1 - q_2) .$$

Therefore the matrix element $\langle q_1| \hat{p} |q_2\rangle \equiv F(q_1, q_2)$ must satisfy the equation

$$i\hbar\delta (q_1 - q_2) = (q_1 - q_2) F(q_1, q_2) . \tag{2.28}$$

To solve this equation for $F(q_1, q_2)$ we cannot simply divide both sides by $q_1 - q_2$ because the expression obtained, $x^{-1}\delta(x)$, is undefined. Therefore we first apply the Fourier transform to (2.28). Introducing the variable $q \equiv q_1 - q_2$ and taking into account that

$$\int \delta(q) e^{-ipq} dq = 1,$$

we obtain

$$i\hbar = \int qF(q_1, q_1 - q) e^{-ipq} dq = i\frac{\partial}{\partial p} \int F(q_1, q_1 - q) e^{-ipq} dq.$$

Integration over p gives

$$\hbar p + C(q_1) = \int F(q_1, q_1 - q) e^{-ipq} dq,$$

where $C(q_1)$ is an undetermined function. The inverse Fourier transform yields

$$F(q_1, q_2) = \frac{1}{2\pi} \int (\hbar p + C) e^{ipq} dp = \left[-i\hbar \frac{\partial}{\partial q_1} + C(q_1) \right] \delta(q_1 - q_2),$$

and hence

$$\langle q_1|\,\hat{p}\,|q_2\rangle = -i\hbar\delta'(q_1-q_2) + C(q_1)\,\delta(q_1-q_2)\,, \tag{2.29}$$

where the prime denotes the derivative of $\delta(q)$ with respect to $q = q_1 - q_2$. The function $C(q_1)$ cannot be determined from the commutation relation alone because the replacement of the operator $\hat{p}$ by $\hat{p} + c(\hat{q})$, where c is an arbitrary function, does not destroy the commutation relation. The above transformation changes the matrix element $\langle q_1|\,\hat{p}\,|q_2\rangle$ by the term $c(q_1)\,\delta(q_1-q_2)$ and therefore the term proportional to $\delta(q_1-q_2)$ in (2.29) can always be removed by redefinition of the operator $\hat{p}$. Thus the final result for $\langle q_1|\,\hat{p}\,|q_2\rangle$ is

$$\langle q_1|\,\hat{p}\,|q_2\rangle = -i\hbar\delta'(q_1-q_2)\,. \tag{2.30}$$

Remark Note that for a given $\hat{p}$ the term $C(q_1)\delta(q_1-q_2)$ in (2.29) can be removed by redefining the basis vectors $|q\rangle$ themselves. Multiplying each $|q\rangle$ by a q-dependent phase,

$$|\tilde{q}\rangle \equiv e^{-ic(q)}\,|q\rangle\,,$$

we obtain

$$\langle \tilde{q}_1|\,\hat{p}\,|\tilde{q}_2\rangle = \hbar c'(q)\delta(q_1-q_2) - i\hbar\delta'(q_1-q_2) + C(q_1)\,\delta(q_1-q_2)\,.$$

The function $c(q)$ can always be chosen such that $C(q_1)\delta(q_1-q_2)$ is canceled.

Because $\hat{q}$ and $\hat{p}$ enter the commutation relation on the same "footing" (up to the sign) the matrix element $\langle p_1|\,\hat{q}\,|p_2\rangle$ can be immediately inferred by interchanging $q \longleftrightarrow p$ in (2.30) and changing the sign:

$$\langle p_1|\,\hat{q}\,|p_2\rangle = i\hbar\delta'(p_1-p_2)\,. \tag{2.31}$$

The matrix elements $\langle p|q\rangle$ To calculate $\langle p|q\rangle$, we consider the matrix element

$$\langle p|\,\hat{p}\,|q\rangle = p\,\langle p|q\rangle\,. \tag{2.32}$$

On the other hand, we have

$$\langle p|\,\hat{p}\,|q\rangle = \langle p|\left[\int dq_1\,|q_1\rangle\,\langle q_1|\right]\hat{p}\,|q\rangle = \int dq_1\,\langle p|q_1\rangle\,\langle q_1|\,\hat{p}\,|q\rangle = i\hbar\frac{\partial}{\partial q}\,\langle p|q\rangle\,, \tag{2.33}$$

where we have substituted the expression in (2.30) for $\langle q_1|\,\hat{p}\,|q\rangle$. Comparing (2.32) and (2.33) we obtain

$$p\,\langle p|q\rangle = i\hbar\frac{\partial}{\partial q}\,\langle p|q\rangle\,.$$

Similarly, considering the matrix element $\langle p|\,\hat{q}\,|q\rangle$, one derives

$$q\,\langle p|q\rangle = i\hbar\frac{\partial}{\partial p}\,\langle p|q\rangle\,.$$

Integrating these equations, we find

$$\langle p|q\rangle = C_1(p)\exp\left[-\frac{ipq}{\hbar}\right], \quad \langle p|q\rangle = C_2(q)\exp\left[-\frac{ipq}{\hbar}\right],$$

respectively. The above solutions are compatible only if $C_1(p) = C_2(q) = \text{const}$, and thus

$$\langle p|q\rangle = C\exp\left[-\frac{ipq}{\hbar}\right], \tag{2.34}$$

where the constant of integration C is determined (up to an irrelevant phase factor) by the normalization condition $C = (2\pi\hbar)^{-1/2}$ (see Exercise 2.6). Thus, the final result is

$$(\langle q|p\rangle)^* = \langle p|q\rangle = \frac{1}{\sqrt{2\pi\hbar}}\exp\left[-\frac{ipq}{\hbar}\right]. \tag{2.35}$$

Exercise 2.6
Let $|p\rangle$, $|q\rangle$ be the δ-normalized eigenvectors of the momentum and the position operators, i.e.

$$\hat{p}\,|p_1\rangle = p_1\,|p_1\rangle\,, \quad \langle p_1|p_2\rangle = \delta\,(p_1 - p_2)\,,$$

and the same for $\hat{q}$. Show that the coefficient C in equation (2.34) satisfies $|C| = (2\pi\hbar)^{-1/2}$.

2.7 Evolution in quantum theory

Heisenberg picture In the limit $\hbar \to 0$ the expectation values of the position and momentum operators must satisfy the classical equations of motion. Therefore the simplest way to implement the time evolution in quantum theory is to postulate that the state vector of the system is time-independent, while the operators $\hat{q}\,(t)$ and $\hat{p}\,(t)$ satisfy the "Hamilton equations of motion":

$$\frac{d\hat{q}}{dt} = \frac{\partial H}{\partial p}(\hat{p}, \hat{q}, t) + O(\hbar), \quad \frac{d\hat{p}}{dt} = -\frac{\partial H}{\partial q}(\hat{p}, \hat{q}, t) + O(\hbar), \tag{2.36}$$

where $\hat{p}$, $\hat{q}$ are substituted into $\partial H/\partial q$, $\partial H/\partial p$ *after* taking the derivatives. The expressions on the right-hand side are well-defined if the Hamiltonian H is a polynomial function in p and q. The operator ordering induces the terms which are proportional to $\hbar$ and hence does not influence the classical limit. Most non-polynomial functions can be approximated by polynomials and therefore below we shall not dwell on the mathematical details of defining the operator $\hat{H} = H\,(\hat{p}, \hat{q}, t)$ for a general case.

Of course, ultimately the correct form of the quantum equations of motion is decided by their agreement with experimental data. Presently, in most cases a theory based on the "classical equations of motion" for the $\hat{p}$, $\hat{q}$ operators is in excellent agreement with experiments.

It is convenient to rewrite the equations

$$\frac{d\hat{q}}{dt} = \frac{\partial H}{\partial p}(\hat{p}, \hat{q}, t), \quad \frac{d\hat{p}}{dt} = -\frac{\partial H}{\partial q}(\hat{p}, \hat{q}, t) \tag{2.37}$$

in a purely algebraic form. Using the identities (see Exercise 2.7):

$$[\hat{q}, f(\hat{p}, \hat{q})] = i\hbar\frac{\partial f}{\partial p}(\hat{p}, \hat{q}), \quad [\hat{p}, f(\hat{p}, \hat{q})] = -i\hbar\frac{\partial f}{\partial q}(\hat{p}, \hat{q}),$$

the equations (2.37) become:

$$\frac{d\hat{q}}{dt} = -\frac{i}{\hbar}\left[\hat{q}, \hat{H}\right], \quad \frac{d\hat{p}}{dt} = -\frac{i}{\hbar}\left[\hat{p}, \hat{H}\right]. \tag{2.38}$$

They are called the Heisenberg equations of motion.

Exercise 2.7

(a) Using the canonical commutation relation, prove that

$$[\hat{q}, \hat{q}^m\hat{p}^n] = i\hbar n\hat{q}^m\hat{p}^{n-1}.$$

This relation can symbolically be written as

$$[\hat{q}, \hat{q}^m\hat{p}^n] = i\hbar\frac{\partial}{\partial\hat{p}}(\hat{q}^m\hat{p}^n).$$

Derive the similar relation for $\hat{p}$,

$$[\hat{p}, \hat{p}^m\hat{q}^n] = -i\hbar\frac{\partial}{\partial\hat{q}}(\hat{p}^m\hat{q}^n).$$

(b) Suppose that $f(p, q)$ is an analytic function in p, q given by a series expansion that converges for all p and q. The operator $f(\hat{p}, \hat{q})$ is defined by substituting the operators $\hat{p}$, $\hat{q}$ into that expansion (the ordering of $\hat{q}$ and $\hat{p}$ must be somehow fixed). Show that

$$[\hat{q}, f(\hat{p}, \hat{q})] = i\hbar\frac{\partial}{\partial\hat{p}}f(\hat{p}, \hat{q}). \tag{2.39}$$

Here it is implied that the derivative $\partial/\partial\hat{p}$ acts on each $\hat{p}$ with no change to the operator ordering, e.g.

$$\frac{\partial}{\partial\hat{p}}(\hat{p}^3\hat{q}\hat{p}^2\hat{q}) = 3\hat{p}^2\hat{q}\hat{p}^2\hat{q} + 2\hat{p}^3\hat{q}\hat{p}\hat{q}.$$

Exercise 2.8

Show that an operator $\hat{A} = f(\hat{p}, \hat{q}, t)$, where $f(p, q, t)$ is an analytic function, satisfies the equation

$$\frac{d}{dt}\hat{A} = -\frac{i}{\hbar}\left[\hat{A}, \hat{H}\right] + \frac{\partial\hat{A}}{\partial t}. \tag{2.40}$$

So far we have considered the time-dependent operators $\hat{q}(t)$, $\hat{p}(t)$ that act on fixed state vectors $|\psi\rangle$; this description of quantized systems is called the *Heisenberg picture.*

Schrödinger picture An alternative way to describe the time evolution in quantum theory is to refer the time dependence to the state vector assuming that the operators are time-independent. In fact, what is relevant for the measurements is not the operators themselves, but only their eigenvalues and the expectation values. It turns out that the time evolution of the expectation values can be entirely encoded in $|\psi(t)\rangle$. Let us consider for simplicity an operator which does not depend on time explicitly, that is, $\hat{A} = f(\hat{p}, \hat{q})$. The general solution of equation (2.40) is then

$$\hat{A}(t) = \exp\left[\frac{i}{\hbar}(t-t_0)\hat{H}\right]\hat{A}_0 \exp\left[-\frac{i}{\hbar}(t-t_0)\hat{H}\right], \tag{2.41}$$

where $\hat{A}_0 \equiv A(t_0) = \text{const}$, and for an arbitrary quantum state $|\psi_0\rangle$ the time-dependent expectation value of $\hat{A}(t)$ is

$$\langle A(t)\rangle \equiv \langle\psi_0|\hat{A}(t)|\psi_0\rangle = \langle\psi_0|e^{\frac{i}{\hbar}\hat{H}(t-t_0)}\hat{A}_0 e^{-\frac{i}{\hbar}\hat{H}(t-t_0)}|\psi_0\rangle.$$

This relation can be rewritten using a time-dependent state

$$|\psi(t)\rangle \equiv e^{-\frac{i}{\hbar}\hat{H}(t-t_0)}|\psi_0\rangle \tag{2.42}$$

and the time-independent operator $\hat{A}_0$ as

$$\langle A(t)\rangle = \langle\psi(t)|\hat{A}_0|\psi(t)\rangle.$$

The description of dynamics using evolving state vectors and time-independent operators is called the *Schrödinger picture.*

Taking the time derivative of (2.42), we find that the state vector $|\psi(t)\rangle$ satisfies the *Schrödinger equation,*

$$i\hbar\frac{\partial|\psi(t)\rangle}{\partial t} = \hat{H}|\psi(t)\rangle. \tag{2.43}$$

The above quantization procedure is equally well applicable to nonrelativistic mechanics, to solid state physics (a very large but finite number of degrees of freedom), and to relativistic field theory (infinitely many degrees of freedom). In the case of a system with local symmetries, some complications (mainly of technical nature) arise, but the general idea of quantization remains the same. It is clear that the Schrödinger equation (2.43) is simply a way to implement the Hamiltonian dynamics in quantum theory and it can be either relativistically invariant or not, depending on the Hamiltonian of the physical system. If the Hamiltonian $\hat{H}$ describes a relativistic system, then the corresponding Schrödinger equation is also relativistically invariant. For example, in field theory the Hamiltonian depends on infinitely many degrees of freedom and, as we will see later, equation (2.43) becomes a functional differential equation. The relativistic invariance of this equation is not manifest and is only revealed with extra effort. This is related to the fact that the quantization procedure is based on the commutation relations which are naturally implemented only in the Hamiltonian approach, where Lorentz invariance is also not manifest.

Remark: Schrödinger equations The use of a Schrödinger equation does *not* necessarily imply nonrelativistic physics. There is a widespread confusion about the role of the Schrödinger equation vs. that of the basic relativistic field equations: the Klein–Gordon equation, the Dirac equation, or the Maxwell equations. It would be a mistake to think that the Dirac equation and the Klein–Gordon equation are "relativistic forms" of the Schrödinger equation (although some textbooks say that). This was how the Dirac and the Klein–Gordon equations were discovered, but their actual place in quantum theory is quite different. The three field equations named above describe *classical* relativistic fields of spin 0, $1/2$ and 1 respectively. These equations need to be quantized to obtain a quantum field theory. Their role is analogous to that of the harmonic oscillator equation: they provide a classical Hamiltonian for quantization. The Schrödinger equations corresponding to the Klein–Gordon, the Dirac and the Maxwell equations describe quantum theories of these classical fields. (In practice, Schrödinger equations are rarely used in quantum field theory because in most cases it is much easier to work in the Heisenberg picture.)

Remark: second quantization The term "second quantization" is frequently used to refer to quantum field theory, whereas "first quantization" means ordinary quantum mechanics. However, this is obsolete terminology originating from the historical development of QFT as a relativistic extension of quantum mechanics. In fact, a quantization procedure can only be applied to a *classical* theory and yields the corresponding quantum theory. One does not quantize a *quantum* theory for a second time. It is more logical to say "quantization of fields" instead of "second quantization."

Historically it was not immediately realized that relativistic particles can be described only by quantized fields. At first, fields were regarded as wave functions of point particles. Old QFT textbooks present the picture of (1) "quantizing" a relativistic point particle to obtain the Klein–Gordon or Dirac equations, which are sometimes mistakenly identified with the relativistic generalization of the Schrödinger equation; and (2) "second-quantizing" the "relativistic Schrödinger wave function" to obtain a quantum field theory. The confusion between Schrödinger equations and relativistic wave equations has been cleared, but the old illogical terminology of "first" and "second" quantization persists. It is unnecessary to talk about a "second-quantized Dirac equation" if the Dirac equation is actually quantized only once.

The modern view is that one must describe relativistic particles by fields. Therefore one starts right away with a classical relativistic field equation, such as the Dirac equation (for the electron field) and the Maxwell equations (for the photon field), and applies the quantization procedure (only once) to obtain the relativistic quantum theory of photons and electrons.

3

Driven harmonic oscillator

Summary Quantization of harmonic oscillator driven by external classical force. "In" and "out" states. Matrix elements and Green's functions.

A quantum harmonic oscillator driven by an external classical force is a simple physical system which allows us to introduce several important concepts, such as Green's functions, "in" and "out" states, and to formulate the problem of particle production.

3.1 Quantizing an oscillator

A harmonic oscillator driven by a given external force $J(t)$ satisfies the classical equation of motion

$$\ddot{q} = -\omega^2 q + J(t)$$

which follows from the Lagrangian

$$L(t, q, \dot{q}) = \frac{1}{2}\dot{q}^2 - \frac{1}{2}\omega^2 q^2 + J(t) q.$$

The corresponding Hamiltonian is

$$H(p, q) = \frac{p^2}{2} + \frac{\omega^2 q^2}{2} - J(t) q, \tag{3.1}$$

and the Hamilton equations of motion are

$$\dot{q} = p, \quad \dot{p} = -\omega^2 q + J(t). \tag{3.2}$$

Note that the Hamiltonian depends explicitly on the time t and therefore the energy of the harmonic oscillator is not conserved.

In quantum theory the coordinate q and the momentum p become the operators $\hat{q}(t)$ and $\hat{p}(t)$ satisfying the equal time commutation relation $[\hat{q}, \hat{p}] = i$. (From now on, we use the units where $\hbar = 1$.) In the Heisenberg picture these operators satisfy the Heisenberg equations which are obtained by replacing q and p in (3.2) by the corresponding operators:

$$\frac{d\hat{q}}{dt} = \hat{p}, \quad \frac{d\hat{p}}{dt} = -\omega^2 \hat{q} + J(t). \tag{3.3}$$

It is convenient to introduce two Hermitian conjugated operators $\hat{a}^-(t)$ and $\hat{a}^+(t)$ instead of $\hat{p}(t)$ and $\hat{q}(t)$:

$$\hat{a}^-(t) \equiv \sqrt{\frac{\omega}{2}} \left[\hat{q}(t) + \frac{i}{\omega} \hat{p}(t) \right], \quad \hat{a}^+(t) \equiv \sqrt{\frac{\omega}{2}} \left[\hat{q}(t) - \frac{i}{\omega} \hat{p}(t) \right].$$

These operators, called the *annihilation* and *creation* operators respectively, satisfy the commutation relation

$$\left[\hat{a}^-(t), \hat{a}^+(t) \right] = 1 \tag{3.4}$$

at every moment of time t.

Exercise 3.1
Using the commutation relation $[\hat{q}, \hat{p}] = i$, verify that $[\hat{a}^-(t), \hat{a}^+(t)] = 1$.

The equations of motion for the operators $\hat{a}^\mp(t)$ follow immediately from (3.3):

$$\frac{d}{dt} \hat{a}^- = -i\omega \hat{a}^- + \frac{i}{\sqrt{2\omega}} J(t), \tag{3.5}$$

$$\frac{d}{dt} \hat{a}^+ = i\omega \hat{a}^+ - \frac{i}{\sqrt{2\omega}} J(t). \tag{3.6}$$

They are readily integrated to give

$$\hat{a}^-(t) = \left[\hat{a}^-_{\text{in}} + \frac{i}{\sqrt{2\omega}} \int_0^t e^{i\omega t'} J(t') dt' \right] e^{-i\omega t}, \tag{3.7}$$

$$\hat{a}^+(t) = \left[\hat{a}^+_{\text{in}} - \frac{i}{\sqrt{2\omega}} \int_0^t e^{-i\omega t'} J(t') dt' \right] e^{i\omega t}, \tag{3.8}$$

where $\hat{a}^\mp_{in} = a^\mp (t = 0)$ are the operator-valued constants of integration.

Exercise 3.2
Derive solutions (3.7) and (3.8).

Substituting

$$\hat{q} = \frac{1}{\sqrt{2\omega}} \left(\hat{a}^- + \hat{a}^+ \right), \quad \hat{p} = \frac{\sqrt{\omega}}{i\sqrt{2}} \left(\hat{a}^- - \hat{a}^+ \right) \tag{3.9}$$

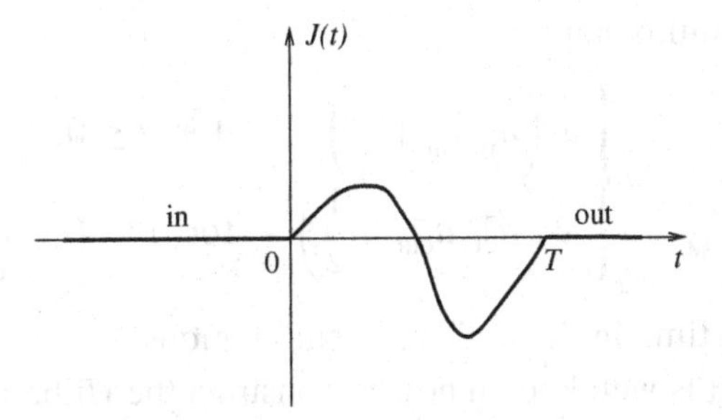

Fig. 3.1 The external force $J(t)$ and the “in”/“out” regions.

into (3.1), we find the following expression for the Hamiltonian in terms of the creation and annihilation operators $\hat{a}^{\pm}$,

$$\hat{H} = \frac{\omega}{2}\left(\hat{a}^{+}\hat{a}^{-} + \hat{a}^{-}\hat{a}^{+}\right) - \frac{\hat{a}^{+} + \hat{a}^{-}}{\sqrt{2\omega}} J(t)$$
$$= \frac{\omega}{2}\left(2\hat{a}^{+}\hat{a}^{-} + 1\right) - \frac{\hat{a}^{+} + \hat{a}^{-}}{\sqrt{2\omega}} J(t). \quad (3.10)$$

3.2 The “in” and “out” states

To simplify the calculations, we assume that the external force $J(t)$ is nonzero only during a time interval $T > t > 0$. The regions $t \leq 0$ and $t \geq T$, where the oscillator is unperturbed, are called the “in” and “out” regions respectively (see Fig. 3.1). Our purpose is to determine the relation between the states of the oscillator in these regions.

It follows from (3.7) and (3.8) that

$$\hat{a}^{-}(t) = \hat{a}^{-}_{\text{in}} e^{-i\omega t}, \quad \hat{a}^{+}(t) = \hat{a}^{+}_{\text{in}} e^{i\omega t} \quad (3.11)$$

in the “in” region. Correspondingly in the “out” region we have

$$\hat{a}^{-}(t) = \hat{a}^{-}_{\text{out}} e^{-i\omega t}, \quad \hat{a}^{+}(t) = \hat{a}^{+}_{\text{out}} e^{i\omega t} \quad (3.12)$$

where

$$\hat{a}^{-}_{\text{out}} \equiv \hat{a}^{-}_{\text{in}} + \frac{i}{\sqrt{2\omega}} \int_0^T e^{i\omega t'} J(t') dt' \equiv \hat{a}^{-}_{\text{in}} + J_0, \quad \hat{a}^{+}_{\text{out}} = \hat{a}^{+}_{\text{in}} + J_0^{*}. \quad (3.13)$$

It is obvious that both pairs of the time-independent operators $\hat{a}^{\pm}_{\text{in}}$ and $\hat{a}^{\pm}_{\text{out}}$ satisfy the commutation relation $\left[\hat{a}^{-}, \hat{a}^{+}\right] = 1$. Substituting (3.11) and (3.12) into (3.10), we find that the Hamiltonian

$$\hat{H} = \begin{cases} \omega\left(\hat{a}_{\text{in}}^{+}\hat{a}_{\text{in}}^{-} + \frac{1}{2}\right) & \text{for } t \le 0, \\ \omega\left(\hat{a}_{\text{out}}^{+}\hat{a}_{\text{out}}^{-} + \frac{1}{2}\right) & \text{for } t \ge T, \end{cases} \tag{3.14}$$

does not depend on time in the "in" and "out" regions.

Hilbert space It is well known how to construct the Hilbert space of quantum states for the unperturbed oscillator. We *assume* the existence of the unique normalized vector $|0\rangle$ for which $\hat{a}^{-}|0\rangle = 0$. This vector corresponds to the ground (vacuum) state. One can then prove that the orthogonal normalized vectors

$$|n\rangle = \frac{1}{\sqrt{n!}}(\hat{a}^{+})^{n}|0\rangle, \tag{3.15}$$

describing the excited states for $n \ge 1$, form a complete basis in the Hilbert space. In other words, all possible quantum states of the oscillator are of the form

$$|\psi\rangle = \sum_{n=0}^{\infty}\psi_n|n\rangle, \quad \sum_{n=0}^{\infty}|\psi_n|^2 = 1. \tag{3.16}$$

This is a standard result and we omit its proof. Details can be found e.g. in the book by P. A. M. Dirac, *Principles of Quantum Mechanics* (Oxford, 1948). Ultimately, the agreement between the theory and experiment determines whether a particular Hilbert space is suitable for describing a physical system; for a harmonic oscillator, the space spanned by the orthonormal basis $\{|n\rangle\}$, where $n = 0, 1, \ldots$, is adequate.

In the case under consideration there are two regions where the external classical force vanishes. Therefore, two annihilation operators, $\hat{a}_{\text{in}}^{-}$ and $\hat{a}_{\text{out}}^{-}$, define two different vacuum states – the "in" vacuum $|0_{\text{in}}\rangle$ and the "out" vacuum $|0_{\text{out}}\rangle$:

$$\hat{a}_{\text{in}}^{-}|0_{\text{in}}\rangle = 0, \quad \hat{a}_{\text{out}}^{-}|0_{\text{out}}\rangle = 0.$$

It follows from (3.14) that the vectors $|0_{\text{in}}\rangle$ and $|0_{\text{out}}\rangle$ are the lowest-energy states for $t \le 0$ and for $t \ge T$ respectively. One can easily see that the vectors $|0_{\text{in}}\rangle$ and $|0_{\text{out}}\rangle$ are different: The state $|0_{\text{out}}\rangle$ is an eigenstate of the operator $\hat{a}_{\text{out}}^{-}$ with zero eigenvalue, while

$$\hat{a}_{\text{out}}^{-}|0_{\text{in}}\rangle = (\hat{a}_{\text{in}}^{-} + J_0)|0_{\text{in}}\rangle = J_0|0_{\text{in}}\rangle,$$

that is, the vector $|0_{\text{in}}\rangle$ is an eigenstate of the same operator with the eigenvalue J_0. (We recall that the eigenstates of the annihilation operator with nonzero eigenvalues are called the *coherent states*.) Conversely, $\hat{a}_{\text{in}}^{-}|0_{\text{in}}\rangle = 0$ and $\hat{a}_{\text{in}}^{-}|0_{\text{out}}\rangle = -J_0|0_{\text{out}}\rangle$.

Using the creation operators $\hat{a}^+_{\text{in}}$ and $\hat{a}^+_{\text{out}}$, we can build two complete orthonormal sets of excited states,

$$|n_{\text{in}}\rangle = \frac{1}{\sqrt{n!}}\left(\hat{a}^+_{\text{in}}\right)^n |0_{\text{in}}\rangle, \quad |n_{\text{out}}\rangle = \frac{1}{\sqrt{n!}}\left(\hat{a}^+_{\text{out}}\right)^n |0_{\text{out}}\rangle, \quad n = 1, 2, \ldots$$

It is easy to verify that

$$\hat{H}(t)\,|n_{\text{in}}\rangle = \omega\left(n + \frac{1}{2}\right)|n_{\text{in}}\rangle, \quad t \leq 0;$$

$$\hat{H}(t)\,|n_{\text{out}}\rangle = \omega\left(n + \frac{1}{2}\right)|n_{\text{out}}\rangle, \quad t \geq T.$$

Hence, the vectors $|n_{\text{in}}\rangle$ are the eigenstates of the Hamiltonian (3.14) for $t \leq 0$ (but not for $t \geq T$), and $|n_{\text{out}}\rangle$ are its eigenstates for $t \geq T$. For this reason it is natural to interpret $|n_{\text{in}}\rangle$ as "n-particle states" of the oscillator for $t \leq 0$, while for $t \geq T$ the n-particle states are $|n_{\text{out}}\rangle$.

Remark: interpretation of the "in" and "out" states We work in the Heisenberg picture where quantum states are time-independent and operators change in time. In this picture the physical interpretation of a constant state vector $|\psi\rangle$ depends on time. For example, we have found that the vector $|0_{\text{in}}\rangle$ is no longer the lowest-energy state for $t \geq T$. This happens because the energy of the system changes due to the external force $J(t)$. In the absence of this force, $\hat{a}^-_{\text{in}} = \hat{a}^-_{\text{out}}$ and the state $|0_{\text{in}}\rangle$ describes the vacuum state at all times.

Relation between "in" and "out" states Both sets of states $\{|n_{\text{in}}\rangle\}$ and $\{|n_{\text{out}}\rangle\}$ form separately a complete basis in the Hilbert space. Therefore the vector $|0_{\text{in}}\rangle$ can be written as a linear combination of the "out" states,

$$|0_{in}\rangle = \sum_{n=0}^{\infty} \Lambda_n \,|n_{\text{out}}\rangle. \tag{3.17}$$

The coefficients Λ_n satisfy the recurrence relation

$$\Lambda_{n+1} = \frac{J_0}{\sqrt{n+1}}\Lambda_n. \tag{3.18}$$

Exercise 3.3
Derive the recurrence relation (3.18) from (3.13).

It is easy to see that the solution of (3.18) is

$$\Lambda_n = \frac{J_0^n}{\sqrt{n!}}\Lambda_0.$$

The constant Λ_0 is fixed by the normalization condition $\langle 0_{\text{in}} | 0_{\text{in}} \rangle = 1$. To compute Λ_0, we consider the normalization

$$\langle 0_{\text{in}} | 0_{\text{in}} \rangle = \sum_{n=0}^{\infty} |\Lambda_n|^2 = |\Lambda_0|^2 \sum_{n=0}^{\infty} \frac{|J_0|^{2n}}{n!} = |\Lambda_0|^2 e^{|J_0|^2} = 1$$

and hence

$$|\Lambda_0| = \exp\left[-\frac{1}{2} |J_0|^2 \right].$$

The unimportant phase of Λ_0 remains undetermined.

Thus, the state $|0_{\text{in}}\rangle$ can be expressed in terms of the "out" states as

$$|0_{\text{in}}\rangle = \exp\left[-\frac{1}{2} |J_0|^2 \right] \sum_{n=0}^{\infty} \frac{J_0^n}{\sqrt{n!}} |n_{\text{out}}\rangle, \tag{3.19}$$

or, equivalently,

$$|0_{\text{in}}\rangle = \exp\left[-\frac{1}{2} |J_0|^2 + J_0 \hat{a}_{\text{out}}^+ \right] |0_{\text{out}}\rangle.$$

Hence for $t > T$ the state vector $|0_{\text{in}}\rangle$ describes the coherent state of the harmonic oscillator, which is an eigenstate of $\hat{a}_{\text{out}}^-$ with eigenvalue J_0.

It follows from (3.19) that the initial vacuum state is a superposition of excited states for $t > T$. In particular, the probability of detecting the oscillator in an excited state n is

$$|\Lambda_n|^2 = e^{-|J_0|^2} \frac{|J_0|^{2n}}{n!}.$$

If we interpret a state with the occupation number n as describing n particles, then one concludes that the presence of the external force $J(t)$ leads to the particle production.

3.3 Matrix elements and Green's functions

The experimentally measurable quantities are expectation values of various Hermitian operators such as $\langle 0_{\text{in}} | \hat{q}(t) | 0_{\text{in}} \rangle$. Unlike the expectation values, "in-out" matrix elements, e.g. $\langle 0_{\text{out}} | \hat{q}(t) | 0_{\text{in}} \rangle$, cannot be directly measured and they are in general complex numbers. However, as we shall see in Chapter 12, such matrix elements are sometimes useful in intermediate calculations. Therefore we shall now calculate the expectation values and the matrix elements for various operators for $t \le 0$ (the "in" region) and $t \ge T$ (the "out" region).

Example 3.1 First we compute the expectation value of the Hamiltonian $\hat{H}(t)$ for the "in" vacuum state $|0_{\text{in}}\rangle$. It follows from (3.14) that

$$\langle 0_{\text{in}}| \hat{H}(t) |0_{\text{in}}\rangle = \frac{\omega}{2}$$

for $t \leq 0$ and

$$\langle 0_{\text{in}}| \hat{H}(t) |0_{\text{in}}\rangle = \langle 0_{\text{in}}| \omega \left(\frac{1}{2} + \hat{a}^{+}_{\text{out}}\hat{a}^{-}_{\text{out}}\right) |0_{\text{in}}\rangle = \left(\frac{1}{2} + |J_0|^2\right) \omega$$

for $t \geq T$, where the relations in (3.13) have been used to express $\hat{a}^{\pm}_{\text{out}}$ in terms of $\hat{a}^{\pm}_{\text{in}}$.

It is now apparent that the energy of the oscillator becomes larger than the zero-point energy $\frac{1}{2}\omega$ after applying the force $J(t)$. The constant $|J_0|^2$ is expressed in terms of $J(t)$ as

$$|J_0|^2 = \frac{1}{2\omega} \int_0^T \int_0^T dt_1 dt_2 e^{i\omega(t_1 - t_2)} J(t_1) J(t_2).$$

Example 3.2 The occupation number operator, defined as

$$\hat{N}(t) \equiv \hat{a}^{+}(t)\hat{a}^{-}(t) = \begin{cases} \hat{a}^{+}_{\text{in}}\hat{a}^{-}_{\text{in}} & \text{for} \quad t \leq 0, \\ \hat{a}^{+}_{\text{out}}\hat{a}^{-}_{\text{out}} & \text{for} \quad t \geq T, \end{cases}$$

has the expectation value

$$\langle 0_{\text{in}}| \hat{N}(t) |0_{\text{in}}\rangle = \begin{cases} 0 & \text{for} \quad t \leq 0, \\ |J_0|^2 & \text{for} \quad t \geq T. \end{cases} \tag{3.20}$$

The in-out matrix element of $\hat{N}(t)$ is

$$\langle 0_{\text{out}}| \hat{N}(t) |0_{\text{in}}\rangle = 0$$

for $t \leq 0$ and $t \geq T$.

Example 3.3 The expectation value of the position operator,

$$\hat{q}(t) = \frac{1}{\sqrt{2\omega}} \left(\hat{a}^{-}(t) + \hat{a}^{+}(t)\right), \tag{3.21}$$

is equal to zero,

$$\langle 0_{\text{in}}| \hat{q}(t \leq 0) |0_{\text{in}}\rangle = 0, \tag{3.22}$$

for $t \le T$. It follows from (3.12) and (3.13) that

$$\hat{a}^-(t \ge T) = \hat{a}^-_{\text{out}} e^{-i\omega t} = \left(\hat{a}^-_{\text{in}} + J_0\right) e^{-i\omega t},$$
$$\hat{a}^+(t \ge T) = \left(\hat{a}^+_{\text{in}} + J_0^*\right) e^{i\omega t},$$

and therefore

$$\langle 0_{\text{in}} | \hat{q}(t) | 0_{\text{in}} \rangle = \frac{1}{\sqrt{2\omega}} \left(J_0 e^{-i\omega t} + J_0^* e^{i\omega t} \right) = \int_0^T \frac{\sin \omega(t-t')}{\omega} J(t') dt' \tag{3.23}$$

for $t \ge T$.

Introducing the *retarded Green's function* for the harmonic oscillator,

$$G_{\text{ret}}(t, t') \equiv \frac{\sin \omega(t-t')}{\omega} \theta(t-t'), \tag{3.24}$$

results (3.22), (3.23) can be rewritten as

$$q(t) = \int_{-\infty}^{+\infty} J(t') G_{\text{ret}}(t, t') dt'. \tag{3.25}$$

Example 3.4 The in-out matrix element of the position operator $\hat{q}$ is

$$\begin{aligned} \frac{\langle 0_{\text{out}} | \hat{q}(t \le 0) | 0_{\text{in}} \rangle}{\langle 0_{\text{out}} | 0_{\text{in}} \rangle} &= \frac{e^{-i\omega t}}{\sqrt{2\omega}} \frac{\langle 0_{\text{out}} | \hat{a}^+_{\text{in}} | 0_{\text{in}} \rangle}{\langle 0_{\text{out}} | 0_{\text{in}} \rangle} = -J_0 \frac{e^{-i\omega t}}{\sqrt{2\omega}}, \\ \frac{\langle 0_{\text{out}} | \hat{q}(t \ge T) | 0_{\text{in}} \rangle}{\langle 0_{\text{out}} | 0_{\text{in}} \rangle} &= \frac{e^{-i\omega t}}{\sqrt{2\omega}} \frac{\langle 0_{\text{out}} | \hat{a}^-_{\text{out}} | 0_{\text{in}} \rangle}{\langle 0_{\text{out}} | 0_{\text{in}} \rangle} = J_0 \frac{e^{-i\omega t}}{\sqrt{2\omega}}. \end{aligned} \tag{3.26}$$

In general, these matrix elements are complex numbers. Noting that

$$\frac{1}{\sqrt{2\omega}} J_0 e^{-i\omega t} = \frac{i}{2\omega} \int_0^T e^{-i\omega(t-t')} J(t') dt',$$

and introducing the *Feynman Green's function*

$$G_F(t, t') \equiv \frac{i e^{-i\omega|t-t'|}}{2\omega} \tag{3.27}$$

result (3.26) can be rewritten as

$$\frac{\langle 0_{\text{out}} | \hat{q}(t) | 0_{\text{in}} \rangle}{\langle 0_{\text{out}} | 0_{\text{in}} \rangle} = \int_{-\infty}^{+\infty} J(t') G_F(t, t') dt'. \tag{3.28}$$

Other matrix elements, such as $\langle 0_{\text{in}} | \hat{q}(t_1) \hat{q}(t_2) | 0_{\text{in}} \rangle$, can be calculated similarly.

Exercise 3.4

Verify that for $t_1,\ t_2 \geq T$

$$\langle 0_{\text{in}}|\,\hat{q}\,(t_2)\,\hat{q}\,(t_1)\,|0_{\text{in}}\rangle = \frac{1}{2\omega}e^{-i\omega(t_2-t_1)} + \int_0^T dt_1' \int_0^T dt_2' J\,(t_1')\,J\,(t_2')\,G_{\text{ret}}\,(t_1,t_1')\,G_{\text{ret}}\,(t_2,t_2')$$

and

$$\frac{\langle 0_{\text{out}}|\,\hat{q}\,(t_2)\,\hat{q}\,(t_1)\,|0_{\text{in}}\rangle}{\langle 0_{\text{out}}|\,0_{\text{in}}\rangle} = \frac{1}{2\omega}e^{-i\omega(t_2-t_1)} + \int_0^T dt_1' \int_0^T dt_2' J\,(t_1')\,J\,(t_2')\,G_{\text{F}}\,(t_1,t_1')\,G_{\text{F}}\,(t_2,t_2')\,.$$

4

From harmonic oscillators to fields

Summary Ensemble of harmonic oscillators. Field quantization and mode expansion. Vacuum energy. Schrödinger equation for quantum fields.

4.1 Quantum harmonic oscillators

A free field can be treated as a set of infinitely many harmonic oscillators. Therefore we quantize a scalar field by simply generalizing the method used to describe a finite set of oscillators.

The most general classical action describing N harmonic oscillators with generalized coordinates $q_1, \ldots, q_N$ is

$$S[q_i] = \frac{1}{2} \int \left[\sum_{i=1}^{N} \dot{q}_i^2 - \sum_{i,j=1}^{N} M_{ij} q_i q_j \right] dt, \tag{4.1}$$

where the matrix M_{ij} is symmetric and positive-definite.

By choosing an appropriate set of normal coordinates

$$\tilde{q}_\alpha = \sum_{i=1}^{N} C_{\alpha i} q_i$$

the matrix M_{ij} can be diagonalized, $M_{ij} \to M_{\alpha\beta} = \delta_{\alpha\beta} \omega_\alpha^2$, and the oscillators "decoupled" from each other. In terms of these new coordinates the action (4.1) reduces to

$$S[\tilde{q}_\alpha] = \frac{1}{2} \int \sum_{\alpha=1}^{N} \left(\dot{\tilde{q}}_\alpha^2 - \omega_\alpha^2 \tilde{q}_\alpha^2 \right) dt,$$

where ω_α are the eigenfrequencies.

Exercise 4.1
Find a linear transformation which "decouples" the oscillators.

For brevity, we shall omit the tilde and write q_α instead of $\tilde{q}_\alpha$ below. The normal modes q_α are quantized (in the Heisenberg picture) by introducing the operators $\hat{q}_\alpha(t)$, $\hat{p}_\alpha(t)$ which satisfy the standard equal-time commutation relations:

$$[\hat{q}_\alpha, \hat{p}_\beta] = i\delta_{\alpha\beta}, \quad [\hat{q}_\alpha, \hat{q}_\beta] = [\hat{p}_\alpha, \hat{p}_\beta] = 0.$$

In turn, the creation and annihilation operators $\hat{a}^\pm_\alpha(t)$, defined as

$$\hat{a}^\pm_\alpha(t) = \sqrt{\frac{\omega_\alpha}{2}} \left(\hat{q}_\alpha(t) \mp \frac{i}{\omega_\alpha} \hat{p}_\alpha(t) \right),$$

obey the equations similar to (3.5) and (3.6):

$$\frac{d}{dt} \hat{a}^\pm_\alpha(t) = \pm i\omega_\alpha \hat{a}^\pm_\alpha(t).$$

The general solution of these equations for each oscillator is

$$\hat{a}^\pm_\alpha(t) = {}^{(0)}\hat{a}^\pm_\alpha e^{\pm i\omega_\alpha t},$$

where ${}^{(0)}\hat{a}^\pm_\alpha$ are operator-valued constants of integration which satisfy the commutation relations

$$\left[{}^{(0)}\hat{a}^-_\alpha, {}^{(0)}\hat{a}^+_\beta \right] = \delta_{\alpha\beta}.$$

Below we shall use mostly *time-independent* operators ${}^{(0)}\hat{a}$, so we skip the cumbersome superscript ${}^{(0)}$ and denote them simply by $\hat{a}^\pm_\alpha$.

The Hilbert space for the system of oscillators is constructed, as usual, with the help of the operators $\hat{a}^\pm_\alpha$. In particular, the vacuum state $|0, \ldots, 0\rangle$ is the unique eigenvector of all annihilation operators $\hat{a}^-_\alpha$ with eigenvalue 0:

$$\hat{a}^-_\alpha |0, \ldots, 0\rangle = 0 \text{ for } \alpha = 1, \ldots, N.$$

The state $|n_1, n_2, \ldots, n_N\rangle$ with occupation numbers n_α for each oscillator q_α is defined by

$$|n_1, \ldots, n_N\rangle = \left[\prod_{\alpha=1}^{N} \frac{\left(\hat{a}^+_\alpha\right)^{n_\alpha}}{\sqrt{n_\alpha!}} \right] |0, 0, \ldots, 0\rangle, \tag{4.2}$$

and the vectors $|n_1, \ldots, n_N\rangle$, with all possible choices of occupation numbers n_α, span the whole Hilbert space.

4.2 From oscillators to fields

A *classical field* is described by a function, $\phi(\mathbf{x}, t)$, characterizing the field strength at every moment t and at each point $\mathbf{x}$ in space. One can interpret a field as an infinite set of oscillators $q_i(t) \Longleftrightarrow \phi_\mathbf{x}(t)$ "attached" to each point $\mathbf{x}$. Note that the oscillator "position" $\phi_\mathbf{x}(t)$ "takes its values" in the configuration space, i.e. in the space of the field strength. The spatial coordinate $\mathbf{x}$ plays the role of index labeling the oscillators, similarly to the discrete index i for the oscillators q_i.

Using this analogy, we treat the scalar field $\phi(\mathbf{x}, t)$ as an infinite collection of oscillators. Replacing the sums over the discrete indices i by the integrals over the continuous indices $\mathbf{x}$ in (4.1) we find that the action for the scalar field must be of the form

$$S[\phi] = \frac{1}{2}\int dt \left[\int d^3\mathbf{x}\, \dot{\phi}^2(\mathbf{x}, t) - \int d^3\mathbf{x}\, d^3\mathbf{y}\, \phi(\mathbf{x}, t)\, \phi(\mathbf{y}, t)\, M(\mathbf{x}, \mathbf{y})\right], \quad (4.3)$$

where the function M is yet to be determined. This action must be invariant with respect to the Lorentz transformations and to the spacetime translations (the *Poincaré group*). The simplest Poincaré-invariant action for a real scalar field $\phi(\mathbf{x}, t)$ is obtained if we set

$$M(\mathbf{x}, \mathbf{y}) = \left[-\Delta_{\mathbf{x}} + m^2\right] \delta(\mathbf{x} - \mathbf{y}). \quad (4.4)$$

Then action (4.3) becomes

$$\begin{aligned} S[\phi] &= \frac{1}{2}\int d^4x \left[\eta^{\mu\nu}(\partial_\mu \phi)(\partial_\nu \phi) - m^2\phi^2\right] \\ &= \frac{1}{2}\int d^3\mathbf{x}\, dt \left[\dot{\phi}^2 - (\nabla\phi)^2 - m^2\phi^2\right], \end{aligned} \quad (4.5)$$

where $\eta^{\mu\nu} = \mathrm{diag}(1, -1, -1, -1)$, $x^0 \equiv t$ and $(x^1, x^2, x^3) \equiv \mathbf{x}$. It is obviously translationally invariant, and Lorentz invariance is the subject of the next exercise.

Exercise 4.2

Verify that the action (4.5) does not change under the Lorentz transformations:

$$x^\mu \to \tilde{x}^\mu = \Lambda^\mu_\nu x^\nu, \quad \phi(\mathbf{x}, t) \to \tilde{\phi}(\mathbf{x}, t) = \phi(\tilde{\mathbf{x}}, \tilde{t}), \quad (4.6)$$

where the matrix Λ^μ_ν satisfies

$$\eta_{\mu\nu}\Lambda^\mu_\alpha \Lambda^\nu_\beta = \eta_{\alpha\beta}.$$

Calculating the functional derivative, we find that the scalar field satisfies the equation of motion

$$\frac{\delta S}{\delta\phi(\mathbf{x}, t)} = \ddot{\phi}(\mathbf{x}, t) - \Delta\phi(\mathbf{x}, t) + m^2\phi(\mathbf{x}, t) = 0. \quad (4.7)$$

Exercise 4.3

Derive equation (4.7).

It is clear from (4.7) that the "oscillators" $\phi(\mathbf{x}, t) \equiv \phi_{\mathbf{x}}(t)$ are "coupled." To see this, note that the Laplacian $\Delta\phi$ contains the second derivatives of ϕ with respect to the spatial coordinates and, in particular,

$$\frac{\partial^2 \phi_x}{\partial x^2} \approx \frac{\phi_{x+\delta x} - 2\phi_x + \phi_{x-\delta x}}{(\delta x)^2}.$$

Therefore the behavior of the oscillator ϕ_x depends on the nearby oscillators at points $x \pm \delta x$.

To decouple the oscillators, we use the Fourier transform,

$$\phi(\mathbf{x}, t) \equiv \int \frac{d^3\mathbf{k}}{(2\pi)^{3/2}} e^{i\mathbf{k}\cdot\mathbf{x}} \phi_{\mathbf{k}}(t). \tag{4.8}$$

Substituting (4.8) into (4.7), we find that the complex functions $\phi_{\mathbf{k}}(t)$ satisfy the following ordinary differential equations,

$$\frac{d^2}{dt^2}\phi_{\mathbf{k}}(t) + \left(k^2 + m^2\right)\phi_{\mathbf{k}}(t) = 0, \tag{4.9}$$

and thus describe an infinite set of decoupled harmonic oscillators with frequencies

$$\omega_k \equiv \sqrt{k^2 + m^2}.$$

It is easy to verify that in terms of $\phi_{\mathbf{k}}(t)$ the action (4.5) takes the following form

$$S = \frac{1}{2}\int dt\, d^3\mathbf{k} \left(\dot{\phi}_{\mathbf{k}}\dot{\phi}_{-\mathbf{k}} - \omega_k^2 \phi_{\mathbf{k}}\phi_{-\mathbf{k}}\right). \tag{4.10}$$

Exercise 4.4
Show that for a real scalar field $\phi(\mathbf{x}, t)$ the relation $(\phi_{\mathbf{k}})^* = \phi_{-\mathbf{k}}$ must be satisfied.

4.3 Quantizing fields in a flat spacetime

To quantize the scalar field, we first need to reformulate the classical theory in the Hamiltonian formalism. Noting that the action is an integral of the Lagrangian only over the *time* (but not over space), we conclude that for the system described by the action (4.5) the Lagrangian is

$$L[\phi] = \int \mathcal{L} d^3\mathbf{x}; \quad \mathcal{L} \equiv \frac{1}{2}\eta^{\mu\nu}\phi_{,\mu}\phi_{,\nu} - \frac{1}{2}m^2\phi^2,$$

where $\mathcal{L}$ is called the Lagrangian density. Taken at a given moment of time, the Lagrangian is a functional that depends on the field configuration. Therefore, the canonical momenta are defined as the functional derivatives of the Lagrangian with respect to the generalized "velocities" $\dot{\phi} \equiv \partial\phi/\partial t$,

$$\pi(\mathbf{x}, t) \equiv \frac{\delta L[\phi]}{\delta\dot{\phi}(\mathbf{x}, t)} = \dot{\phi}(\mathbf{x}, t).$$

The classical Hamiltonian is then

$$H = \int \pi(\mathbf{x}, t)\,\dot{\phi}(\mathbf{x}, t)\,d^3\mathbf{x} - L = \frac{1}{2}\int d^3\mathbf{x}\left[\pi^2 + (\nabla\phi)^2 + m^2\phi^2\right], \qquad (4.11)$$

and the Hamilton equations of motion are

$$\frac{\partial\phi(\mathbf{x}, t)}{\partial t} = \frac{\delta H}{\delta\pi(\mathbf{x}, t)} = \pi(\mathbf{x}, t),$$

$$\frac{\partial\pi(\mathbf{x}, t)}{\partial t} = -\frac{\delta H}{\delta\phi(\mathbf{x}, t)} = \Delta\phi(\mathbf{x}, t) - m^2\phi(\mathbf{x}, t). \qquad (4.12)$$

Remark: Lorentz invariance We have noted previously that the Hamiltonian formalism is better suited for quantization. Although the Lorentz invariance is not manifest in the Hamiltonian formalism, this does not mean that the resulting quantum theory breaks this invariance. If the classical theory is relativistically invariant, then the resulting quantum theory is also relativistically invariant.

To quantize the scalar field, we introduce the operators $\hat{\phi}(\mathbf{x}, t)$ and $\hat{\pi}(\mathbf{x}, t)$ and postulate the standard commutation relations

$$\left[\hat{\phi}(\mathbf{x}, t),\, \hat{\pi}(\mathbf{y}, t)\right] = i\delta(\mathbf{x} - \mathbf{y}); \qquad (4.13)$$

$$\left[\hat{\phi}(\mathbf{x}, t),\, \hat{\phi}(\mathbf{y}, t)\right] = \left[\hat{\pi}(\mathbf{x}, t),\, \hat{\pi}(\mathbf{y}, t)\right] = 0.$$

Substituting

$$\hat{\phi}(\mathbf{x}, t) = \int \frac{d^3\mathbf{k}}{(2\pi)^{3/2}} e^{i\mathbf{k}\cdot\mathbf{x}}\hat{\phi}_{\mathbf{k}}(t), \quad \hat{\pi}(\mathbf{y}, t) = \int \frac{d^3\mathbf{k}'}{(2\pi)^{3/2}} e^{i\mathbf{k}\cdot\mathbf{y}}\hat{\pi}_{\mathbf{k}'}(t) \qquad (4.14)$$

into these commutation relations, after some algebra we find the following result for the mode operators:

$$\left[\hat{\phi}_{\mathbf{k}}(t), \hat{\pi}_{\mathbf{k}'}(t)\right] = i\delta\left(\mathbf{k} + \mathbf{k}'\right).$$

Note that the plus sign in $\delta(\mathbf{k} + \mathbf{k}')$ indicates that the variable conjugate to $\hat{\phi}_{\mathbf{k}}$ is $\hat{\pi}_{-\mathbf{k}} = (\hat{\pi}_{\mathbf{k}})^\dagger$. This is also evident from the action (4.10).

Remark: complex oscillators For the real scalar field ϕ, the variables $\phi_{\mathbf{k}}(t)$ are complex and each $\phi_{\mathbf{k}}$ may be thought of as a pair of real-valued functions, $\phi_{\mathbf{k}} = \phi_{\mathbf{k}}^{(1)} + i\phi_{\mathbf{k}}^{(2)}$, satisfying the constraints $\phi_{-\mathbf{k}}^{(1)} = \phi_{\mathbf{k}}^{(1)}$ and $\phi_{-\mathbf{k}}^{(2)} = -\phi_{\mathbf{k}}^{(2)}$. Accordingly, the operators $\hat{\phi}_{\mathbf{k}}$ are not Hermitian and $(\hat{\phi}_{\mathbf{k}})^\dagger = \hat{\phi}_{-\mathbf{k}}$. In principle, one could rewrite the theory in terms of the Hermitian variables, but it is more convenient to keep the complex $\phi_{\mathbf{k}}$.

Substituting (4.14) into the Hamilton equations (4.12) we obtain

$$\frac{d\hat{\phi}_{\mathbf{k}}}{dt} = \hat{\pi}_{\mathbf{k}}, \quad \frac{d\hat{\pi}_{\mathbf{k}}}{dt} = -\omega_k^2\hat{\phi}_{\mathbf{k}}. \qquad (4.15)$$

Similarly to Section 4.1, it is convenient to introduce the creation and annihilation operators,

$$\hat{a}_{\mathbf{k}}^{-}(t) \equiv \sqrt{\frac{\omega_k}{2}}\left(\hat{\phi}_{\mathbf{k}} + \frac{i\hat{\pi}_{\mathbf{k}}}{\omega_k}\right); \; \hat{a}_{\mathbf{k}}^{+}(t) \equiv (\hat{a}_{\mathbf{k}}^{-}(t))^{\dagger} = \sqrt{\frac{\omega_k}{2}}\left(\hat{\phi}_{-\mathbf{k}} - \frac{i\hat{\pi}_{-\mathbf{k}}}{\omega_k}\right), \quad (4.16)$$

which satisfy the commutation relations

$$\left[\hat{a}_{\mathbf{k}}^{-}(t), \hat{a}_{\mathbf{k}'}^{+}(t)\right] = \delta\left(\mathbf{k}-\mathbf{k}'\right), \quad \left[\hat{a}_{\mathbf{k}}^{-}(t), \hat{a}_{\mathbf{k}'}^{-}(t)\right] = \left[\hat{a}_{\mathbf{k}}^{+}(t), \hat{a}_{\mathbf{k}'}^{+}(t)\right] = 0, \quad (4.17)$$

and obey the equations

$$\frac{d}{dt}\hat{a}_{\mathbf{k}}^{\pm}(t) = \pm i\omega_k \hat{a}_{\mathbf{k}}^{\pm}(t).$$

The general solution of these equations is

$$\hat{a}_{\mathbf{k}}^{\pm}(t) = {}^{(0)}\hat{a}_{\mathbf{k}}^{\pm} e^{\pm i\omega_k t}, \quad (4.18)$$

where the time-independent operators ${}^{(0)}\hat{a}_{\mathbf{k}}^{\pm}$ obviously also satisfy the commutation relations (4.17). Below we will mainly use these time-independent operators and will omit the superscript $^{(0)}$ for brevity.

The Hilbert space is built in the standard way. We postulate the existence of the vacuum state $|0\rangle$ which is annihilated by all operators $\hat{a}_{\mathbf{k}}^{-}$, that is, $\hat{a}_{\mathbf{k}}^{-}|0\rangle = 0$ for all $\mathbf{k}$. The state with the occupation numbers n_s in every mode $\mathbf{k}_s$ is

$$|n_1, n_2, \ldots\rangle = \left[\prod_s \frac{\left(\hat{a}_{\mathbf{k}_s}^{+}\right)^{n_s}}{\sqrt{n_s!}}\right]|0\rangle, \quad (4.19)$$

where $s = 1, 2, \ldots$ enumerates the excited modes and $|0\rangle \equiv |0, 0, \ldots\rangle$. The vector (4.19) corresponds to the quantum state in which n_1 particles have momentum $\mathbf{k}_1$, n_2 particles have momentum $\mathbf{k}_2$, etc. The vectors $|n_1, n_2, \ldots\rangle$ with all possible choices of n_s form the complete orthonormal basis in the Hilbert space.

The Hamiltonian (4.11) can be rewritten as

$$\hat{H} = \frac{1}{2}\int d^3\mathbf{k}\left[\hat{\pi}_{\mathbf{k}}\hat{\pi}_{-\mathbf{k}} + \omega_k^2\hat{\phi}_{\mathbf{k}}\hat{\phi}_{-\mathbf{k}}\right]. \quad (4.20)$$

Equivalently, we may express $\hat{H}$ in terms of the creation and annihilation operators:

$$\hat{H} = \int d^3\mathbf{k}\frac{\omega_k}{2}\left(\hat{a}_{\mathbf{k}}^{-}\hat{a}_{\mathbf{k}}^{+} + \hat{a}_{\mathbf{k}}^{+}\hat{a}_{\mathbf{k}}^{-}\right) = \int d^3\mathbf{k}\,\omega_k[\hat{a}_{\mathbf{k}}^{+}\hat{a}_{\mathbf{k}}^{-} + \frac{1}{2}\delta^{(3)}(0)]. \quad (4.21)$$

Exercise 4.5
Derive equations (4.20) and (4.21).

4.4 The mode expansion

Taking into account (4.18) we find from (4.16) that

$$\hat{\phi}_{\mathbf{k}}(t) = \frac{1}{\sqrt{2\omega_k}}\left(\hat{a}_{\mathbf{k}}^{-} e^{-i\omega_k t} + \hat{a}_{-\mathbf{k}}^{+} e^{i\omega_k t}\right).$$

Substituting this expression into (4.14) gives the following expansion of the field operator $\hat{\phi}(\mathbf{x}, t)$ in terms of the time-independent creation and annihilation operators,

$$\hat{\phi}(\mathbf{x}, t) = \int \frac{d^3\mathbf{k}}{(2\pi)^{3/2}} \frac{1}{\sqrt{2\omega_k}} \left[e^{-i\omega_k t + i\mathbf{k}\cdot\mathbf{x}} \hat{a}_{\mathbf{k}}^{-} + e^{i\omega_k t - i\mathbf{k}\cdot\mathbf{x}} \hat{a}_{\mathbf{k}}^{+}\right], \tag{4.22}$$

where we have replaced $-\mathbf{k}$ by $\mathbf{k}$ in the second term. The obtained expression is called the *mode expansion* of the quantum field $\hat{\phi}(\mathbf{x}, t)$.

This observation suggests an alternative quantization procedure without explicitly introducing the operators $\hat{\phi}_{\mathbf{k}}$ and $\hat{\pi}_{\mathbf{k}}$. We can begin immediately with the field operator expansion

$$\hat{\phi}(\mathbf{x}, t) = \int \frac{d^3\mathbf{k}}{(2\pi)^{3/2}} \frac{1}{\sqrt{2}} \left[v_{\mathbf{k}}^{*}(t) e^{i\mathbf{k}\cdot\mathbf{x}} \hat{a}_{\mathbf{k}}^{-} + v_{\mathbf{k}}(t) e^{-i\mathbf{k}\cdot\mathbf{x}} \hat{a}_{\mathbf{k}}^{+}\right], \tag{4.23}$$

and postulate the commutation relations for the time-independent operators $\hat{a}_{\mathbf{k}}^{-}$ and $\hat{a}_{\mathbf{k}}^{+}$:

$$\left[\hat{a}_{\mathbf{k}}^{-}, \hat{a}_{\mathbf{k}'}^{+}\right] = \delta\left(\mathbf{k} - \mathbf{k}'\right); \quad \left[\hat{a}_{\mathbf{k}}^{-}, \hat{a}_{\mathbf{k}'}^{-}\right] = \left[\hat{a}_{\mathbf{k}}^{+}, \hat{a}_{\mathbf{k}'}^{+}\right] = 0. \tag{4.24}$$

Because the field operator satisfies (4.7), the mode functions $v_{\mathbf{k}}(t)$ must obey the equation

$$\ddot{v}_{\mathbf{k}} + \omega_k^2 v_{\mathbf{k}} = 0, \tag{4.25}$$

where $\omega_k^2 = k^2 + m^2$. Substituting (4.23) together with

$$\hat{\pi}(\mathbf{y}, t) = \frac{\partial \hat{\phi}(\mathbf{y}, t)}{\partial t} = \int \frac{d^3\mathbf{k}}{(2\pi)^{3/2}} \frac{1}{\sqrt{2}} \left[\dot{v}_{\mathbf{k}}^{*}(t) e^{i\mathbf{k}\cdot\mathbf{y}} \hat{a}_{\mathbf{k}}^{-} + \dot{v}_{\mathbf{k}}(t) e^{-i\mathbf{k}\cdot\mathbf{y}} \hat{a}_{\mathbf{k}}^{+}\right] \tag{4.26}$$

into (4.13), we find that the canonical commutation relations are compatible with (4.24) only if the normalization conditions,

$$\dot{v}_{\mathbf{k}}(t) v_{\mathbf{k}}^{*}(t) - v_{\mathbf{k}}(t) \dot{v}_{\mathbf{k}}^{*}(t) = 2i, \tag{4.27}$$

are satisfied. The expression on the left-hand side is the Wronskian of the two independent complex solutions $v_{\mathbf{k}}(t)$ and $v_{\mathbf{k}}^{*}(t)$ of equation (4.25), and therefore does not depend on time. Substituting the general solution of equation (4.25),

$$v_{\mathbf{k}}(t) = \frac{1}{\sqrt{\omega_k}}\left(\alpha_{\mathbf{k}} e^{i\omega_k t} + \beta_{\mathbf{k}} e^{-i\omega_k t}\right), \tag{4.28}$$

into (4.27), we find that the constants of integration $\alpha_{\mathbf{k}}$ and $\beta_{\mathbf{k}}$ must obey

$$|\alpha_{\mathbf{k}}|^2 - |\beta_{\mathbf{k}}|^2 = 1. \tag{4.29}$$

This condition does not suffice to determine the two constants of integration, $\alpha_{\mathbf{k}}$ and $\beta_{\mathbf{k}}$. Therefore the operators $\hat{a}^-_{\mathbf{k}}$ and $\hat{a}^+_{\mathbf{k}}$ are not yet unambiguously defined. To resolve this ambiguity, let us calculate the Hamiltonian. Substituting (4.23) and (4.26) into (4.11), and using $v_{\mathbf{k}}(t)$ given in (4.28), we obtain

$$\hat{H} = \int d^3\mathbf{k}\, \omega_k \Big[\alpha^*_{\mathbf{k}}\beta^*_{\mathbf{k}}\hat{a}^-_{\mathbf{k}}\hat{a}^-_{-\mathbf{k}} + \alpha_{\mathbf{k}}\beta_{\mathbf{k}}\hat{a}^+_{\mathbf{k}}\hat{a}^+_{-\mathbf{k}} + \left(|\alpha_{\mathbf{k}}|^2 + |\beta_{\mathbf{k}}|^2\right)\left(\hat{a}^+_{\mathbf{k}}\hat{a}^-_{\mathbf{k}} + \tfrac{1}{2}\delta^{(3)}(0)\right)\Big], \tag{4.30}$$

where we have used $\alpha_{\mathbf{k}} = \alpha_{-\mathbf{k}}$ and $\beta_{\mathbf{k}} = \beta_{-\mathbf{k}}$. It is obvious from (4.30) that the vector $|0\rangle$, defined by the conditions $\hat{a}^-_{\mathbf{k}}|0\rangle = 0$ for all $\mathbf{k}$, is an eigenvector of the Hamiltonian only if $\alpha_{\mathbf{k}}\beta_{\mathbf{k}} = 0$. Combined with (4.29) this condition tells us that

$$\alpha_{\mathbf{k}} = e^{i\delta_{\mathbf{k}}}, \quad \beta_{\mathbf{k}} = 0,$$

where $\delta_{\mathbf{k}}$ are the irrelevant constant phase factors which we set to zero. Thus, the vector $|0\rangle$ can be interpreted as the vacuum state with the minimal energy only if the mode functions in the expansion (4.23) are taken as

$$v_k(t) = \frac{1}{\sqrt{\omega_k}} e^{i\omega_k t}. \tag{4.31}$$

The obtained result is in complete agreement with (4.22).

Remark: positive and negative frequency modes The modes $v^*_k(t) \propto e^{-i\omega_k t}$ and $v_k(t) \propto e^{i\omega_k t}$ are usually referred to as the positive and negative frequency modes respectively. Alternatively, these solutions are called positive and negative energy solutions. This rather confusing terminology has no particular meaning and is of historical origin. In "first quantized relativistic theory" the field was interpreted as the wave function of a relativistic particle. Therefore the solution $v^*_k(t) \propto e^{-i\omega_k t}$ was regarded as describing particles with positive energy,

$$\hat{H} v^*_k(t) = i\hbar \frac{\partial v^*_k(t)}{\partial t} = \hbar\omega_k v^*_k(t),$$

and the solution $v_k(t) \propto e^{i\omega_k t}$ as corresponding to negative-energy states. However, this interpretation does not make sense in a quantum field theory where particles always have positive energy.

4.5 Vacuum energy and vacuum fluctuations

Vacuum energy It is easy to see from (4.21) that the total energy of the field in the vacuum state $|0\rangle$ is

$$E_0 = \langle 0|\hat{H}|0\rangle = \frac{1}{2}\delta^{(3)}(0)\int d^3\mathbf{k}\,\omega_k. \qquad (4.32)$$

This energy is divergent: there is an infinite multiplicative factor $\delta^{(3)}(0)$ in (4.32) and in addition the integral

$$\int d^3\mathbf{k}\,\omega_k = \int_0^\infty 4\pi k^2\sqrt{m^2+k^2}\,dk$$

diverges as k^4 at the upper limit of integration.

The origin of the divergent factor $\delta^{(3)}(0)$ is easy to understand: It is simply the infinite volume of space. Indeed, the factor $\delta^{(3)}(0)$ arises from the commutation relation (4.17) when we evaluate $\delta^{(3)}(\mathbf{k}-\mathbf{k}')$ at $\mathbf{k}=\mathbf{k}'$ (note that $\delta^{(3)}(\mathbf{k})$ has the dimension of 3-volume). For a field in a finite box of volume V, the vacuum energy is (see (1.11)):

$$E_0 = \frac{1}{2}\sum_{\mathbf{k}}\omega_k \approx \frac{1}{2}\frac{V}{(2\pi)^3}\int d^3\mathbf{k}\,\omega_k.$$

Comparing this expression with (4.32), we find that the formally infinite factor $\delta^{(3)}(0)$ arises when the box volume V goes to infinity. In this limit, the vacuum energy density is equal to

$$\lim_{V\to\infty}\frac{E_0}{V} = \frac{1}{2}\int\frac{d^3\mathbf{k}}{(2\pi)^3}\omega_k, \qquad (4.33)$$

and it also diverges at $|\mathbf{k}|\to\infty$. This *ultraviolet divergence* arises because the number of "oscillators" with large momentum grows as k^3 and each of them has zero-point energy $\omega_k/2 \approx k/2$.

In a flat spacetime there is a simple recipe to circumvent the problem of the vacuum energy divergence. The energy of an excited state $|n_1, n_2, \ldots\rangle$ can be computed using equations (4.17), (4.21) and taking into account that

$$\left[\hat{a}_{\mathbf{k}}^-, (\hat{a}_{\mathbf{k}'}^+)^n\right] = n(\hat{a}_{\mathbf{k}}^+)^{n-1}\delta\left(\mathbf{k}-\mathbf{k}'\right).$$

As a result we obtain

$$E(n_1, n_2, \ldots) = E_0 + \int d^3\mathbf{k}\left(\sum_s n_s\delta\left(\mathbf{k}-\mathbf{k}_s\right)\right)\omega_k = E_0 + \sum_s n_s\omega_{k_s}.$$

Thus the total energy is always a sum of the divergent vacuum energy E_0 and a finite state-dependent contribution. The absolute value of the energy is relevant only for the gravitational field. If we neglect gravity, then the presence of the vacuum energy cannot be detected in experiments involving only transitions between the excited states. Therefore, we can postulate that the vacuum energy E_0 does not contribute to the gravitational field and simply subtract E_0 from the Hamiltonian. After this subtraction, the modified Hamiltonian becomes

$$\hat{H} \equiv \int d^3\mathbf{k}\, \omega_k \hat{a}^+_{\mathbf{k}} \hat{a}^-_{\mathbf{k}},$$

and the vacuum is an eigenstate of the Hamiltonian with *zero* eigenvalue:

$$\hat{H}\,|0\rangle = 0.$$

Vacuum fluctuations The subtraction of the vacuum energy density does not, however, remove the vacuum fluctuations of the quantum fields. To estimate the magnitude of fluctuations, let us calculate the correlation function

$$\xi_\phi\left(|\mathbf{x}-\mathbf{y}|\right) \equiv \langle 0|\,\hat{\phi}\left(\mathbf{x},t\right)\hat{\phi}\left(\mathbf{y},t\right)|0\rangle\,.$$

Substituting the mode expansion (4.22), we obtain

$$\xi_\phi\left(|\mathbf{x}-\mathbf{y}|\right) = \frac{1}{4\pi^2}\int \frac{k^3}{\omega_k}\frac{\sin\left(k\,|\mathbf{x}-\mathbf{y}|\right)}{k\,|\mathbf{x}-\mathbf{y}|}\frac{dk}{k},$$

where the integration over the angles was performed. It follows that the typical squared amplitude of the scalar field fluctuations on scales $L \sim 1/k$ is of order

$$\delta\phi_L^2 = \frac{1}{4\pi^2}\left.\frac{k^3}{\omega_k}\right|_{k\sim L^{-1}} \sim \frac{1}{L^2\sqrt{1+(mL)^2}}. \tag{4.34}$$

Thus, the amplitude of the vacuum fluctuations $\delta\phi_L$ decays as L^{-1} for $L < m^{-1}$ and as $L^{-3/2}$ for $L > m^{-1}$. This result is already familiar to us from Section 1.4 (see (1.13)).

4.6 The Schrödinger equation for a quantum field

So far we have been working in the Heisenberg picture. However, the fields can also be quantized using the Schrödinger picture. Let us begin with the Schrödinger equation for an ensemble of harmonic oscillators. Their Hamiltonian is obtained from action (4.1) in the standard way:

$$H = \frac{1}{2}\sum_i p_i^2 + \frac{1}{2}\sum_{i,j} M_{ij} q_i q_j.$$

In the Schrödinger picture, the operators $\hat{p}_i$, $\hat{q}_i$ are time-*independent* and act on the time-dependent wave function $|\psi(t)\rangle$. The Hilbert space is spanned by the basis vectors $|q_1, \ldots, q_N\rangle$ which are the generalized eigenvectors of the coordinate operators $\hat{q}_i$. In this basis a state vector $|\psi(t)\rangle$ can then be decomposed as

$$|\psi(t)\rangle = \int dq_1 \ldots dq_N \psi(q_1, \ldots, q_N, t) |q_1, \ldots, q_N\rangle ,$$

where

$$\psi(q_1, \ldots, q_N, t) = \langle q_1, \ldots, q_N | \psi(t)\rangle .$$

The momentum operators $\hat{p}_i$ act on the wave function $\psi(q_1, \ldots, q_N, t)$ as derivatives $-i\partial/\partial q_i$. The Schrödinger equation then takes the form

$$i\frac{\partial \psi}{\partial t} = \hat{H}\psi = \frac{1}{2}\sum_{i,j}\left(-\delta_{ij}\frac{\partial^2}{\partial q_i \partial q_j} + M_{ij}q_i q_j\right)\psi. \tag{4.35}$$

To generalize this Schrödinger equation to the case of quantum scalar field, we replace the oscillator coordinates q_i by field values $\phi_{\mathbf{x}} \equiv \phi(\mathbf{x})$ and the wave function $\psi(q_1, \ldots, q_N, t)$ becomes a wave functional $\Psi[\phi(\mathbf{x}), t]$ defined in the space of the field configurations $\phi(\mathbf{x})$ (recall that the spatial coordinate $\mathbf{x}$ plays here the role of the index i). The wave functional has a simple physical interpretation, namely, the probability of measuring a particular spatial field configuration $\phi(\mathbf{x})$ at time t is proportional to $|\Psi[\phi(\mathbf{x}), t]|^2$.

The Schrödinger equation for the scalar field is obtained from (4.35) by replacing the partial derivatives $\partial/\partial q_i$ by functional derivatives $\delta/\delta\phi(\mathbf{x})$ and the sums over discrete indices i by an integral over the spatial coordinate $\mathbf{x}$:

$$i\frac{\partial \Psi[\phi, t]}{\partial t} = -\frac{1}{2}\int d^3\mathbf{x}\frac{\delta^2 \Psi[\phi, t]}{\delta\phi(\mathbf{x})\,\delta\phi(\mathbf{x})} + \frac{1}{2}\int d^3\mathbf{x}\,d^3\mathbf{y}\,M(\mathbf{x}, \mathbf{y})\,\phi(\mathbf{x})\,\phi(\mathbf{y})\,\Psi[\phi, t].$$

For the kernel $M(\mathbf{x}, \mathbf{y})$ given in (4.4), this equation is relativistically invariant although this invariance is not immediately manifest from the form of the equation.

In quantum field theory, the wave functionals and the functional Schrödinger equation are rarely used. We wrote this equation here mainly for illustrative purposes and will proceed to use the Heisenberg picture and the basis of the Hamiltonian eigenstates in the following chapters.

Remark: canonical quantum gravity The use of the wave functional is inevitable in canonical nonperturbative quantum gravity. In this theory the metric is quantized and the role of generalized coordinates of the gravitational field is played by the spatial part of the metric $g_{ik}(\mathbf{x})$. In the coordinate basis, the wave functional Ψ depends on $g_{ik}(\mathbf{x})$ and the matter field configuration $\phi(\mathbf{x})$, that is, $\Psi[g_{ik}, \phi]$. This wave functional is constrained to respect three-dimensional diffeomorphism invariance. In addition, it obeys the Schrödinger-like equation which takes an unusual form,

$$\hat{H}\Psi\left[g_{ik},\phi\right]=0, \tag{4.36}$$

where $\hat{H}$ is a constraint quadratic in momenta conjugate to g_{ik}. This constraint generates the dynamics in classical gravity and therefore it plays the role of the Hamiltonian. One can crudely understand why the "Hamiltonian" vanishes by considering a closed universe where the positive energy of matter is exactly compensated by the "negative energy" of the gravitational field. Equation (4.36) is called the Wheeler–DeWitt equation.

The most remarkable feature of canonical quantum gravity is the disappearance of time in the fundamental theory and hence an explicit breaking of the spacetime diffeomorphism invariance. This invariance and the concept of time are recovered only at the quasiclassical level. Note that the time-dependent Schrödinger equation for the matter fields can be obtained from (4.36) assuming a quasi-classical background metric. The wave functional $\Psi\left[g_{ik},\phi\right]$ describes the availability of the corresponding three-geometries, which are the "building blocks" for the four-dimensional quasiclassical spacetimes.

Unfortunately, at present the canonical quantum gravity remains only a formal scheme. In spite of many years of effort, this scheme has not yielded reliable physical results which could highlight the nonperturbative aspects of quantum geometry.

5

Reminder: classical fields

Summary Action functional in general. Minimal and conformal coupling of scalar field to gravity. Internal symmetries and gauge fields. Gravitational field. The energy-momentum tensor. Conservation laws.

5.1 The action functional

In this book we mainly consider a quantum scalar field interacting with the classical gravitational or electromagnetic fields. To determine the admissible form of their couplings we have to recall the basic principles of classical field theory.

A theory of a classical field $\phi_i(x)$ is based on the action

$$S[\phi] = \int d^4x \, \mathcal{L}(\phi_i, \partial_\mu \phi_i, \ldots) \tag{5.1}$$

where i is an "internal index," $\partial_\mu \phi_i \equiv \partial \phi_i / \partial x^\mu$, and the Lagrangian density $\mathcal{L}$ depends on the field strength and its derivatives. The variable ϕ_i can designate a real or complex scalar field, a vector field, or a tensor field. For instance, in the case of the gravitational field $\phi_i \equiv g_{\alpha\beta}(x^\gamma)$, where $g_{\alpha\beta}$ is the metric. Consideration of fermionic fields is beyond the scope of this book.

Choosing the action functional The action functional is usually chosen in accordance with the following guiding principles:

(i) The action must be real-valued because otherwise the total probability is not conserved in the corresponding quantum theory.

(ii) Local theories have so far been successful in describing experiments and therefore the action is usually taken as a *local functional* of the fields and their derivatives. The Lagrangian density $\mathcal{L}$ is then a *function* of the fields and their derivatives. Otherwise, as for example in the case

$$\int d^4x \, d^4x' \, \phi^{,\mu}(x - x') \phi_{,\mu}(x'),$$

the Lagrangian density is also a functional and the values of the field at separated points x and x' are "coupled."

(iii) Usually it is sufficient to specify the initial conditions for the field itself and at most for its first derivatives in order to unambiguously predict the subsequent evolution of the field. This means that the equations of motion contain derivatives of at most second order and hence we can restrict ourselves to actions which depend only on the field strength and its first derivatives, that is, $\mathcal{L} = \mathcal{L}(\phi_i, \phi_{i,\mu})$.

(iv) In the absence of gravity, the action must be Poincaré-invariant in order to respect the translational and Lorentz invariance of the flat spacetime. This requirement strongly constrains possible Lagrangians. In particular, the Lagrangian density $\mathcal{L}$ cannot depend explicitly on $\mathbf{x}$ or t.

(v) In an arbitrary curved spacetime, the action must be invariant with respect to general coordinate transformations since physical properties of the system are independent of the coordinate system used ("geometrical character" of nature).

(vi) The conservation of different charges is usually related to the existence of some internal symmetries, among which gauge symmetry is of particular importance. In such cases the action must respect these symmetries. For example, the Lagrangian describing an electrically charged complex scalar field must be invariant with respect to $U(1)$ local gauge transformations because otherwise the electric charge would not be conserved. The Standard Model of particle physics is based on $SU(3) \times SU(2) \times U(1)$ group of local gauge transformations.

As we will see shortly, the above requirements usually suffice to fix the coupling of the different fields almost unambiguously.

Equations of motion The requirement that the action takes an extremum value for a classically allowed field configuration $\phi_i(x)$ leads to the Euler–Lagrange equations of motion,

$$\frac{\delta S}{\delta \phi_i(x)} = 0.$$

For a Lagrangian density $\mathcal{L} = \mathcal{L}(\phi_i, \phi_{i,\mu})$, the variation of the action is

$$\delta S = \int d^4x \left(\frac{\partial \mathcal{L}}{\partial \phi_i} - \frac{\partial}{\partial x^\mu} \frac{\partial \mathcal{L}}{\partial \phi_{i,\mu}} \right) \delta \phi_i(x) + O\left([\delta \phi]^2 \right),$$

and so the equations of motion are

$$\frac{\delta S[\phi]}{\delta \phi_i(x)} = \frac{\partial \mathcal{L}}{\partial \phi_i} - \frac{\partial}{\partial x^\mu} \frac{\partial \mathcal{L}}{\partial \phi_{i,\mu}} = 0. \tag{5.2}$$

Here summation over μ is implied and we have assumed that the boundary terms vanish sufficiently rapidly as $|\mathbf{x}| \to \infty$, $|t| \to \infty$.

The formula (5.2) holds for all Lagrangians that depend on the field strength and at most on its first derivatives. If the Lagrangian contains second-order derivatives such as $\phi_{,\mu\nu}$, the corresponding equations of motion are generally of third or fourth order.

5.2 Real scalar field and its coupling to the gravity

The simplest relativistically invariant Lagrangian density for a real scalar field $\phi(x)$ in a flat spacetime takes the form

$$\mathcal{L}\left(\phi, \partial_\mu \phi\right) = \frac{1}{2}\eta^{\mu\nu}\phi_{,\mu}\phi_{,\nu} - V(\phi), \tag{5.3}$$

where $\eta^{\mu\nu} \equiv \mathrm{diag}(1, -1, -1, -1)$ is the Minkowski metric and $V(\phi)$ is a potential that describes the self-interaction of the field. For a *free* (i.e. noninteracting) field of mass m the potential is

$$V(\phi) = \frac{1}{2}m^2\phi^2.$$

Note that a linear term $A\phi$ in the potential can be always removed by a field redefinition $\phi(x) = \tilde{\phi}(x) + \phi_0$.

To generalize the Lagrangian (5.3) to the case of a curved spacetime with an arbitrary metric $g_{\mu\nu}$, we have to:

(i) replace $\eta_{\mu\nu}$ with the metric $g_{\mu\nu}$;
(ii) replace ordinary derivatives by covariant derivatives (note that the first covariant derivative of a scalar function coincides with the ordinary derivative);
(iii) use the covariant volume element $d^4x\sqrt{-g}$, where $g \equiv \det g_{\mu\nu}$, instead of the usual volume element $d^3\mathbf{x}\, dt$.

The resulting action,

$$S = \int d^4x\sqrt{-g}\left[\frac{1}{2}g^{\mu\nu}\phi_{,\mu}\phi_{,\nu} - V(\phi)\right], \tag{5.4}$$

explicitly depends on $g_{\mu\nu}$ and describes a scalar field *minimally coupled* to the gravity. This coupling necessarily follows from the requirement of general covariance.

Remark: covariant volume element To understand the appearance of the factor $\sqrt{|g|}$ in the volume element, let us consider a two-dimensional Euclidean plane covered by curvilinear coordinates $\tilde{x}$, $\tilde{y}$. In these coordinates, the metric $g_{ij}(\tilde{x})$, where $i, j = 1, 2$, is generally different from the Euclidean metric δ_{ij}. Infinitesimal coordinate increments $d\tilde{x}$, $d\tilde{y}$ define an area element corresponding to the infinitesimal parallelogram spanned by the vectors $\mathbf{l}_1 = (d\tilde{x}, 0)$ and $\mathbf{l}_2 = (0, d\tilde{y})$. The length of the vector $\mathbf{l}_1$ is given by

$$|\mathbf{l}_1| = \sqrt{g_{ij}l_1^i l_1^j} = \sqrt{g_{11}}d\tilde{x}.$$

Similarly, we find $|\mathbf{l}_2| = \sqrt{g_{22}}d\tilde{y}$. The scalar product of the vectors $\mathbf{l}_1$ and $\mathbf{l}_2$ is

$$\mathbf{l}_1 \cdot \mathbf{l}_2 = g_{ij} l_1^i l_2^j = g_{12} d\tilde{x} d\tilde{y}.$$

On the other hand, one can express the scalar product $\mathbf{l}_1 \cdot \mathbf{l}_2$ through the angle θ between the vectors (according to the cosine theorem),

$$\mathbf{l}_1 \cdot \mathbf{l}_2 = |\mathbf{l}_1|\,|\mathbf{l}_2| \cos\theta = \sqrt{g_{11} g_{22}}\, d\tilde{x} d\tilde{y} \cos\theta.$$

Comparing the two expressions for $\mathbf{l}_1 \cdot \mathbf{l}_2$, we find that

$$\cos\theta = \frac{g_{12}}{\sqrt{g_{11} g_{22}}},$$

and the infinitesimal area dA of the parallelogram is

$$dA = |\mathbf{l}_1|\,|\mathbf{l}_2| \sin\theta = \sqrt{g_{11} g_{22} - (g_{12})^2}\, d\tilde{x} d\tilde{y} = \sqrt{\det g_{ij}}\, d\tilde{x} d\tilde{y}.$$

In n dimensions this formula is generalized to $dV = d^n x \sqrt{|g(x)|}$. In a four-dimensional spacetime where the metric has the signature $(+,-,-,-)$, the determinant g is always negative and hence the volume element is $d^4 x \sqrt{-g}$.

Nonminimal and conformal couplings The action can in principle contain additional terms which directly couple the fields to the curvature tensor $R_{\mu\nu\rho\sigma}$. Such couplings to the gravity are called *nonminimal* and they violate the strong equivalence principle, which states that *all local* effects of gravity must disappear in the local inertial frame. However, curvature does not vanish in a local inertial frame and hence influences the behavior of fields in theories with nonminimal coupling. However, the only criterion for the legitimacy of a theory is agreement with experiment. Theories violating the strong equivalence principle are allowed as long as they agree with available experiments.

The simplest action for a nonminimally coupled scalar field is

$$S = \int d^4 x \sqrt{-g} \left[\frac{1}{2} g^{\mu\nu} \phi_{,\mu} \phi_{,\nu} - V(\phi) - \frac{\xi}{2} R \phi^2 \right], \tag{5.5}$$

where R is the Ricci curvature scalar and ξ is a constant parameter. The additional term induces a "mass" correction which is proportional to the scalar curvature. With $V = 0$ and $\xi = 1/6$, this theory has an additional symmetry, namely, the action (5.5) is invariant under conformal transformations,

$$g_{\mu\nu} \rightarrow \tilde{g}_{\mu\nu} = \Omega^2(x) g_{\mu\nu}, \tag{5.6}$$

where the conformal factor $\Omega^2(x)$ is an arbitrary function.[1]

[1] Verifying the conformal invariance of the above action takes a fair amount of algebra. We omit the details of this calculation which can be found in Chapter 6 of the book by S. Fulling, *Aspects of Quantum Field Theory in Curved Space-Time* (Cambridge, 1989).

As we will see later, the energy-momentum tensor of a conformally invariant field is traceless in the classical theory. Conformal invariance has also another important aspect: In *conformally flat spacetimes*, where the metric can be written down as $g_{\mu\nu} = \Omega^2(x)\eta_{\mu\nu}$, the influence of the gravitational field on the behavior of the conformally invariant quantum field is greatly simplified. An important example of a conformally flat spacetime is the Friedmann universe.

The equation of motion for the real scalar field ϕ coupled to the gravity follows from the action (5.5),

$$\partial_\alpha \frac{\partial \mathcal{L}}{\partial \phi_{,\alpha}} - \frac{\partial \mathcal{L}}{\partial \phi} = \left(\sqrt{-g} g^{\alpha\beta} \phi_{,\beta}\right)_{,\alpha} + \left(\frac{\partial V}{\partial \phi} + \xi R \phi\right)\sqrt{-g} = 0. \tag{5.7}$$

For the minimally coupled scalar field we have $\xi = 0$.

Equation (5.7) can also be rewritten in a manifestly covariant form

$$\phi_{;\alpha}^{;\alpha} + \frac{\partial V}{\partial \phi} + \xi R \phi = 0, \tag{5.8}$$

where $;\alpha$ denotes the covariant derivative with respect to the coordinate x^α.

5.3 Gauge invariance and coupling to the electromagnetic field

A real scalar field is electrically neutral and does not couple to the electromagnetic field. The conservation of the electric charge is related to the existence of internal symmetry associated with $U(1)$ gauge transformations. These transformations are implemented as multiplication of the field by $\exp(i\alpha)$, where α is an arbitrary real constant. Therefore they cannot be realized for the real scalar field and the electrically charged scalar field is necessarily described by the complex variable φ, or equivalently, by two real scalar fields.

The action

$$S[\phi] = \int d^4x \sqrt{-g}\left[\frac{1}{2} g^{\alpha\beta} \varphi_{,\alpha}\varphi^*_{,\beta} - V(\varphi\varphi^*)\right] \tag{5.9}$$

is clearly invariant with respect to the *global gauge transformation*

$$\varphi(x) \to \tilde{\varphi}(x) = e^{i\alpha}\varphi(x), \tag{5.10}$$

where α is a real constant, i.e. α is the same number for all points in the spacetime. This explains why transformation (5.10) is called global.

The minimal coupling of the charged scalar field with the electromagnetic field is unambiguously determined by the requirement of the *local* gauge invariance. Generalizing (5.10) to a *local gauge transformation*,

$$\varphi(x) \to \tilde{\varphi}(x) = e^{i\alpha(x)}\varphi(x), \tag{5.11}$$

where $\alpha(x)$ an arbitrary function of the spacetime coordinates, we find that

$$\varphi_{,\mu}(x) \to \tilde{\varphi}_{,\mu}(x) = e^{i\alpha(x)}\left(\varphi_{,\mu} + i\alpha_{,\mu}\varphi\right).$$

Hence, action (5.9) is not invariant under local gauge transformations. To regain its local gauge invariance we are forced to introduce an additional vector field A_μ called the *gauge field*, which compensates the extra term in the derivative $\varphi_{,\mu}$. (In the present case the field A_μ is interpreted as the electromagnetic field.) Then, replacing the ordinary derivatives ∂_μ by the gauge covariant derivatives D_μ,

$$\varphi_{,\mu} \to D_\mu \varphi \equiv \varphi_{,\mu} + iA_\mu \varphi, \tag{5.12}$$

and postulating for the field A^μ the transformation law

$$A_\mu \to \tilde{A}_\mu \equiv A_\mu - \alpha_{,\mu}, \tag{5.13}$$

we find

$$\tilde{D}_\mu \tilde{\varphi} = \left(\partial_\mu + i\tilde{A}_\mu\right)\left(e^{i\alpha(x)}\varphi\right) = e^{i\alpha(x)} D_\mu \varphi.$$

It follows that the action

$$S = \int d^4x \sqrt{-g}\left[\frac{1}{2} g^{\alpha\beta} \left(D_\alpha \varphi\right)\left(D_\beta \varphi\right)^* - V\left(\varphi\varphi^*\right)\right] \tag{5.14}$$

$$= \int d^4x \sqrt{-g}\left[\frac{1}{2} g^{\alpha\beta} \varphi_{,\alpha}\varphi^*_{,\beta} - V\left(\varphi\varphi^*\right) + \frac{1}{2} g^{\alpha\beta} iA_\alpha \left(\varphi\varphi^*_{,\beta} - \varphi^*\varphi_{,\beta}\right) + \frac{1}{2} g^{\alpha\beta} A_\alpha A_\beta \varphi\varphi^*\right] \tag{5.15}$$

is invariant under local gauge transformations (5.11)–(5.13). This action describes the minimal coupling of the charged scalar field to the electromagnetic field and to gravity. In the following chapters we consider the behavior of a quantum scalar field in the presence of the classical gravitational and electromagnetic fields.

5.4 Action for the gravitational and gauge fields

In the second part of the book we will need the action for the classical gravitational and electromagnetic fields themselves. The simplest possible action for the gravitational field is the *Einstein–Hilbert action*:

$$S^{\text{grav}} = -\frac{1}{16\pi G} \int d^4x \sqrt{-g}(R + 2\Lambda), \tag{5.16}$$

where G is Newton's gravitational constant, R is the Ricci scalar curvature and Λ is a constant parameter (the cosmological constant). The Einstein equations are obtained by extremizing this action with respect to $g^{\alpha\beta}$.

Exercise 5.1
Derive the Einstein equations in the absence of matter and cosmological constant using the *Palatini method:* assume that the metric $g_{\mu\nu}$ and the Christoffel symbol $\Gamma^{\mu}_{\alpha\beta}$ are independent and vary the action with respect to both of them.

Hint: The expression for the Ricci tensor in terms of $g_{\mu\nu}$ and $\Gamma^{\mu}_{\alpha\beta}$ is

$$R = g^{\alpha\beta} R_{\alpha\beta} = g^{\alpha\beta}\left(\partial_{\mu}\Gamma^{\mu}_{\alpha\beta} - \partial_{\beta}\Gamma^{\mu}_{\alpha\mu} + \Gamma^{\mu}_{\alpha\beta}\Gamma^{\nu}_{\mu\nu} - \Gamma^{\nu}_{\alpha\mu}\Gamma^{\mu}_{\beta\nu}\right). \tag{5.17}$$

First find the variation of $R\sqrt{-g}$ with respect to Γ and assuming that $\Gamma^{\mu}_{\alpha\beta}$ is symmetric in α,β establish the standard relation between Γ and g,

$$\Gamma^{\mu}_{\alpha\beta} = \frac{1}{2} g^{\mu\nu}\left(g_{\alpha\nu,\beta} + g_{\beta\nu,\alpha} - g_{\alpha\beta,\nu}\right). \tag{5.18}$$

Then compute the variation of $R\sqrt{-g}$ with respect to $g^{\alpha\beta}$ and obtain the vacuum Einstein equation

$$\frac{\delta S^{\mathrm{grav}}}{\delta g^{\alpha\beta}} = -\frac{\sqrt{-g}}{16\pi G}\left(R_{\alpha\beta} - \frac{1}{2} g_{\alpha\beta} R\right) = 0. \tag{5.19}$$

Remark: higher-derivative gravity The Einstein equations are the only possible second-order covariant equations for the gravitational field. Any modification of the Einstein–Hilbert action in four dimensions leads to a higher-derivative gravity. At present, the Einstein theory is in a very good agreement with experiments. However, it is likely that this theory breaks down in regions with an extremely strong gravitational field, where the curvature radius is comparable to the Planck length. In this case, the correct gravitational action must contain additional terms quadratic in the curvature, such as R^2, $R_{\mu\nu}R^{\mu\nu}$, or $R_{\mu\nu\rho\sigma}R^{\mu\nu\rho\sigma}$, and the Einstein equations are modified by higher derivative terms. The R^2 terms can be of fundamental origin or, as we will see later, can arise due to vacuum polarization effects.

The simplest action describing the dynamics of the electromagnetic field A_{μ} itself can be easily built out of the gauge invariant *field strength*

$$F_{\mu\nu} \equiv A_{\nu;\mu} - A_{\mu;\nu} = A_{\nu,\mu} - A_{\mu,\nu}, \tag{5.20}$$

and the result is

$$S[A_{\mu}] = -\frac{1}{16\pi}\int d^4x \sqrt{-g} g^{\alpha\beta} g^{\mu\nu} F_{\alpha\mu} F_{\beta\nu}. \tag{5.21}$$

It also describes the coupling of the electromagnetic field to gravity. Action (5.21) is conformally invariant: under the conformal transformation (5.6) the factors $\sqrt{-g}$ and $g^{\alpha\beta}$ are multiplied by Ω^4 and Ω^{-2} respectively, resulting in no net change in the action. Therefore the evolution of the electromagnetic field in conformally flat spacetimes is greatly simplified. In particular, the gravitational field in the Friedmann universe does not produce any photons.

5.5 Energy-momentum tensor

The total action describing gravity coupled to the matter fields ϕ_i can be written as

$$S\left[\phi_i, g_{\mu\nu}\right] = S^{\text{grav}}\left[g_{\mu\nu}\right] + S^m\left[\phi_i, g_{\mu\nu}\right].$$

The equations of motion for the gravitational field are obtained by varying this action with respect to the metric:

$$\frac{\delta S\left[\phi_i, g_{\mu\nu}\right]}{\delta g^{\alpha\beta}} = \frac{\delta}{\delta g^{\alpha\beta}} S^{\text{grav}}\left[g_{\mu\nu}\right] + \frac{\delta}{\delta g^{\alpha\beta}} S^m\left[\phi_i, g_{\mu\nu}\right] = 0. \tag{5.22}$$

These equations must coincide with the Einstein equations,

$$G_{\alpha\beta} \equiv R_{\alpha\beta} - \frac{1}{2} g_{\alpha\beta} R = 8\pi G T_{\alpha\beta}, \tag{5.23}$$

where $T_{\alpha\beta}$ is the energy-momentum tensor of the matter fields ϕ_i. Taking into account (5.23) we find that the equations (5.22) and (5.23) are consistent only if

$$T_{\alpha\beta} = \frac{2}{\sqrt{-g}} \frac{\delta S^m}{\delta g^{\alpha\beta}}. \tag{5.24}$$

Thus equation (5.24) can be viewed as a *definition* of the energy-momentum tensor (EMT) for the matter fields. The resulting tensor $T_{\alpha\beta}$ is automatically symmetric and covariantly conserved,

$$T^{\alpha}_{\beta;\alpha} = 0. \tag{5.25}$$

Example 5.1 The energy-momentum tensor of the minimally coupled scalar field ϕ with the action (5.4) is

$$T_{\alpha\beta}(x) = \frac{2}{\sqrt{-g}} \frac{\delta S}{\delta g^{\alpha\beta}(x)} = \phi_{,\alpha}\phi_{,\beta} - g_{\alpha\beta}\left[\frac{1}{2} g^{\mu\nu}\phi_{,\mu}\phi_{,\nu} - V(\phi)\right]. \tag{5.26}$$

Conservation of the EMT The covariant conservation of the EMT is a consequence of the invariance of the action with respect to general coordinate transformations. Considering an infinitesimal coordinate transformation

$$x^\alpha \to \tilde{x}^\alpha = x^\alpha + \xi^\alpha(x), \tag{5.27}$$

we find that the metric transforms as

$$g^{\alpha\beta}(x) \to \tilde{g}^{\alpha\beta}(x) = g^{\alpha\beta}(x) + \xi^{\alpha;\beta} + \xi^{\beta;\alpha} + O\left(|\xi|^2\right).$$

Note that $g^{\alpha\beta}(x)$ and $\tilde{g}^{\alpha\beta}(x)$ refer to different points of the manifold which have the same coordinate values in two different coordinate systems x and $\tilde{x}$. The specific form of the transformation law for the matter field,

$$\phi_i(x) \to \tilde{\phi}_i(x) = \phi_i(x) + \delta\phi_i(x),$$

depends on the type of the field. For example, for the real scalar field

$$\phi(x) \to \tilde{\phi}(x) = \phi(x) - \phi_{,\beta}\xi^{\beta}. \tag{5.28}$$

The action does not depend on the coordinate system and hence its variation under the coordinate transformation (5.27) must vanish, that is,

$$\delta S^m = \int \frac{\delta S^m}{\delta g^{\alpha\beta}(x)} \left(\xi^{\alpha;\beta} + \xi^{\beta;\alpha}\right) d^4x + \int \frac{\delta S^m}{\delta \phi_i(x)} \delta\phi_i(x) d^4x = 0. \tag{5.29}$$

Taking into account (5.24), one obtains

$$\int \frac{\delta S^m}{\delta g^{\alpha\beta}(x)} \left(\xi^{\alpha;\beta} + \xi^{\beta;\alpha}\right) d^4x = \int T_{\alpha\beta}\xi^{\beta;\alpha}\sqrt{-g}d^4x$$

$$= \int \left[\left(T^{\alpha}_{\beta}\xi^{\beta}\right)_{;\alpha} - T^{\alpha}_{\beta;\alpha}\xi^{\beta}\right]\sqrt{-g}d^4x = -\int T^{\alpha}_{\beta;\alpha}\xi^{\beta}\sqrt{-g}d^4x, \tag{5.30}$$

where we have assumed that ξ^{α} vanishes sufficiently quickly at infinity. Considering the real scalar field ϕ and substituting (5.30) and (5.28) into (5.29), we find that $\delta S^m = 0$ only if

$$T^{\alpha}_{\beta;\alpha} + \frac{\delta S^m}{\delta \phi}\phi_{,\beta} = 0. \tag{5.31}$$

It follows that the energy-momentum tensor is covariantly conserved, $T^{\alpha}_{\beta;\alpha} = 0$, if the equations of motion

$$\frac{\delta S^m}{\delta \phi} = 0 \tag{5.32}$$

are satisfied.

Moreover, one can easily see from (5.31) that the covariant conservation of T^{α}_{β} is equivalent to the equations of motion (5.32). The Einstein tensor $G_{\alpha\beta}$ defined in (5.23) satisfies the Bianchi identities $G^{\alpha}_{\beta;\alpha} = 0$ and hence the conservation of the energy-momentum tensor is incorporated in the Einstein equations. Therefore the equations of motion for matter do not need to be postulated separately.

If the action for the matter field is invariant with respect to the conformal transformations (5.6) then the trace of the corresponding EMT vanishes. In fact, considering an infinitesimal conformal variation of the metric,

$$\delta g^{\alpha\beta} = g^{\alpha\beta}\delta\Theta, \tag{5.33}$$

where $\delta\Theta(x)$ is an arbitrary function, we find that

$$\delta S^m = \int \frac{\delta S^m}{\delta g^{\alpha\beta}} \delta g^{\alpha\beta} d^4x = \int \frac{\sqrt{-g}}{2} T_{\alpha\beta} g^{\alpha\beta} \delta\Theta d^4x = \int \frac{\sqrt{-g}}{2} T^\alpha_\alpha \delta\Theta d^4x,$$

and hence $T^\alpha_\alpha = 0$ if S^m is invariant with respect to transformation (5.33). We will show in the second part of the book that the vacuum polarization effects generally spoil conformal invariance of the original classical theory and generate a nonzero value of T^α_α. This phenomenon is called the *trace anomaly.*

6

Quantum fields in expanding universe

Summary Quantization of a scalar field in a Friedmann universe. Bogolyubov transformations. Choice of the vacuum state. Particle production. Correlation functions. Amplitude of quantum fluctuations.

Scalar field quantization is tremendously simplified in an isotropic and homogeneous expanding universe. Homogeneity and isotropy dictate a preferable choice of the spacetime foliation and a preferable time parametrization. In this chapter we will consider a minimally coupled quantum scalar field and study how its state changes in a homogeneous and isotropic gravitational background.

6.1 Classical scalar field in expanding background

For simplicity, we consider only the case of the spatially flat Friedmann universe. In the preferred coordinate system where the symmetries of the spacetime are manifest, the metric takes the form

$$ds^2 = dt^2 - a^2(t)\delta_{ik}dx^i dx^k. \tag{6.1}$$

Note that although the three-dimensional surfaces of constant time are flat, the spacetime is nevertheless curved. It is convenient to introduce the *conformal time*

$$\eta(t) \equiv \int^t \frac{dt}{a(t)},$$

instead of the physical time t. With this new coordinate interval (6.1) is

$$ds^2 = a^2(\eta)\left[d\eta^2 - \delta_{ik}dx^i dx^k\right] = a^2(\eta)\eta_{\mu\nu}dx^\mu dx^\nu, \tag{6.2}$$

and it is obvious that the metric is conformally equivalent to the Minkowski metric $\eta_{\mu\nu}$.

A real minimally coupled massive scalar field $\phi(x)$ in a curved spacetime is described by the action

$$S=\frac{1}{2}\int\sqrt{-g}d^4x\left[g^{\alpha\beta}\phi_{,\alpha}\phi_{,\beta}-m^2\phi^2\right]. \quad (6.3)$$

With the substitution $g^{\alpha\beta}=a^{-2}\eta^{\alpha\beta}$ and $\sqrt{-g}=a^4$ this action becomes

$$S=\frac{1}{2}\int d^3\mathbf{x}\,d\eta\,a^2\left[\phi'^2-(\nabla\phi)^2-m^2a^2\phi^2\right], \quad (6.4)$$

where the prime $'$ denotes derivatives with respect to the conformal time η. Moreover, introducing the auxiliary field

$$\chi\equiv a(\eta)\phi, \quad (6.5)$$

we can rewrite action (6.4) in terms of χ as

$$S=\frac{1}{2}\int d^3\mathbf{x}\,d\eta\left[\chi'^2-(\nabla\chi)^2-\left(m^2a^2-\frac{a''}{a}\right)\chi^2\right], \quad (6.6)$$

where the total derivative terms were omitted.

Exercise 6.1
Derive (6.6) from (6.4).

Variation of the action (6.6) with respect to χ gives the following equation of motion,

$$\chi''-\Delta\chi+\left(m^2a^2-\frac{a''}{a}\right)\chi=0. \quad (6.7)$$

Comparing (6.7) with (4.7), we find that the field χ obeys the same equation of motion as a massive scalar field in Minkowski spacetime, except that the *effective mass*,

$$m^2_{\text{eff}}(\eta)\equiv m^2a^2-\frac{a''}{a}, \quad (6.8)$$

becomes time-dependent. This time dependence of the effective mass accounts for the interaction of the scalar field with the gravitational background.

Thus, the problem of quantization of a scalar field ϕ in a flat Friedmann universe is reduced to the mathematically equivalent problem of quantization of a free scalar field χ in Minkowski spacetime. All information about the influence of the gravitational field on ϕ is encapsulated in the time-dependent mass $m_{\text{eff}}(\eta)$. Note that action (6.6) is explicitly time-dependent, so the energy of the scalar field χ is not conserved. In quantum theory this leads to particle creation; the energy for new particles is supplied by the classical gravitational field.

6.1.1 Mode expansion

Expanding the field χ in Fourier modes,

$$\chi(\mathbf{x}, \eta) = \int \frac{d^3\mathbf{k}}{(2\pi)^{3/2}} \chi_{\mathbf{k}}(\eta) e^{i\mathbf{k}\cdot\mathbf{x}}, \tag{6.9}$$

and substituting this expansion into (6.7) we find that the Fourier modes $\chi_{\mathbf{k}}(\eta)$ satisfy a set of decoupled ordinary differential equations

$$\chi''_{\mathbf{k}} + \omega_k^2(\eta)\chi_{\mathbf{k}} = 0, \tag{6.10}$$

where

$$\omega_k^2(\eta) \equiv k^2 + m_{\text{eff}}^2(\eta) = k^2 + m^2 a^2(\eta) - \frac{a''}{a}. \tag{6.11}$$

Because the field $\chi(\mathbf{x}, \eta)$ is real, i.e. $\chi^*(\mathbf{x}, \eta) = \chi(\mathbf{x}, \eta)$, the complex Fourier modes $\chi_{\mathbf{k}}(\eta)$ must satisfy the condition

$$\chi_{\mathbf{k}}^*(\eta) = \chi_{-\mathbf{k}}(\eta). \tag{6.12}$$

Since $\omega_k^2(\eta)$ in equation (6.10) depends only on $k \equiv |\mathbf{k}|$, its general solution can be written as

$$\chi_{\mathbf{k}}(\eta) = \frac{1}{\sqrt{2}} \left[a_{\mathbf{k}}^- v_k^*(\eta) + a_{-\mathbf{k}}^+ v_k(\eta) \right]. \tag{6.13}$$

Here $v_k(\eta)$ and its complex conjugate $v_k^*(\eta)$ are two linearly independent solutions of (6.10) which are the same for all Fourier modes with the given magnitude of the wavevector $k \equiv |\mathbf{k}|$, and $a_{\mathbf{k}}^\pm$ are two complex constants of integration that can depend also on the direction of vector $\mathbf{k}$. The index $-\mathbf{k}$ in the second term and the factor $1/\sqrt{2}$ are chosen for later convenience. Since χ is real, it follows from (6.12) that $a_{\mathbf{k}}^+ = (a_{\mathbf{k}}^-)^*$.

Exercise 6.2

Verify that if v_k satisfies (6.10), then the *Wronskian*

$$W[v_k, v_k^*] \equiv v_k' v_k^* - v_k v_k^{*\prime} = 2i \operatorname{Im}(v' v^*) \tag{6.14}$$

is time-independent. Show that $W[v_k, v_k^*] \neq 0$ if and only if v_k and v_k^* are linearly independent solutions. Verify that the coefficients $a_{\mathbf{k}}^\pm$ can be expressed in terms of $\chi_{\mathbf{k}}(\eta)$ and $v_k(\eta)$ as:

$$a_{\mathbf{k}}^- = \sqrt{2} \frac{v_k' \chi_{\mathbf{k}} - v_k \chi_{\mathbf{k}}'}{v_k' v_k^* - v_k v_k^{*\prime}} = \sqrt{2} \frac{W[v_k, \chi_{\mathbf{k}}]}{W[v_k, v_k^*]}; \quad a_{\mathbf{k}}^+ = (a_{\mathbf{k}}^-)^*. \tag{6.15}$$

Note that the numerators and denominators in (6.15) are time-independent.

When $v(\eta)$ is multiplied by a constant, $v(\eta) \to \lambda v(\eta)$, the Wronskian $W[v, v^*]$ scales as $|\lambda|^2$. Therefore if $W \neq 0$ we can always normalize v_k in such a way that

$\text{Im}(\dot{v}v^*) = 1$. In this case a complex solution $v_k(\eta)$ is called a *mode function.* It follows from the results of Exercise 6.2 that $v_k(\eta)$ and $v_k^*(\eta)$ are linearly independent.

Substituting (6.13) into (6.9) we find

$$\chi(\mathbf{x},\eta) = \frac{1}{\sqrt{2}}\int \frac{d^3\mathbf{k}}{(2\pi)^{3/2}}\left[a_{\mathbf{k}}^- v_k^*(\eta) + a_{-\mathbf{k}}^+ v_k(\eta)\right] e^{i\mathbf{k}\cdot\mathbf{x}}$$

$$= \frac{1}{\sqrt{2}}\int \frac{d^3\mathbf{k}}{(2\pi)^{3/2}}\left[a_{\mathbf{k}}^- v_k^*(\eta) e^{i\mathbf{k}\cdot\mathbf{x}} + a_{\mathbf{k}}^+ v_k(\eta) e^{-i\mathbf{k}\cdot\mathbf{x}}\right], \tag{6.16}$$

where in the second term the integration variable $\mathbf{k}$ was changed from $\mathbf{k}$ to $-\mathbf{k}$ to make the integrand a manifestly real expression.

Remark: isotropic mode functions In equation (6.13) we expressed all Fourier modes $\chi_{\mathbf{k}}(\eta)$ with a given $|\mathbf{k}| = k$ in terms of the *same* mode function $v_k(\eta)$. This *isotropic* choice of the mode functions is possible because of the isotropy of the Friedmann universe where ω_k depends only on $k = |\mathbf{k}|$.

6.2 Quantization

The field χ is quantized by imposing the standard equal-time commutation relations on the field operator $\hat{\chi}$ and its canonically conjugate momentum $\hat{\pi} \equiv \hat{\chi}'$,

$$[\hat{\chi}(\mathbf{x},\eta), \hat{\pi}(\mathbf{y},\eta)] = i\delta(\mathbf{x}-\mathbf{y}); \tag{6.17}$$

$$[\hat{\chi}(\mathbf{x},t), \hat{\chi}(\mathbf{y},t)] = [\hat{\pi}(\mathbf{x},t), \hat{\pi}(\mathbf{y},t)] = 0. \tag{6.18}$$

The Hamiltonian for the quantum field $\hat{\chi}$ is

$$\hat{H}(\eta) = \frac{1}{2}\int d^3\mathbf{x}\left[\hat{\pi}^2 + (\nabla\hat{\chi})^2 + m_{\text{eff}}^2(\eta)\hat{\chi}^2\right]. \tag{6.19}$$

The creation and annihilation operators $\hat{a}_{\mathbf{k}}^\pm$ can be introduced via the mode operators $\hat{\chi}_{\mathbf{k}}$ as in Chapter 4. However, a quicker way is to begin directly with the mode expansion (6.16) considering the constants of integration $a_{\mathbf{k}}^\pm$ as operators $\hat{a}_{\mathbf{k}}^\pm$. Then the field operator $\hat{\chi}$ is expanded as

$$\hat{\chi}(\mathbf{x},\eta) = \frac{1}{\sqrt{2}}\int \frac{d^3\mathbf{k}}{(2\pi)^{3/2}}\left(e^{i\mathbf{k}\cdot\mathbf{x}} v_k^*(\eta)\hat{a}_{\mathbf{k}}^- + e^{-i\mathbf{k}\cdot\mathbf{x}} v_k(\eta)\hat{a}_{\mathbf{k}}^+\right), \tag{6.20}$$

where the mode functions $v_k(\eta)$ obey the equations

$$v_k'' + \omega_k^2(\eta) v_k = 0, \quad \omega_k(\eta) \equiv \sqrt{k^2 + m_{\text{eff}}^2(\eta)}, \tag{6.21}$$

and satisfy the normalization condition

$$\text{Im}\left(v_k' v_k^*\right) = \frac{v_k' v_k^* - v_k v_k^{*\prime}}{2i} = 1. \tag{6.22}$$

Substituting (6.20) into (6.17) and taking into account (6.22) we find that the operators $\hat{a}_{\mathbf{k}}^{\pm}$ satisfy the commutation relations

$$\left[\hat{a}_{\mathbf{k}}^{-}, \hat{a}_{\mathbf{k}'}^{+}\right] = \delta\left(\mathbf{k} - \mathbf{k}'\right), \quad \left[\hat{a}_{\mathbf{k}}^{-}, \hat{a}_{\mathbf{k}'}^{-}\right] = \left[\hat{a}_{\mathbf{k}}^{+}, \hat{a}_{\mathbf{k}'}^{+}\right] = 0, \tag{6.23}$$

and thus can be interpreted as the creation and annihilation operators.

Exercise 6.3
Verify this last statement.

Remark: complex scalar field If χ were a complex field, then in general $\chi_{\mathbf{k}}^{*} \neq \chi_{-\mathbf{k}}$ and the mode expansion would be written as

$$\hat{\chi}(\mathbf{x}, \eta) = \frac{1}{\sqrt{2}} \int \frac{d^3\mathbf{k}}{(2\pi)^{3/2}} \left(e^{i\mathbf{k}\cdot\mathbf{x}} v_k^{*}(\eta) \hat{a}_{\mathbf{k}}^{-} + e^{-i\mathbf{k}\cdot\mathbf{x}} v_k(\eta) \hat{b}_{\mathbf{k}}^{+} \right),$$

where the operators $\hat{a}_{\mathbf{k}}^{-}$ and $\hat{b}_{\mathbf{k}}^{+}$ are independent. In this case we have two sets of creation and annihilation operators, $\hat{a}_{\mathbf{k}}^{\pm}$ and $\hat{b}_{\mathbf{k}}^{\pm}$, satisfying $(\hat{a}_{\mathbf{k}}^{-})^{\dagger} = \hat{a}_{\mathbf{k}}^{+}$ and $(\hat{b}_{\mathbf{k}}^{-})^{\dagger} = \hat{b}_{\mathbf{k}}^{+}$. The operators $\hat{a}_{\mathbf{k}}^{+}$ and $\hat{b}_{\mathbf{k}}^{+}$ create the particles and antiparticles respectively. This agrees with the picture that a complex scalar field describes particles which are different from their antiparticles. For a real scalar field particles are their own antiparticles.

6.3 Bogolyubov transformations

The operators $\hat{a}_{\mathbf{k}}^{\pm}$ can be used to construct the basis of quantum states in the Hilbert space. However, the corresponding states acquire an unambiguous physical interpretation only after the particular mode functions $v_k(\eta)$ are selected. The normalization condition (6.22) is not enough to completely specify the complex solutions $v_k(\eta)$ of the second-order differential equation (6.21).

In fact, the functions

$$u_k(\eta) = \alpha_k v_k(\eta) + \beta_k v_k^{*}(\eta), \tag{6.24}$$

where α_k and β_k are time-independent complex coefficients, also satisfy equation (6.21). Moreover, if the coefficients α_k and β_k obey the condition

$$|\alpha_k|^2 - |\beta_k|^2 = 1, \tag{6.25}$$

then the functions $u_k(\eta)$ satisfy the normalization condition (6.22) and therefore they can be used as the mode functions instead of $v_k(\eta)$.

Exercise 6.4
Verify that if (6.25) is satisfied then $\mathrm{Im}(u_k' u_k^{*}) = 1$.

In terms of the mode functions $u_k(\eta)$ the field operator expansion takes the following form,

$$\hat{\chi}(\mathbf{x}, \eta) = \frac{1}{\sqrt{2}} \int \frac{d^3\mathbf{k}}{(2\pi)^{3/2}} \left[e^{i\mathbf{k}\cdot\mathbf{x}} u_k^{*}(\eta) \hat{b}_{\mathbf{k}}^{-} + e^{-i\mathbf{k}\cdot\mathbf{x}} u_k(\eta) \hat{b}_{\mathbf{k}}^{+} \right], \tag{6.26}$$

where $\hat{b}_{\mathbf{k}}^{\pm}$ is another set of the creation and annihilation operators satisfying the standard commutation relations (6.23). To determine how the operators $\hat{b}_{\mathbf{k}}^{\pm}$ are related to $\hat{a}_{\mathbf{k}}^{\pm}$, we note that the two expressions (6.20) and (6.26) for the same field operator $\hat{\chi}(\mathbf{x}, \eta)$ agree only if

$$e^{i\mathbf{k}\cdot\mathbf{x}}\left[u_k^*(\eta)\hat{b}_{\mathbf{k}}^- + u_k(\eta)\hat{b}_{-\mathbf{k}}^+\right] = e^{i\mathbf{k}\cdot\mathbf{x}}\left[v_k^*(\eta)\hat{a}_{\mathbf{k}}^- + v_k(\eta)\hat{a}_{-\mathbf{k}}^+\right].$$

Substituting here the expression for u_k in terms of v_k from (6.24), we find the following relation between the operators $\hat{a}_{\mathbf{k}}^{\pm}$ and $\hat{b}_{\mathbf{k}}^{\pm}$:

$$\hat{a}_{\mathbf{k}}^- = \alpha_k^*\hat{b}_{\mathbf{k}}^- + \beta_k\hat{b}_{-\mathbf{k}}^+, \quad \hat{a}_{\mathbf{k}}^+ = \alpha_k\hat{b}_{\mathbf{k}}^+ + \beta_k^*\hat{b}_{-\mathbf{k}}^-. \tag{6.27}$$

The above relations are called the *Bogolyubov transformation*. One can reverse (6.27) to obtain

$$\hat{b}_{\mathbf{k}}^- = \alpha_k\hat{a}_{\mathbf{k}}^- - \beta_k\hat{a}_{-\mathbf{k}}^+, \quad \hat{b}_{\mathbf{k}}^+ = \alpha_k^*\hat{a}_{\mathbf{k}}^+ - \beta_k^*\hat{a}_{-\mathbf{k}}^-. \tag{6.28}$$

The *Bogolyubov coefficients* α_k and β_k can in turn be expressed in terms of the mode functions $v_k(\eta)$ and $u_k(\eta)$. From the relations

$$u_k(\eta) = \alpha_k v_k(\eta) + \beta_k v_k^*(\eta),$$
$$u_k'(\eta) = \alpha_k v_k'(\eta) + \beta_k v_k^{*\prime}(\eta),$$

we find

$$\alpha_k = \frac{W(u_k, v_k^*)}{2i}, \quad \beta_k = \frac{W(v_k, u_k)}{2i}, \tag{6.29}$$

where W is the Wronskian.

6.4 Hilbert space; "*a*- and *b*-particles"

Both sets of the operators $\hat{a}_{\mathbf{k}}^{\pm}$ and $\hat{b}_{\mathbf{k}}^{\pm}$ can be used to build orthonormal bases in the Hilbert space. The two different "vacuum vectors" $\left|_{(a)}0\right\rangle$ and $\left|_{(b)}0\right\rangle$ can be defined in the standard way,

$$\hat{a}_{\mathbf{k}}^-\left|_{(a)}0\right\rangle = 0, \quad \hat{b}_{\mathbf{k}}^-\left|_{(b)}0\right\rangle = 0,$$

for all $\mathbf{k}$. We call them the "*a*-vacuum" and the "*b*-vacuum" respectively. Two sets of excited states,

$$\left|_{(a)}m_{\mathbf{k}_1}, n_{\mathbf{k}_2}, \ldots\right\rangle \equiv \frac{1}{\sqrt{m!n!\ldots}}\left[\left(\hat{a}_{\mathbf{k}_1}^+\right)^m\left(\hat{a}_{\mathbf{k}_2}^+\right)^n\ldots\right]\left|_{(a)}0\right\rangle \tag{6.30}$$

and

$$\left|_{(b)}m_{\mathbf{k}_1}, n_{\mathbf{k}_2}, \ldots\right\rangle \equiv \frac{1}{\sqrt{m!n!\ldots}}\left[\left(\hat{b}_{\mathbf{k}_1}^+\right)^m\left(\hat{b}_{\mathbf{k}_2}^+\right)^n\ldots\right]\left|_{(b)}0\right\rangle, \tag{6.31}$$

describe the "a- and b-particles" respectively. An arbitrary quantum state $|\psi\rangle$ can be written as a linear combination of the excited states,

$$|\psi\rangle = \sum_{m,n,\ldots} C^{(a)}_{mn\ldots} \left|{}_{(a)}m_{\mathbf{k}_1}, n_{\mathbf{k}_2}, \ldots\right\rangle = \sum_{m,n,\ldots} C^{(b)}_{mn\ldots} \left|{}_{(b)}m_{\mathbf{k}_1}, n_{\mathbf{k}_2}, \ldots\right\rangle,$$

and the probability to find m "a-particles" in the mode $\mathbf{k}_1$, etc. is $\left|C^{(a)}_{mn\ldots}\right|^2$. For the "$b$-particles" the corresponding probabilities are given by $\left|C^{(b)}_{mn\ldots}\right|^2$.

The b-states are in general different from the a-states. In particular, if $\beta_k \neq 0$ then the "b-vacuum" contains "a-particles." To verify this let us calculate the expectation value of the a-particle number operator $\hat{N}^{(a)}_{\mathbf{k}} = \hat{a}^+_{\mathbf{k}} \hat{a}^-_{\mathbf{k}}$ in the state $\left|{}_{(b)}0\right\rangle$. Using (6.27) we obtain

$$\begin{aligned}\left\langle{}_{(b)}0\right| \hat{N}^{(a)}_{\mathbf{k}} \left|{}_{(b)}0\right\rangle &= \left\langle{}_{(b)}0\right| \hat{a}^+_{\mathbf{k}} \hat{a}^-_{\mathbf{k}} \left|{}_{(b)}0\right\rangle \\ &= \left\langle{}_{(b)}0\right| \left(\alpha_k \hat{b}^+_{\mathbf{k}} + \beta^*_k \hat{b}^-_{-\mathbf{k}}\right)\left(\alpha^*_k \hat{b}^-_{\mathbf{k}} + \beta_k \hat{b}^+_{-\mathbf{k}}\right)\left|{}_{(b)}0\right\rangle \\ &= \left\langle{}_{(b)}0\right| \left(\beta^*_k \hat{b}^-_{-\mathbf{k}}\right)\left(\beta_k \hat{b}^+_{-\mathbf{k}}\right)\left|{}_{(b)}0\right\rangle = |\beta_k|^2\, \delta^{(3)}(0). \end{aligned} \tag{6.32}$$

The divergent factor $\delta^{(3)}(0)$ accounts for an infinite spatial volume and hence the mean *density* of the a-particles in the mode $\mathbf{k}$ is

$$n_{\mathbf{k}} = |\beta_k|^2 .$$

The total mean density of all particles,

$$n = \int d^3\mathbf{k}\, |\beta_k|^2 ,$$

is finite only if $|\beta_k|^2$ decays faster than k^{-3} for large k.

The "b-vacuum" can be expressed as a superposition of excited a-particle states as (see Exercise 6.5)

$$\begin{aligned}\left|{}_{(b)}0\right\rangle &= \left[\prod_{\mathbf{k}} \frac{1}{|\alpha_k|^{1/2}} \exp\left(\frac{\beta_k}{2\alpha_k} \hat{a}^+_{\mathbf{k}} \hat{a}^+_{-\mathbf{k}}\right)\right] \left|{}_{(a)}0\right\rangle \\ &= \prod_{\mathbf{k}} \frac{1}{|\alpha_k|^{1/2}} \left(\sum_{n=0}^{\infty} \left(\frac{\beta_k}{2\alpha_k}\right)^n \left|{}_{(a)}n_{\mathbf{k}}, n_{-\mathbf{k}}\right\rangle\right). \end{aligned} \tag{6.33}$$

Because of the isotropy the particles come in pairs with opposite momenta $\mathbf{k}$ and $-\mathbf{k}$.

Exercise 6.5
Derive (6.33).

Quantum states defined by an exponential of a quadratic combination of creation operators acting on the vacuum, as in (6.33), are called *squeezed* states. The "b-vacuum" is therefore a squeezed state with respect to the "a-vacuum." Similarly, the "a-vacuum" is a squeezed state with respect to the "b-vacuum."

The "b-vacuum" is normalized by the infinite product $\prod_{\mathbf{k}} |\alpha_k|$. This product converges only if $|\alpha_k|$ tends to unity rapidly enough as $k \to \infty$ or, more precisely, if $|\beta_k|^2$ vanishes faster than k^{-3} for large k. Otherwise, the vacuum state $\left|_{(b)}0\right\rangle$ is not expressible as a normalized combination of a-states, and the Bogolyubov transformation is not well-defined. Note that the same condition guarantees that the total mean number density is finite.

6.5 Choice of the physical vacuum

It is clear from the above consideration that the particle interpretation of the theory depends on the choice of the mode functions. We have seen that the a-vacuum $\left|_{(a)}0\right\rangle$, being a state without a-particles, nevertheless contains b-particles. A natural question is whether the a- or b-particles correspond to the observable particles. So far, all mode functions related by linear transformations (6.24) are on the same footing and our problem here is to determine the preferable set of the mode functions that describe the "actual" physical vacuum and particles.

6.5.1 The instantaneous lowest-energy state

In Chapter 4 the vacuum was defined as an eigenstate of the Hamiltonian with the lowest possible energy. This allowed us to choose the preferable set of mode functions and thus to unambiguously determine the physical vacuum. However, in the case under consideration the Hamiltonian (6.19) depends explicitly on time and thus does not possess time-independent eigenvectors that could serve as a vacuum. Nevertheless, given a particular moment of time η_0 we can still define the *instantaneous* vacuum $\left|_{\eta_0}0\right\rangle$ as the lowest-energy state of the Hamiltonian $\hat{H}(\eta_0)$.

To find a set of mode functions that determine $\left|_{\eta_0}0\right\rangle$, we first compute the expectation value $\left\langle_{(v)}0\right| \hat{H}(\eta_0) \left|_{(v)}0\right\rangle$ for the "vacuum" state $\left|_{(v)}0\right\rangle$ determined by arbitrarily chosen mode functions $v_k(\eta)$. Then we shall minimize this expectation value with respect to $v_k(\eta)$. (A standard result of the linear algebra is that the minimization of $\langle x| \hat{A} |x\rangle$ with respect to normalized vectors $|x\rangle$ is equivalent to finding the eigenvector $|x\rangle$ of the operator $\hat{A}$ with the smallest eigenvalue.)

Taking into account that $\hat{\pi} \equiv \hat{\chi}'$ and substituting the mode expansion (6.20) into (6.19) we find:

$$\hat{H}(\eta) = \frac{1}{4} \int d^3\mathbf{k} \left[\hat{a}_{\mathbf{k}}^- \hat{a}_{-\mathbf{k}}^- F_k^* + \hat{a}_{\mathbf{k}}^+ \hat{a}_{-\mathbf{k}}^+ F_k + \left(2\hat{a}_{\mathbf{k}}^+ \hat{a}_{\mathbf{k}}^- + \delta^{(3)}(0) \right) E_k \right], \quad (6.34)$$

where

$$E_k(\eta) \equiv \left|v_k'\right|^2 + \omega_k^2(\eta) \left|v_k\right|^2, \quad (6.35)$$

$$F_k(\eta) \equiv v_k'^2 + \omega_k^2(\eta) v_k^2. \quad (6.36)$$

Exercise 6.6
Derive (6.34).

Since $\hat{a}_{\mathbf{k}}^- \left|_{(v)}0\right\rangle = 0$, the expectation value of $\hat{H}(\eta_0)$ in the state $\left|_{(v)}0\right\rangle$ is

$$\left\langle_{(v)}0\right| \hat{H}(\eta_0) \left|_{(v)}0\right\rangle = \frac{1}{4} \delta^{(3)}(0) \int d^3\mathbf{k} E_k(\eta_0).$$

As discussed above, the divergent factor $\delta^{(3)}(0)$ is a harmless manifestation of the infinite total volume of space. The energy density is then

$$\varepsilon(\eta_0) = \frac{1}{4} \int d^3\mathbf{k} E_k(\eta_0) = \frac{1}{4} \int d^3\mathbf{k} \left(\left|v_k'(\eta_0)\right|^2 + \omega_k^2(\eta_0) \left|v_k(\eta_0)\right|^2 \right), \quad (6.37)$$

and our task is to determine which mode functions $v_k(\eta)$ minimize $\varepsilon(\eta_0)$. It is clear that for each mode $\mathbf{k}$ its contribution to the energy must be minimized separately. Thus, for a given k we have to determine $v_k(\eta_0)$ and $v_k'(\eta_0)$ which minimize the expression

$$E_k(\eta_0) = \left|v_k'(\eta_0)\right|^2 + \omega_k^2(\eta_0) \left|v_k(\eta_0)\right|^2 \quad (6.38)$$

while obeying the normalization condition (6.22),

$$v_k'(\eta_0) v_k^*(\eta_0) - v_k(\eta_0) v_k^{*\prime}(\eta_0) = 2i. \quad (6.39)$$

Substituting

$$v_k = r_k \exp(i\alpha_k)$$

into (6.39), we infer that the real functions r_k and α_k obey

$$r_k^2 \alpha_k' = 1. \quad (6.40)$$

With this relation we find that

$$E_k(\eta_0) = \left|v_k'\right|^2 + \omega_k^2 \left|v_k\right|^2$$
$$= r_k'^2 + r_k^2 \alpha_k'^2 + \omega_k^2 r_k^2 = r_k'^2 + \frac{1}{r_k^2} + \omega_k^2 r_k^2 \quad (6.41)$$

is minimized when $r_k'(\eta_0) = 0$ and $r_k(\eta_0) = \omega_k^{-1/2}(\eta_0)$. We thus find that the initial conditions

$$v_k(\eta_0) = \frac{1}{\sqrt{\omega_k(\eta_0)}} e^{i\alpha_k(\eta_0)}, \quad v_k'(\eta_0) = i\sqrt{\omega_k(\eta_0)} e^{i\alpha_k(\eta_0)} = i\omega_k v_k(\eta_0), \quad (6.42)$$

select the preferred mode functions which determine the vacuum (the lowest-energy state) at a particular moment of time η_0. Although the phase factors $\alpha_k(\eta_0)$ remain undetermined, they are irrelevant and we can set them to zero. Note that the above considerations are valid only if $\omega_k^2 > 0$. For $\omega_k^2(\eta_0) < 0$ the function E_k has no minimum. In this case the instantaneous lowest-energy vacuum does not exist.

Remark: Hamiltonian diagonalization The mode functions satisfying the conditions (6.42) define a certain set of operators $\hat{a}_{\mathbf{k}}^{\pm}$ and the corresponding vacuum $\left|_{\eta_0} 0\right\rangle$. For these mode functions one finds $E_k(\eta_0) = 2\omega_k(\eta_0)$ and $F_k(\eta_0) = 0$, so the Hamiltonian at time η_0 is

$$\hat{H}(\eta_0) = \int d^3\mathbf{k}\, \omega_k(\eta_0) \left[\hat{a}_{\mathbf{k}}^{+} \hat{a}_{\mathbf{k}}^{-} + \frac{1}{2} \delta^{(3)}(0) \right]. \quad (6.43)$$

At $\eta = \eta_0$ this Hamiltonian is diagonal in the eigenbasis of the occupation number operators $\hat{N}_{\mathbf{k}} = \hat{a}_{\mathbf{k}}^{+} \hat{a}_{\mathbf{k}}^{-}$, which consists of the vacuum state $\left|_{\eta_0} 0\right\rangle$ and the corresponding excited states. Accordingly, the state $\left|_{\eta_0} 0\right\rangle$ is sometimes called the *vacuum of instantaneous Hamiltonian diagonalization*. The vacuum states at two different moments of time are related by Bogolyubov coefficients α_k and β_k, so particles are produced if $\beta_k \neq 0$.

Remark: zero-point energy As before, the zero-point energy density of the quantum field in the vacuum state $\left|_{\eta_0} 0\right\rangle$ is divergent,

$$\frac{1}{4} \int d^3\mathbf{k}\, E_k(\eta_0) = \frac{1}{2} \int d^3\mathbf{k}\, \omega_k(\eta_0).$$

This quantity is time-dependent and cannot be simply subtracted. A more sophisticated renormalization procedure (developed in Part II of this book) is needed to obtain the correct value of the energy density.

For a scalar field in Minkowski spacetime, ω_k is time-independent and the prescription (6.42) yields the standard mode functions (4.31), which determine the time-independent vacuum state. But if ω_k changes with time then the mode functions satisfying (6.42) at $\eta = \eta_0$ will generally differ from the mode functions that satisfy the same conditions at a different time $\eta_1 \neq \eta_0$. In other words, the vacua $\left|_{\eta_0} 0\right\rangle$ and $\left|_{\eta_1} 0\right\rangle$ are different and the state $\left|_{\eta_0} 0\right\rangle$ is not the lowest-energy state at a later moment of time η_1. In this case there are no states which remain eigenstates of the Hamiltonian at all times. This can be easily seen from (6.34).

A vacuum state could remain an eigenstate of the Hamiltonian only if $F_k = 0$ for all η, i.e.

$$F_k(\eta) = \left(v_k'\right)^2 + \omega_k^2(\eta) v_k^2 = 0.$$

This differential equation has the exact solutions,

$$v_k(\eta) = C \exp\left[\pm i \int \omega_k(\eta) d\eta\right],$$

which do not satisfy the mode function equation (6.21) if $\omega_k(\eta)$ depends on time.

The operators $\hat{a}_{\mathbf{k}}^{\pm}(\eta_0)$ and $\hat{a}_{\mathbf{k}}^{\pm}(\eta_1)$ defining the instantaneous vacuum states $\left|_{\eta_0}0\right\rangle$ and $\left|_{\eta_1}0\right\rangle$ at two different moments of time are related by a Bogolyubov transformation. The expectation value of the Hamiltonian $\hat{H}(\eta_1)$ in the vacuum state $\left|_{\eta_0}0\right\rangle$ is

$$\left\langle_{\eta_0}0\right| \hat{H}(\eta_1) \left|_{\eta_0}0\right\rangle = \left\langle_{\eta_0}0\right| \int d^3\mathbf{k}\, \omega_k(\eta_1) \left[\hat{a}_{\mathbf{k}}^{+}(\eta_1)\hat{a}_{\mathbf{k}}^{-}(\eta_1) + \tfrac{1}{2}\delta^{(3)}(0)\right] \left|_{\eta_0}0\right\rangle$$

$$= \delta^3(0) \int d^3\mathbf{k}\, \omega_k(\eta_1) \left[\tfrac{1}{2} + |\beta_k|^2\right],$$

where β_k is the corresponding Bogolyubov coefficient. Unless $\beta_k = 0$ for all k, this energy is larger than the minimum possible value and hence the state $\left|_{\eta_0}0\right\rangle$ contains particles at time η_1.

Remark: minimal fluctuations The value of the mode function $v_k(\eta_0)$ can be chosen to be arbitrarily small without violating the normalization condition (6.22). However, in this case $v_k'(\eta_0)$ must be very large. This is the consequence of the Heisenberg uncertainty relation. In this case v_k would acquire large values within a short time, leading to large field fluctuations. In the lowest-energy state both $v_k(\eta_0)$ and $v_k'(\eta_0)$ are optimized so that the generation of large fluctuations within a short time is avoided. In this sense the vacuum state is the state with the minimal quantum fluctuations.

6.5.2 Ambiguity of the vacuum state

Minimization of the instantaneous energy is not the only possible way to define the "vacuum state" and there is no unique "best" physical prescription for choosing the vacuum state of a field in a general curved spacetime. The physical reason for this ambiguity is easy to understand. The usual definitions of the vacuum and particle states in Minkowski spacetime are based on a decomposition of fields into plane waves $\exp(i\mathbf{kx} - i\omega_k t)$. A localized particle with momentum k is described by a wavepacket with a momentum spread Δk and the particle momentum is well-defined only if $\Delta k \ll k$. The spatial size λ of the wavepacket is inversely proportional to Δk, so that $\lambda \sim 1/\Delta k$, and therefore $\lambda \gg 1/k$. However, when the geometry of a curved spacetime varies significantly across a region of size λ, the plane waves are not a good approximation to the solution of the wave equation and the usual definition of a particle with momentum k fails. This definition is meaningful only if the curvature scale (the distance below which the spacetime can be well approximated by Minkowski space) exceeds k^{-1}. Note that the relevant quantity is the four-dimensional curvature. Therefore, even in a spatially flat Friedmann universe the vacuum and particle states are not always well-defined for some modes. For example, those modes of the scalar field for which the squared frequency,

$$\omega_k^2(\eta) = k^2 + m^2 a^2 - \frac{a''}{a},$$

is negative, do not oscillate and for them the analogy with a harmonic oscillator breaks down. Formally, even for $\omega_k^2 < 0$ mode expansions still make sense, but the interpretation of the corresponding states in Hilbert space in terms of physical particles is problematic. In particular, the "excited" states can have negative energy. Moreover, an eigenstate with the lowest instantaneous energy does not exist for such modes: For $\omega_k^2 < 0$ the condition $F_k(\eta_0) = 0$ contradicts the normalization condition (6.22) and hence the lowest energy eigenstate cannot be defined.

Further complications arise in curved spacetimes without symmetries. We shall see in Chapter 8 that an accelerated detector in a flat spacetime registers particles even when the field is in the true Minkowski vacuum state. Thus the definition of particles depends in general on the coordinate system which is preferred "from the point of view" of a detector. In a curved spacetime there is no a priori preferable coordinate system. Moreover, in the presence of gravity the energy is not necessarily bounded from below and the definition of the "true" vacuum state as the lowest energy state can therefore also fail.

Remark: short distances We have seen that the minimal energy state does not exist for modes with $\omega_k^2 < 0$. However, because $\omega_k^2 = k^2 + m_{\text{eff}}^2(\eta)$, modes with large enough k, namely,

$$k^2 > k_{\min}^2 \equiv -m_{\text{eff}}^2 = \frac{a''}{a} - m^2 a^2, \tag{6.44}$$

have positive $\omega_k^2 > 0$ even if $m_{\text{eff}}^2 < 0$, and therefore the lowest-energy state is well-defined for these modes. In cosmological applications, a negative m_{eff}^2 can arise because of the field interaction with the gravitational background. In such cases a natural length scale is the radius of curvature and ω_k^2 is negative only for modes exceeding the curvature scale. On much shorter scales, the spacetime can be treated as approximately flat. Therefore the field modes with wavelengths much smaller than the curvature radius are almost unaffected by gravitation. On very small scales, corresponding to large k, we can neglect $|m_{\rm eff}| \ll k$ and set $\omega_k \approx k$. Then the mode functions are those given in (4.31),

$$v_k(\eta) \approx \frac{1}{\sqrt{k}} e^{ik\eta}. \tag{6.45}$$

This leads to a natural definition of the minimal excitation state, which is unambiguous in the leading order and adequate only on small scales, $L \ll L_{\max} \sim k_{\min}^{-1} \sim |m_{\rm eff}|^{-1}$.

Remark: adiabatic vacuum As we have noted above, the notion of the particle in an arbitrary curved spacetime does not have an absolute meaning. Instead one has to consider the detector response which can be unambiguously determined for a given quantum state of the fields. Nevertheless sometimes it is useful to have "an approximate particle definition" which suits our intuition in the best possible way. In spacetimes with slowly changing geometry, the so-called adiabatic vacuum leads sometimes to a more meaningful notion of particles compared to the instantaneous vacuum prescription. In particular, in anisotropic universes a procedure based on the adiabatic vacuum allows one to separate the non-local contribution to the energy-momentum tensor resulting from the "particle production" from the local vacuum polarization effects in a more meaningful manner.

The adiabatic vacuum prescription relies on the WKB approximation for the solution of equation (6.21) in the case of slowly varying $\omega_k^2(\eta)$. Substituting the ansatz

$$v_k(\eta) = \frac{1}{\sqrt{W_k(\eta)}} \exp\left[i \int_{\eta_0}^{\eta} W_k(\eta) d\eta \right] \tag{6.46}$$

into (6.21) we find that the function $W_k(\eta)$ must obey the nonlinear equation

$$W_k^2 = \omega_k^2 - \frac{1}{2}\left[\frac{W_k''}{W_k} - \frac{3}{2}\left(\frac{W_k'}{W_k}\right)^2 \right]. \tag{6.47}$$

Let us consider the case when ω_k is a slowly varying function of time. More precisely, we assume that ω_k and all its derivatives change substantially, i.e. $\Delta\omega_k/\omega_k \sim O(1)$, only during time intervals $T \gg 1/\omega_k$. In this case, equation (6.47) can be used as a recurrence relation which allows us to find a particular solution for W_k in the form of the asymptotic series in the powers of small parameter $(\omega_k T)^{-1}$. For example, to zeroth order in $(\omega_k T)^{-1}$ we have

$$^{(0)}W_k = \omega_k,$$

while to second order

$$^{(2)}W_k = \omega_k \left(1 - \frac{1}{4}\frac{\omega_k''}{\omega_k^3} + \frac{3}{8}\frac{\omega_k'^2}{\omega_k^4} \right).$$

In principle one could find $^{(N)}W_k$ to an arbitrary order N. However the series obtained is asymptotic, and so the accuracy of the approximation reaches an optimum value at a particular N and subsequently becomes worse as N grows. Substituting $^{(N)}W_k$ in (6.46) we obtain an approximate WKB solution $v_k^{(N)}(\eta)$ of the mode equation (6.21) to adiabatic order N. Then the mode functions $v_k(\eta)$ determining the *adiabatic vacuum of order N at a particular time* η_0 are defined by the requirement that the *exact* solution $v_k(\eta)$ of equation (6.21) satisfies the following initial conditions,

$$v_k(\eta_0) = v_k^{(N)}(\eta_0), \quad v_k'(\eta_0) = v_k^{(N)\prime}(\eta_0).$$

All vacuum prescriptions agree if $\omega_k(\eta)$ is exactly constant. In particular, in the case when $\omega_k(\eta)$ tends to a constant both in the remote past ($\eta \ll \eta_1$) and in the future ($\eta \gg \eta_2$) one can unambiguously define "in" and "out" particle states in the past and future respectively. If the frequency $\omega_k(\eta)$ is time-dependent within some interval $\eta_1 < \eta < \eta_2$ then the positive-frequency solution for $\eta \ll \eta_1$ evolves to a mixture of positive and negative frequency solutions for $\eta \gg \eta_2$. As a consequence, particles are produced and the number density of these particles can be unambiguously determined in the "out" region $\eta \gg \eta_2$. On the other hand, the notion of the particle is ambiguous in the intermediate regime, $\eta_1 < \eta < \eta_2$, when ω_k is time-dependent. The reason is that in this case the vacuum fluctuations are not only "excited" but also "deformed" by the external field. This latter effect is called the *vacuum polarization.* There is no unique way to separate the "particles" and the vacuum polarization contributions in the total energy-momentum tensor. However, this does not lead to ambiguities in physical predictions because only the total energy-momentum tensor is relevant as the source of the gravitational field. The response of a specific particle detector can also be unambiguously determined given a quantum state of the field.

Thus, the absence of a generally valid definition of the vacuum and particle states does not impair our ability to make predictions for specific observable quantities in a curved spacetime. All well-posed physical questions can always be unambiguously answered even in the absence of such a definition.

Remark: a quantum-mechanical analogy Note that the stationary Schrödinger equation for a particle in a one-dimensional potential $V(x)$,

$$\frac{d^2\psi}{dx^2} + (E - V(x))\,\psi = 0,$$

coincides with the mode equation (6.21) after we replace the spatial coordinate x by the time η and substitute $\omega_k^2(\eta)$ for $E - V(x)$. The wave function ψ then "plays the role" of the mode function v_k. This allows us to draw a formal mathematical analogy between the problem of particle creation and the problem of a quantum-mechanical penetration through a potential barrier. Considering a plane wave with an amplitude α falling onto the potential barrier from the right (see Fig. 6.1) we find that the incident wave "splits" into reflected and transmitted waves. Normalizing the amplitude of the transmitted wave to unity ($T = 1$) we obtain from the conservation of probability that $|\alpha|^2 = |\beta|^2 + 1$, the condition analogous to (6.25). The transmitted wave in this consideration corresponds to the initial vacuum fluctuations in the problem of particle creation, while the reflected

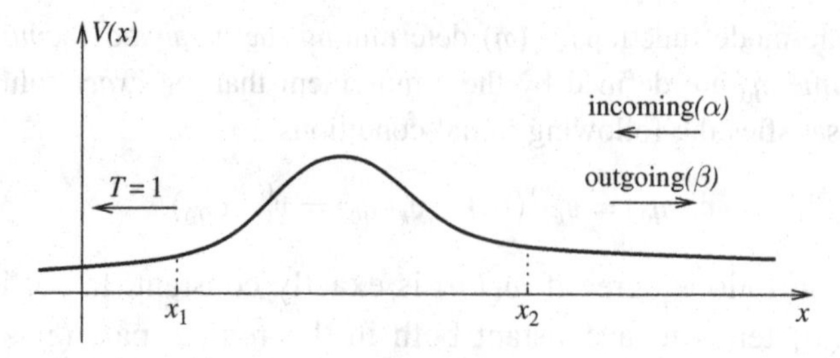

Fig. 6.1 Quantum-mechanical analogy: motion in a potential $V(x)$.

wave "describes the produced particles." We would like to stress once more that this analogy is entirely formal and is useful only to those who have a solid intuition for the corresponding quantum-mechanical problem.

6.6 Amplitude of quantum fluctuations

Correlation function Given a quantum state of the field $|\psi\rangle$, the amplitude of quantum fluctuations is always well-defined irrespective of whether the particle interpretation of the field is available. Let us consider the equal-time correlation function $\langle\psi|\,\hat{\chi}(\mathbf{x},\eta)\,\hat{\chi}(\mathbf{y},\eta)\,|\psi\rangle$. For a "vacuum state" $|0\rangle$ determined by a set of mode functions $v_k(\eta)$, we obtain

$$\langle 0|\,\hat{\chi}(\mathbf{x},\eta)\,\hat{\chi}(\mathbf{y},\eta)\,|0\rangle = \frac{1}{4\pi^2}\int_0^\infty k^2 dk\,|v_k(\eta)|^2\,\frac{\sin kL}{kL}, \tag{6.48}$$

where $L \equiv |\mathbf{x}-\mathbf{y}|$.

Exercise 6.7
Derive (6.48) using the mode expansion (6.20).

Generically the main contribution to the integral in (6.48) comes from wavenumbers $k \sim L^{-1}$, and therefore the magnitude of the correlation function can be estimated as

$$\langle 0|\,\hat{\chi}(\mathbf{x},\eta)\,\hat{\chi}(\mathbf{y},\eta)\,|0\rangle \sim k^3\,|v_k|^2\,, \tag{6.49}$$

with $k \sim L^{-1}$. Note that in the Friedmann universe the comoving coordinate distance $L = |\mathbf{x}-\mathbf{y}|$ is related to the physical distance L_p as $L_p = a(\eta)L$, where $a(\eta)$ is the scale factor. The field χ is related to the original, physical field ϕ by $\phi = \chi/a(\eta)$.

Fluctuations of spatially averaged fields One can consider a field operator averaged over a region of size L (e.g. a cube with sides $L \times L \times L$),

$$\hat{\chi}_L(\eta) \equiv \frac{1}{L^3}\int_{L\times L\times L} \hat{\chi}(\mathbf{x},\eta)\,d^3\mathbf{x},$$

and calculate

$$\delta\chi_L^2(\eta) \equiv \langle\psi| [\hat{\chi}_L(\eta)]^2 |\psi\rangle .$$

This is another way to characterize the typical fluctuations on scales L. Sometimes instead of integrating over the box with the sharp boundaries it is more convenient to define the *window-averaged operator*

$$\hat{\chi}_L(\eta) \equiv \int \hat{\chi}(\mathbf{x}, \eta)\, W_L(\mathbf{x}) d^3\mathbf{x},$$

where $W_L(\mathbf{x})$ is a *window function* $W(\mathbf{x})$ which is of order 1 for $|\mathbf{x}| \lesssim L$ and rapidly decays for $|\mathbf{x}| \gg L$. This function must satisfy the normalization condition

$$\int W(\mathbf{x})\, d^3\mathbf{x} = 1. \tag{6.50}$$

The prototypical example of a window function is the *Gaussian* function

$$W_L(\mathbf{x}) = \frac{1}{(2\pi)^{3/2} L^3} \exp\left(-\frac{|\mathbf{x}|^2}{2L^2}\right)$$

which selects $|\mathbf{x}| \lesssim L$. In the general case it is rather natural to select a window function with the following scaling properties:

$$W_{L'}(\mathbf{x}) \equiv \frac{L^3}{L'^3} W_L\left(\frac{L}{L'}\mathbf{x}\right).$$

In this case

$$\int W_L(\mathbf{x}) e^{-i\mathbf{k}\cdot\mathbf{x}} d^3\mathbf{x} = w(\mathbf{k}L),$$

and the Fourier image $w(\mathbf{k}L)$ satisfies $w|_{\mathbf{k}=0} = 1$ and decays rapidly for $|\mathbf{k}| \gtrsim L^{-1}$.

Given the mode expansion (6.20) for the field operator $\hat{\chi}(\mathbf{x}, \eta)$, after straightforward algebra we find

$$\delta\chi_L^2(\eta) = \langle 0| \left[\int d^3\mathbf{x}\, W_L(\mathbf{x}) \hat{\chi}(\mathbf{x}, \eta)\right]^2 |0\rangle = \frac{1}{2} \int |v_k|^2 |w(\mathbf{k}L)|^2 \frac{d^3\mathbf{k}}{(2\pi)^3}.$$

Since the function $w(\mathbf{k}L)$ is of order unity for $|\mathbf{k}| \lesssim L^{-1}$ and quickly decays for $|\mathbf{k}| \gtrsim L^{-1}$, one can estimate the above integral as

$$\int |v_k|^2 |w(\mathbf{k}L)|^2 \frac{d^3\mathbf{k}}{(2\pi)^3} \sim \int_0^{L^{-1}} k^2 |v_k|^2 dk \sim \frac{1}{L^3} |v_k|^2, \quad k \sim L^{-1}.$$

Thus the amplitude of fluctuations $\delta\chi_L$ is of order

$$\delta\chi_L^2 \sim k^3 |v_k|^2, \tag{6.51}$$

where $k \sim L^{-1}$.

Comparing (6.49) and (6.51), we see that the correlation function and the mean square fluctuation both have the same order of magnitude and both characterize the typical amplitude of quantum fluctuations on scales L. Therefore we refer to

$$\delta(k) \equiv \frac{1}{2\pi} k^{3/2} |v_k| \tag{6.52}$$

as the spectrum of quantum fluctuations.

6.6.1 Comparing fluctuations in the vacuum and excited states

Intuitively one may expect that the fluctuations in an excited state are larger than those in the vacuum state. To demonstrate this, let us compare the fluctuations of a scalar field for the vacuum and excited states in Minkowski spacetime. The spectrum of the vacuum fluctuations in Minkowski spacetime was already calculated in Chapter 4, equation (4.34), and the result is

$$\delta_{\text{vac}}(k) = \frac{1}{2\pi} \frac{k^{3/2}}{\sqrt{\omega_k}} = \frac{1}{2\pi} \frac{k^{3/2}}{\left(k^2 + m^2\right)^{1/4}}. \tag{6.53}$$

This time-independent spectrum is sketched in Fig. 6.2. When measured with a high-resolution device (small L or large k), the field shows large fluctuations. On the other hand, if the field is averaged over a large volume ($L \to \infty$), the amplitude of fluctuations tends to zero.

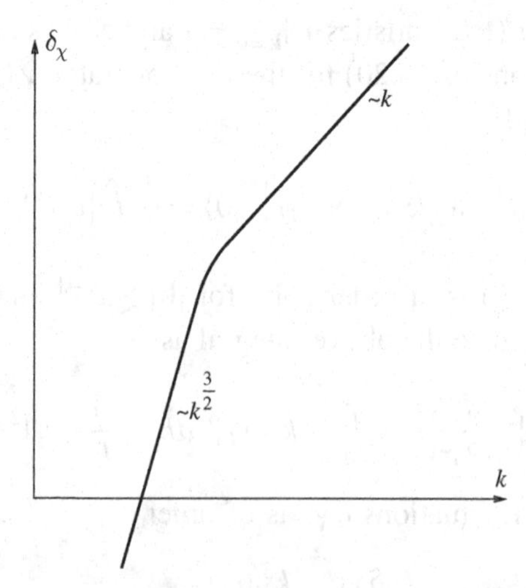

Fig. 6.2 A sketch of the spectrum of fluctuations $\delta\chi_L$ in Minkowski space; $L \equiv 2\pi k^{-1}$. (The logarithmic scaling is used for both axes.)

Let us consider the (nonvacuum) state $|b\rangle$ annihilated by all operators $\hat{b}_{\mathbf{k}}^-$ defined via the field operator expansion

$$\hat{\chi} = \frac{1}{\sqrt{2}} \int \left(e^{i\mathbf{k}\cdot\mathbf{x}} v_k^* \hat{b}_{\mathbf{k}}^- + e^{-i\mathbf{k}\cdot\mathbf{x}} v_k \hat{b}_{\mathbf{k}}^+ \right) \frac{d^3\mathbf{k}}{(2\pi)^{3/2}},$$

where

$$v_k = \frac{1}{\sqrt{\omega_k}} \left(\alpha_k e^{i\omega_k \eta} + \beta_k e^{-i\omega_k \eta} \right). \tag{6.54}$$

Substituting (6.54) in (6.52), we find the spectrum of fluctuations in the state $|b\rangle$:

$$\delta_b(k) = \frac{1}{2\pi} \frac{k^{3/2}}{\sqrt{\omega_k}} \left[|\alpha_k|^2 + |\beta_k|^2 - 2\mathrm{Re}\left(\alpha_k \beta_k e^{2i\omega_k \eta} \right) \right]^{1/2}. \tag{6.55}$$

Thus for the ratio of the amplitudes we obtain

$$\frac{\delta_b^2}{\delta_{\mathrm{vac}}^2} = 1 + 2|\beta_k|^2 - 2\mathrm{Re}\left(\alpha_k \beta_k e^{2i\omega_k \eta} \right). \tag{6.56}$$

After averaging over a sufficiently long time interval, $\Delta\eta \gg \omega_k^{-1}$, the oscillating term $\mathrm{Re}\left(\alpha_k \beta_k e^{2i\omega_k \eta} \right)$ vanishes and the result in (6.56) simply reduces to $1 + 2|\beta_k|^2$.

6.7 An example of particle production

For illustrative purposes we now perform explicit calculations for a rather artificial but simple case when the effective mass of the scalar field changes as follows,

$$m_{\mathrm{eff}}^2(\eta) = \begin{cases} m_0^2, & \eta < 0 \text{ and } \eta > \eta_1; \\ -m_0^2, & 0 < \eta < \eta_1. \end{cases} \tag{6.57}$$

In the regions $\eta < 0$ and $\eta > \eta_1$ the vacuum states are well-defined; they are called the "in" vacuum $|0_{\mathrm{in}}\rangle$ and the "out" vacuum $|0_{\mathrm{out}}\rangle$ respectively. We assume that initially (for $\eta < 0$) the scalar field is the "in" vacuum state and compute (a) the mean particle number, (b) the mean energy of produced particles, and (c) the spectrum of quantum fluctuations for $\eta > \eta_1$.

Mode functions The "in" and "out" vacuum states are entirely determined by specifying the negative frequency mode functions, which are

$$v_k^{(\mathrm{in})}(\eta) = \frac{1}{\sqrt{\omega_k}} e^{i\omega_k \eta}, \tag{6.58}$$

for $\eta < 0$ and

$$v_k^{(\mathrm{out})}(\eta) = \frac{1}{\sqrt{\omega_k}} e^{i(\eta - \eta_1)\omega_k}, \tag{6.59}$$

for $\eta > \eta_1$, where $\omega_k \equiv \sqrt{k^2 + m_0^2}$.

Since $\omega_k^2(\eta) = k^2 + m_{\text{eff}}^2(\eta)$ changes at $\eta = 0$ and $\eta = \eta_1$, the mode functions $v_k^{(\text{in})}(\eta)$ evolve into superposition of the negative and positive frequencies for $\eta > \eta_1$:

$$v_k^{(\text{in})}(\eta) = \frac{1}{\sqrt{\omega_k}} \left[\alpha_k^* e^{i\omega_k(\eta-\eta_1)} + \beta_k^* e^{-i\omega_k(\eta-\eta_1)} \right]. \tag{6.60}$$

The Bogolyubov coefficients α_k, β_k are determined by the requirement that the solution and its first derivative must be continuous at $\eta = 0$ and $\eta = \eta_1$. The result is

$$\alpha_k = \frac{e^{-i\Omega_k\eta_1}}{4} \left(\sqrt{\frac{\omega_k}{\Omega_k}} + \sqrt{\frac{\Omega_k}{\omega_k}} \right)^2 - \frac{e^{i\Omega_k\eta_1}}{4} \left(\sqrt{\frac{\omega_k}{\Omega_k}} - \sqrt{\frac{\Omega_k}{\omega_k}} \right)^2,$$

$$\beta_k = \frac{1}{4} \left(\frac{\Omega_k}{\omega_k} - \frac{\omega_k}{\Omega_k} \right) \left(e^{i\Omega_k\eta_1} - e^{-i\Omega_k\eta_1} \right) = \frac{1}{2} \left(\frac{\Omega_k}{\omega_k} - \frac{\omega_k}{\Omega_k} \right) \sin(\Omega_k\eta_1),$$

where $\omega_k \equiv \sqrt{k^2 + m_0^2}$ and $\Omega_k \equiv \sqrt{k^2 - m_0^2}$.

Exercise 6.8

Derive the above expressions for the Bogolyubov coefficients.

Particle number density For $\eta > \eta_1$ the state $|0_{\text{in}}\rangle$ is different from the true vacuum state $|0_{\text{out}}\rangle$ and so the state $|0_{\text{in}}\rangle$ contains particles. The mean particle number density in a mode **k** is

$$n_k = |\beta_k|^2 = \frac{m_0^4}{|k^4 - m_0^4|} \left| \sin\left(\eta_1 \sqrt{k^2 - m_0^2} \right) \right|^2. \tag{6.61}$$

Note that this expression remains finite as $k \to m_0$. Let us consider separately two limiting cases: $k \gg m_0$ (ultrarelativistic particles) and $k \ll m_0$ (nonrelativistic particles).

For $k \gg m_0$, we have $\omega_k \approx \Omega_k$ and assuming that $m_0\eta_1$ is not too large one can expand (6.61) in powers of the small parameter (m_0/k). After some algebra we obtain

$$n_k = \frac{m_0^4}{k^4} \sin^2(k\eta_1) + O\left(\frac{m_0^5}{k^5} \right). \tag{6.62}$$

It follows that $n_k \ll 1$ or, in other words, very few relativistic particles are created.

The situation is different for $k \ll m_0$. In this case $\sqrt{k^2 - m_0^2} \approx i m_0$ is imaginary and we obtain

$$n_k \sim \sinh^2(m_0\eta_1). \tag{6.63}$$

If $m_0\eta_1 \gg 1$, the density of the produced particles is exponentially large.

Particle energy density Since $n_k \sim k^{-4}$ for $k \to \infty$, the energy density of the produced particles,

$$\varepsilon_0 = \int n_k \omega_k d^3\mathbf{k}, \tag{6.64}$$

logarithmically diverges for large k (ultraviolet limit). This divergence is, however, entirely due to the discontinuous change of $m^2_{\text{eff}}(\eta)$ and it disappears for any smooth function $m^2_{\text{eff}}(\eta)$. Therefore, we ignore this divergence and assume that there is an ultraviolet cutoff at some $k_{\max}$. Then for $m_0\eta_1 \gg 1$, the main contribution to the integral comes from the modes with $k \lesssim m_0$ for which $|\omega_k| \sim m_0$. Using (6.63) we can estimate the energy density of the produced particles as

$$\varepsilon_0 \sim m_0 \int_0^{m_0} dk\, k^2 \exp(2m_0\eta_1) \sim m_0^4 \exp(2m_0\eta_1).$$

Exercise 6.9*

Assuming that the integral in (6.64) is performed over the range $0 < k < k_{\max}$, show that for $m_0\eta_1 \gg 1$ the dominant contribution to the integral comes from $k \approx \sqrt{m_0/\eta_1}$ and derive a more precise estimate for ε_0:

$$\varepsilon_0 \propto \frac{m_0^4}{(m_0\eta_1)^{3/2}} \exp(2m_0\eta_1).$$

Amplitude of fluctuations Neglecting the oscillating term in (6.55) we immediately obtain the following estimate for the amplitude of quantum fluctuations at $\eta > \eta_1$:

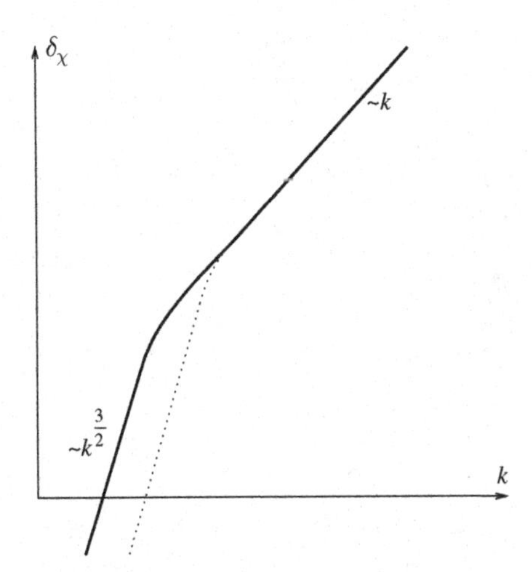

Fig. 6.3 A sketch of the spectrum $\delta\chi_L$ after particle creation; $L \equiv 2\pi k^{-1}$. (The logarithmic scaling is used for both axes.) The dotted line is the spectrum in Minkowski space.

$$\delta \sim \frac{k^{3/2}}{\sqrt{\omega_k}}\left(1+2\left|\beta_k\right|^2\right)^{1/2} \sim \begin{cases} k, & k \gg m_0; \\ k^{3/2} m_0^{-1/2} \exp\left(m_0 \eta_1\right), & k \ll m_0. \end{cases}$$

Thus, we see that on large scales the amplitude of fluctuations is enhanced by the factor $\exp(m_0\eta_1)$ compared to the amplitude of the vacuum fluctuations (see Fig. 6.3).

7

Quantum fields in the de Sitter universe

Summary Field quantization in de Sitter spacetime. Bunch–Davies vacuum. Time evolution of quantum fluctuations.

7.1 De Sitter universe

We now apply the formalism developed in the previous chapter to study the behavior of quantum fluctuations in the *de Sitter universe.* The de Sitter universe is a particular case of a homogeneous and isotropic universe with a positive cosmological constant Λ. Formally this cosmological constant can be thought of as an "ideal hydrodynamical fluid" with the equation of state

$$p_\Lambda = -\varepsilon_\Lambda. \tag{7.1}$$

In this case, the energy-momentum tensor of the perfect fluid becomes

$$T^\mu_\nu = (\varepsilon + p)u^\mu u_\nu - p\delta^\mu_\nu = \varepsilon_\Lambda \delta^\mu_\nu,$$

and it follows from the conservation law $T^\alpha_{\beta;\alpha} = 0$ that $\varepsilon_\Lambda = \text{const}$. This is the energy-momentum tensor corresponding to a cosmological constant. For a flat isotropic universe, the 0-0 component of the Einstein equations (called the Friedmann equation) reduces to

$$H \equiv \left(\frac{\dot{a}}{a}\right)^2 = \frac{8\pi G}{3}\varepsilon_\Lambda, \tag{7.2}$$

where $a(t)$ is the scale factor and the dot denotes the derivative with respect to the physical time t. This equation has the obvious solution

$$a(t) = a_0 e^{H_\Lambda t}, \tag{7.3}$$

which describes a flat de Sitter universe with the time-independent Hubble parameter

$$H_{\Lambda} = \sqrt{\frac{8\pi G}{3}\varepsilon_{\Lambda}}.$$

In this case it is easy to verify that all curvature invariants are constant and therefore the metric

$$ds^2 = dt^2 - H_{\Lambda}^{-2} \exp(2H_{\Lambda}t)\,\delta_{ik}dx^i dx^k \tag{7.4}$$

(where for convenience we set $a_0 = H_{\Lambda}^{-1}$) describes a static maximally symmetric spacetime in expanding coordinates. There exist no static coordinates which can cover the de Sitter spacetime on scales larger that the curvature scale H_{Λ}^{-1}, and even the expanding coordinates in (7.4) are incomplete. To verify this, let us first rewrite metric (7.4) in terms of the conformal time

$$\eta = -\int_t^{\infty} \frac{dt}{a(t)} = -\exp(-H_{\Lambda}t),$$

instead of the physical time t and the spherical coordinates instead of x^i. The result is

$$ds^2 = \frac{1}{H_{\Lambda}^2 \eta^2}\left[d\eta^2 - dr^2 - r^2\left(d\theta^2 + \sin^2\theta d\varphi^2\right)\right], \tag{7.5}$$

where $-\infty < \eta < 0$ and $0 \le r < \infty$. Next we change from η, r to the new coordinates $\tilde{\eta}, \chi$ which are related to the old ones (in the region where both overlap) via

$$\eta = \frac{\sin\tilde{\eta}}{\cos\tilde{\eta} + \cos\chi}, \qquad r = \frac{\sin\chi}{\cos\tilde{\eta} + \cos\chi}. \tag{7.6}$$

Metric (7.5) then takes the form

$$ds^2 = \frac{1}{H_{\Lambda}^2 \sin^2\tilde{\eta}}\left[d\tilde{\eta}^2 - d\chi^2 - \sin^2\chi\left(d\theta^2 + \sin^2\theta d\varphi^2\right)\right], \tag{7.7}$$

and describes a closed de Sitter universe which first contracts for $-\pi < \tilde{\eta} < -\pi/2$, reaches the minimal radius at $\tilde{\eta} = -\pi/2$ and then expands so that $a \to \infty$ as $\tilde{\eta} \to -0$. It is obvious, however, that (7.7) simply corresponds to another coordinate choice for the same de Sitter spacetime. In this sense the de Sitter universe is a very special case of the Friedmann universe. Generally, the energy density in the Friedmann universe is time-dependent and the geometry of hypersurfaces of constant energy density is unambiguously determined; hence the closed and flat universes are physically distinguishable. In the de Sitter universe, however, the energy density is time-independent and therefore any hypersurface is a hypersurface of constant energy. As a consequence, the flat, closed, and open de Sitter universes describe the same spacetime in different coordinate systems. The coordinates in (7.7) span the ranges

$$-\pi < \tilde{\eta} < 0, \quad 0 \leq \chi \leq \pi,$$

covering the entire de Sitter spacetime. The nontrivial time-radial part of this spacetime can be graphically represented by a square in Fig. 7.1, called a *conformal diagram.* Note that each point of the diagram corresponds to a two-dimensional sphere and the radial null geodesics, determined by equation $ds^2 = 0$, are straight lines at $\pm 45°$ angles.

Using relations (7.6) to draw the hypersurfaces $\eta = \text{const}$ and $r = \text{const}$ in the $\tilde{\eta} - \chi$ plane, we find that the coordinates in (7.5) cover only a half of the entire de Sitter spacetime (see Fig. 7.1). Therefore, these coordinates are incomplete. In cosmological applications, however, only a relatively small region of the de Sitter spacetime (shaded in Fig. 7.1) is used to approximate the inflationary epoch in the history of the universe. Within this region the closed and flat coordinates are similar and hence the incompleteness of the flat coordinates is not a problem. On the other hand, the analysis of the behavior of quantum fields is significantly simplified in these coordinates. Therefore, we shall use the flat coordinates and ignore their inability to cover events in the distant past which are irrelevant for physical applications.

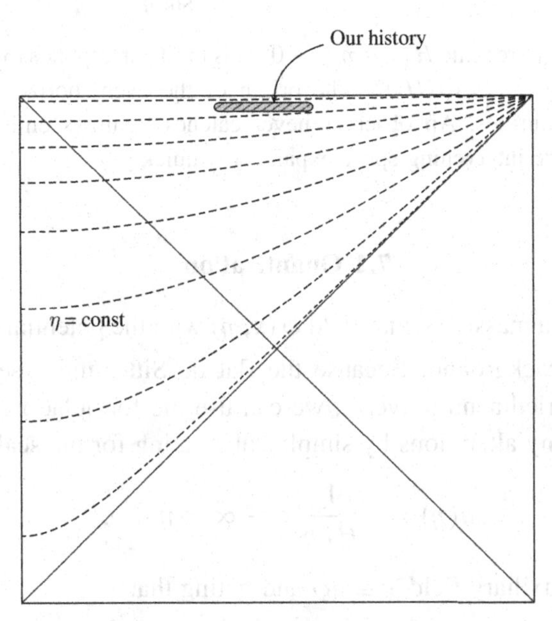

Fig. 7.1 A conformal diagram of de Sitter spacetime. The flat coordinate system covers only the left upper half of the diagram. Dashed lines are surfaces of constant η.

Remark: horizons An interesting feature of de Sitter spacetime is the presence of both particle and event horizons. Given a particular moment of cosmic time, the *particle horizon* is the boundary of the spatial region which consists of the points causally connected to an observer. Let us consider the closed de Sitter universe. Taking into account that the radial light geodesics are described by

$$\chi(\tilde{\eta}) = \pm\tilde{\eta} + \text{const},$$

we find that at time $\tilde{\eta}$ only those points which have comoving coordinates

$$\chi < \chi_{\text{p}}(\tilde{\eta}) = \tilde{\eta} + \pi \tag{7.8}$$

are causally connected to an observer at $\chi = 0$. The physical size of the particle horizon,

$$d_{\text{p}} \equiv a(\tilde{\eta})\,\chi_{\text{p}}(\tilde{\eta}) = -\frac{\tilde{\eta}+\pi}{H_\Lambda \sin\tilde{\eta}}, \tag{7.9}$$

grows as $d_{\text{p}} \propto -1/\tilde{\eta} \propto \exp(H_\Lambda t)$ when $\tilde{\eta} \to -0$.

Any event happening at time $\tilde{\eta}$ at distances

$$\chi > \chi_{\text{e}}(\tilde{\eta}) = -\tilde{\eta} \tag{7.10}$$

will never be seen by an observer at $\chi = 0$. The sphere with comoving radius $\chi_{\text{e}}(\tilde{\eta})$ is called the *event horizon*. Its physical size,

$$d_{\text{e}} \equiv a(\tilde{\eta})\,\chi_{\text{e}}(\tilde{\eta}) = \frac{\tilde{\eta}}{H_\Lambda \sin\tilde{\eta}}, \tag{7.11}$$

approaches the curvature scale H_Λ^{-1} as $\tilde{\eta} \to -0$. This limit corresponds to the exponentially expanding universe: $a \propto \exp(H_\Lambda t)$. The origin of the event horizon in the de Sitter universe is rather curious. An observer never catches lightrays emitted at a distance $d > H_\Lambda^{-1}$ because the intervening space expands too quickly.

7.2 Quantization

Now we quantize a massive scalar field $\phi(\mathbf{x}, \eta)$ with the potential $V(\phi) = \frac{1}{2}m^2\phi^2$ in the de Sitter background. Because the flat de Sitter universe is a particular case of the flat Friedmann universe, we can use the formulae from the previous chapter without any alterations by simply substituting for the scale factor

$$a(\eta) = -\frac{1}{H_\Lambda \eta}, \qquad -\infty < \eta < 0.$$

Introducing the auxiliary field $\chi \equiv a\phi$ and noting that

$$\omega_k^2(\eta) = k^2 + m^2a^2 - \frac{a''}{a} = k^2 + \left(\frac{m^2}{H_\Lambda^2} - 2\right)\frac{1}{\eta^2}, \tag{7.12}$$

we find that the mode function satisfies (see (6.21)):

$$v_k'' + \left[k^2 - \left(2 - \frac{m^2}{H_\Lambda^2} \right) \frac{1}{\eta^2} \right] v_k = 0. \tag{7.13}$$

The general solution of this equation is given in terms of the Bessel functions $J_n(x)$ and $Y_n(x)$:

$$v_k(\eta) = \sqrt{k|\eta|} \left[A_k J_n(k|\eta|) + B_k Y_n(k|\eta|) \right], \quad n \equiv \sqrt{\frac{9}{4} - \frac{m^2}{H_\Lambda^2}} \tag{7.14}$$

(see Exercise 7.1). The normalization condition $\mathrm{Im}\left(v_k^* v_k'\right) = 1$ (see (6.22)) constrains the integration constants A_k and B_k by

$$A_k B_k^* - A_k^* B_k = \frac{i\pi}{k}.$$

Exercise 7.1
Assuming that $m/H_\Lambda < 3/2$, find a change of variables which reduces (7.13) to the Bessel equation

$$s^2 \frac{d^2 f}{ds^2} + s \frac{df}{ds} + \left(s^2 - n^2\right) f = 0$$

with the general solution

$$f(s) = A J_n(s) + B Y_n(s).$$

Using the asymptotics of the Bessel functions, determine the behavior of $v_k(\eta)$ for $k|\eta| \gg 1$ and $k|\eta| \ll 1$.

The asymptotic behavior of the solutions can be found directly from equation (7.13). Given a wavenumber k, let us consider the early time asymptotic $k|\eta| \gg 1$ (which corresponds to large negative η). In this case the physical wavelength,

$$L_\mathrm{p} \sim a(\eta) k^{-1} \simeq \frac{H_\Lambda^{-1}}{k|\eta|}, \tag{7.15}$$

is much smaller than the curvature scale H_Λ^{-1}. Thus we expect that the corresponding mode is not affected by gravity and behaves as in Minkowski space. For $k|\eta| \gg 1$ one can neglect the η^{-2} term compared with k^2 in (7.12) and hence $\omega_k \approx k$. The two independent solutions of (7.13) are then $\propto \exp(\pm ik\eta)$ and we can define the minimal excitation ("vacuum") state for the corresponding modes by choosing the negative-frequency mode as

$$v_k(\eta) \approx \frac{1}{\sqrt{k}} e^{ik\eta}. \tag{7.16}$$

This determines the vacuum state only to the leading order in $|k\eta|^{-1}$, that is with precision which is enough to find the amplitude of the minimal quantum fluctuations.

As the universe expands, the absolute value $|\eta|$ decreases, and so for a given k the value of $k|\eta|$ eventually becomes smaller than unity. It is clear from (7.15) that the physical scale of the mode with a given k becomes of order the curvature scale H_Λ^{-1} at $\eta = \eta_k$ when $k|\eta_k| \sim 1$. We call this time the moment of (event) horizon crossing and refer to the modes with $k|\eta| \gg 1$ and $k|\eta| \ll 1$ as the *subhorizon* and *superhorizon* modes respectively. The subhorizon modes are eventually stretched by expansion and start to feel the curvature of the universe. After horizon crossing, for $k|\eta| \ll 1$, we can neglect the k^2 term in (7.13) and hence

$$v_k'' - \left(2 - \frac{m^2}{H_\Lambda^2}\right)\frac{1}{\eta^2} v_k = 0.$$

The general solution of this equation is

$$v_k(\eta) = A_k |\eta|^{n_1} + B_k |\eta|^{n_2}, \tag{7.17}$$

where

$$n_{1,2} \equiv \frac{1}{2} \pm \sqrt{\frac{9}{4} - \frac{m^2}{H_\Lambda^2}}.$$

At late times ($\eta \to 0$) the term proportional to $B|\eta|^{n_2}$ dominates.

7.2.1 Bunch–Davies vacuum

The superhorizon modes do not oscillate and hence the notion of a particle is not well-defined for $k \ll |\eta|^{-1}$. Moreover, for $m^2 < 2H_\Lambda^2$ the effective mass squared,

$$m_{\text{eff}}^2(\eta) = -\left(2 - \frac{m^2}{H_\Lambda^2}\right)\frac{1}{\eta^2},$$

is negative and the lowest energy state does not exist for the superhorizon modes. However, there exists a preferred quantum state called the Bunch–Davies vacuum. This state is de Sitter invariant and does not change with time. Let us construct the mode functions for the Bunch–Davies vacuum. Considering a mode with a given comoving k, we find that in the far remote past ($\eta \to -\infty$), when $k|\eta| \gg 1$, this mode does not feel the curvature and one can fix the initial conditions by requiring that

$$v_k(\eta) \to \frac{1}{\sqrt{\omega_k}} e^{i\omega_k \eta} \tag{7.18}$$

as $\eta \to -\infty$. In other words, we select the minimal excitation state in the remote past. Assuming that $m < \frac{3}{2}H\Lambda$ and using the results of Exercise 7.1, we find that the functions

$$v_k(\eta) = \sqrt{\frac{\pi\,|\eta|}{2}}\left[J_n(k\,|\eta|) - iY_n(k\,|\eta|)\right], \quad n \equiv \sqrt{\frac{9}{4} - \frac{m^2}{H^2}}, \tag{7.19}$$

have the required asymptotic (7.18). These mode functions determine the *Bunch–Davies vacuum* $|0_{BD}\rangle$ in the standard way: $|0_{BD}\rangle$ is annihilated by all operators $\hat{a}_{\mathbf{k}}^-$ entering expansion (6.20) with v_k given in (7.19). To verify that $|0_{BD}\rangle$ actually describes a time-independent state, let us find how the amplitude of fluctuations of the original field ϕ depends on the physical wavenumber $k_{\text{ph}} = k/a$. The field ϕ is related to the auxiliary field by $\phi = a^{-1}\chi$. Taking into account that $k\,|\eta| = k_{\text{ph}} H_\Lambda^{-1}$, we find (see (6.52))

$$\begin{aligned}\delta_\phi(k_{\text{ph}}) &= \frac{1}{2\pi} a^{-1} k^{3/2}\,|v_k(\eta)| \\ &= \frac{H_\Lambda}{\sqrt{8\pi}}\left(\frac{k_{\text{ph}}}{H_\Lambda}\right)^{3/2}\left[J_n^2\left(\frac{k_{\text{ph}}}{H_\Lambda}\right) + Y_n^2\left(\frac{k_{\text{ph}}}{H_\Lambda}\right)\right]^{1/2},\end{aligned} \tag{7.20}$$

and hence the amplitude of the fluctuations on a given physical scale does not depend on time. Using the asymptotics of the Bessel functions, one obtains from (7.20) that

$$\delta_\phi \simeq \begin{cases} \dfrac{k_{\text{ph}}}{2\pi}, & k_{\text{ph}} \gg H_\Lambda, \\[2ex] \dfrac{2^n \Gamma(n)}{\sqrt{8\pi^3}} H_\Lambda \left(\dfrac{k_{\text{ph}}}{H_\Lambda}\right)^{\frac{3}{2}-n}, & k_{\text{ph}} \ll H_\Lambda. \end{cases} \tag{7.21}$$

In Fig. 7.2 we show the amplitude of quantum fluctuations as a function of the physical wavelength $L_{\text{ph}} = 2\pi/k_{\text{ph}}$. For short-wavelength modes, the Bunch–Davies spectrum is in agreement with the spectrum of fluctuations in Minkowski space (see (6.53)). This confirms our naive expectations that the curvature is not very relevant on subcurvature scales. When $m^2 \ll H_\Lambda^2$, we have

$$\delta_\phi \propto L_{\text{ph}}^{-m^2/3H_\Lambda^2} \tag{7.22}$$

for $L_{\text{ph}} \gg H_\Lambda^{-1}$ and the amplitude of the fluctuations decays only weakly with the scale. In the case of a massless field, this amplitude becomes scale independent on supercurvature scales.

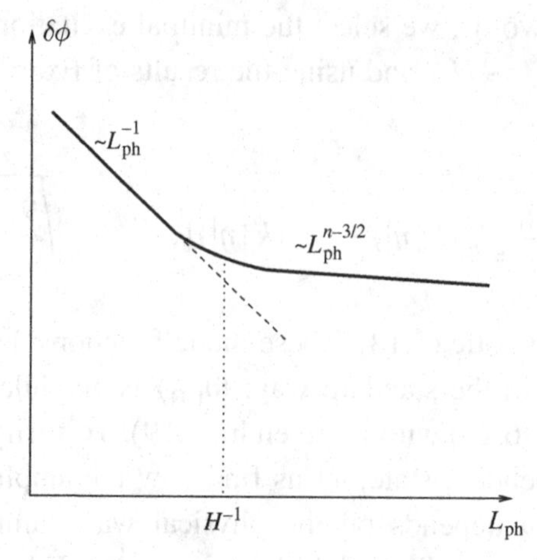

Fig. 7.2 The fluctuation amplitude $\delta\phi_{L_{\mathrm{ph}}}(\eta)$ as function of L_{ph} at fixed time η. The dashed line shows the amplitude of fluctuations in Minkowski spacetime. (Logarithmic scaling is used for both axes.)

7.3 Fluctuations in inflationary universe

The de Sitter universe plays an important role in cosmology. It can be used as a good approximation for the stage of accelerated expansion, known as *inflation*. The inflationary stage has a finite duration and therefore we need only a "piece" of the entire de Sitter spacetime to describe it. Let us assume that inflation begins at time $\eta = \eta_i$ and is over by $\eta = \eta_f$. Within the time interval $\eta_i < \eta < \eta_f$, we approximate the expansion by a flat de Sitter solution. In this case the quantum state of fields at the beginning of inflation depends on the previous history of the universe and can be very different from the Bunch–Davies state. Let us show that regardless of the initial conditions, the spectrum of fluctuations converges to the Bunch–Davies spectrum as the universe expands. To simplify the calculations we assume that at $\eta = \eta_i$ the subhorizon modes are in the state of minimal excitation. This means that for the modes with $k\,|\eta_i| \gg 1$,

$$v_k(\eta) \approx \frac{1}{\sqrt{k}} e^{ik\eta} \tag{7.23}$$

for $\eta_i < \eta < -1/k$ (recall that η is negative). At $\eta = \eta_i$ the minimal excitation state cannot be defined for the modes with $k\,|\eta_i| < 1$ and their spectrum is entirely determined by the unknown preinflationary evolution. Let us see what happens to a subcurvature mode when it crosses the horizon at time $\eta_k \simeq -1/k$. For $\eta > \eta_k$ the asymptotic solution for $v_k(\eta)$ is given in (7.17) where the first term eventually becomes negligible. Ignoring the numerical coefficients of order unity and matching solutions (7.23) and (7.17) at $\eta_k \simeq -1/k$, we obtain

$$v_k(\eta) \sim \frac{1}{\sqrt{k}} |k\eta|^{\frac{1}{2}-n} \tag{7.24}$$

for $\eta > \eta_k$. Thus, after the beginning of inflation we have

$$\delta_\phi(L_{\text{ph}}, \eta) = \frac{k^{3/2}|v_k(\eta)|}{2\pi a(\eta)} \sim \begin{cases} L_{\text{ph}}^{-1}, & L_{\text{ph}} < H_\Lambda^{-1}, \\ H_\Lambda \left[\dfrac{L_{\text{ph}}}{H_\Lambda^{-1}}\right]^{n-\frac{3}{2}}, & H_\Lambda^{-1}\dfrac{\eta_i}{\eta} > L_{\text{ph}} > H_\Lambda^{-1}, \\ \text{unknown}, & L_{\text{ph}} > H_\Lambda^{-1}\dfrac{\eta_i}{\eta}. \end{cases} \tag{7.25}$$

The evolution of the spectrum with time is shown in Fig. 7.3. We see that the perturbations stretched from subhorizon scales build the longwave part of the Bunch–Davies spectrum. The unknown part of the spectrum is redshifted by the expansion to very large physical scales

$$L_{\text{ph}} > H_\Lambda^{-1}(\eta_i/\eta) = H_\Lambda^{-1}\exp\left(H_\Lambda(t-t_i)\right).$$

If inflation continued forever ($\eta \to -0$), then an arbitrary initial state would evolve into the Bunch–Davies vacuum. However, because the duration of inflation is finite, the Bunch–Davies spectrum is formed only on the scales $L_{\text{ph}} < H_\Lambda^{-1}\eta_f/\eta$.

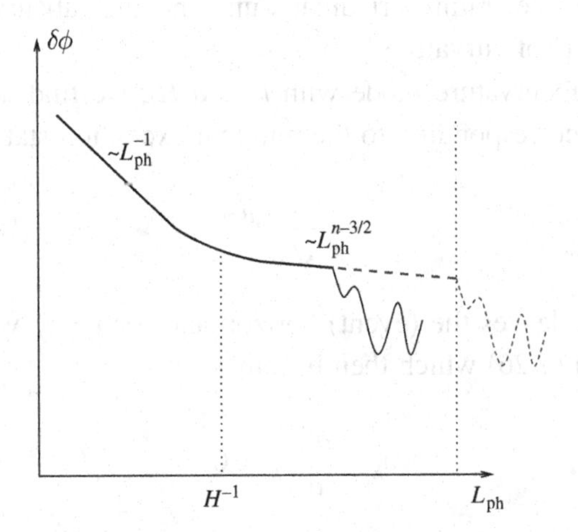

Fig. 7.3 The fluctuation spectrum at time $\eta = \eta_1$ (solid curve) and at later time $\eta = \eta_2$ (dashed curve). The wiggly lines in the infrared region of the spectrum correspond to the scales where the fluctuation amplitude is unknown; this region moves towards very large scales.

In realistic models, the Hubble parameter $H \equiv \dot{a}/a$ changes slightly and decreases towards the end of inflation. Let us find how this influences the spectrum of the generated fluctuations for a massless scalar field. We have found that in this case the amplitude of the large-scale fluctuations must be scale-independent or, in other words, the spectrum is flat, if H is strictly constant. It is clear that the change in H will cause deviations from the flat spectrum. For a massless scalar field ($m = 0$) the mode equation takes the form

$$v_k'' + \left(k^2 - \frac{a''}{a}\right) v_k = 0. \tag{7.26}$$

The effective mass squared is expressed in terms of the Hubble parameter H as

$$m_{\text{eff}}^2 \equiv -\frac{a''}{a} = -a^2 \left(2H^2 + \dot{H}\right),$$

where the dot denotes the derivative with respect to the physical time t. During inflation, the Hubble parameter changes insignificantly during a Hubble time H^{-1} and hence $|\dot{H}| \ll H^2$. Given a mode with a comoving wavenumber $k \gg aH = \dot{a}$, we find that this mode which had originally a physical wavelength $L_{\text{ph}} < H^{-1}$ is stretched to the curvature scale at the moment t_k determined by the condition $k \simeq a_k H_k$. Later on, at $t > t_k$, its physical wavelength exceeds the curvature scale. Note that subcurvature scales can be stretched to supercurvature scales only during inflation when the expansion is accelerating ($\dot{a}$ grows). In a decelerating universe, $\dot{a}$ decreases and the condition $k \gg aH = \dot{a}$ holds at all times if it is satisfied initially. Thus in a decelerating Friedmann universe the subcurvature modes will never feel the effects of curvature.

Considering a subcurvature mode with $k \gg a_i H_i$, we find that for $t_i < t < t_k$ the mode function corresponding to the minimal excitation state is

$$v_k \approx \frac{1}{\sqrt{k}} e^{ik\eta}. \tag{7.27}$$

At $t = t_k$ this mode leaves the (event) horizon and for $t > t_k$ we can neglect the k^2 term in equation (7.26) which then becomes

$$v_k'' - \frac{a''}{a} v_k \simeq 0.$$

The general solution of this equation is

$$v_k = A_k a + B_k a \int \frac{d\eta}{a^2}. \tag{7.28}$$

One can verify that the first term (proportional to a) becomes dominant at late times. Therefore, neglecting the second mode in (7.28) and matching solutions (7.27) and (7.28) at $t = t_k$ (by order of magnitude), we obtain

$$v_k \simeq \frac{1}{\sqrt{k}} \frac{a(t)}{a_k} \simeq \frac{H_k}{k^{3/2}} a(t), \tag{7.29}$$

where a_k and H_k are the values of the scale factor and the Hubble parameter at the moment of horizon crossing determined by the condition $k \simeq a_k H_k$. The spectrum of the fluctuations at $t > t_i$ is then

$$\delta_\phi \sim \frac{k^{3/2} |v_k|}{a(t)} \sim H_k \quad \text{for} \quad H^{-1}(t) < L_{\text{ph}} < H^{-1}\left(\frac{a(t)}{a(t_i)}\right). \tag{7.30}$$

Because the Hubble parameter is decreasing during inflation, the value of H_k is larger for those modes which left the horizon earlier. As a result, the amplitude of fluctuations is slightly higher toward the large scales and the resulting fluctuation spectrum is red-tilted within the corresponding range of scales (see Fig. 7.4).

These results can be directly applied to derive the spectrum of long-wavelength gravitational waves produced during inflation. One can show that the quantization of gravitational waves in an expanding universe can be reduced to the problem of quantization of a massless scalar field. Therefore the spectrum of gravitational waves produced in an accelerated universe also deviates from the flat spectrum. Since the Hubble parameter changes very slowly, the amplitude of fluctuations depends on the scale only logarithmically. The observed structure of the universe

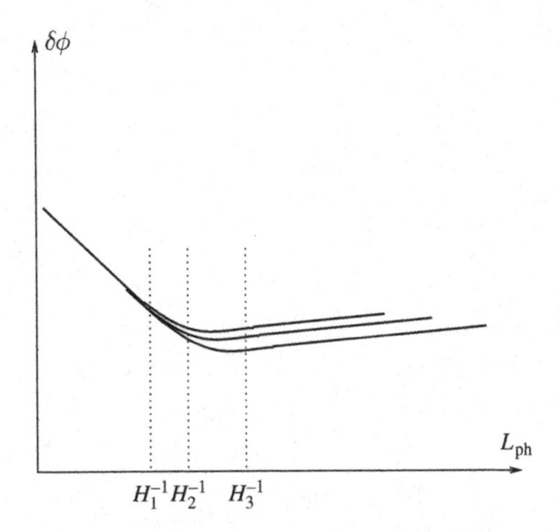

Fig. 7.4 The fluctuation spectrum resulting from inflation at three consecutive moments of time where the Hubble parameter has values H_1^{-1}, H_2^{-1}, H_3^{-1}. The spectrum is red-tilted towards large scales in the region $L_{\text{ph}} > H^{-1}$.

can be explained by considering the quantum scalar metric perturbations during inflation. This is a technically much more involved problem. However, the physical explanation of the production of the primordial inhomogeneities with a nearly scale-invariant spectrum is not very different from that presented above. In the case of matter inhomogeneities the "backreaction" of the gravitational field potential makes the m^2 term in (7.22) negative and this adds to the red tilt of the spectrum of scalar metric perturbations when compared to the spectrum of gravity waves.[1]

[1] For a detailed treatment of the quantum theory of cosmological perturbations, we refer the reader to the book *Physical Foundations of Cosmology* by V. Mukhanov (Cambridge University Press, 2005).

8

Unruh effect

Summary Uniformly accelerated observer. The Rindler spacetime. The Rindler and Minkowski vacua. The Unruh temperature.

W. G. Unruh discovered that an observer accelerating in the Minkowski vacuum sees particles which have a thermal spectrum, with the temperature being proportional to the acceleration. This effect is called the *Unruh effect*, and in this chapter we shall derive it in a simplified case. We shall consider a massless scalar field and assume that the observer moves with a constant acceleration in a 1+1-dimensional spacetime.

8.1 Accelerated motion

The metric of the two-dimensional Minkowski spacetime is

$$ds^2 = dt^2 - dx^2 = \eta_{\alpha\beta} dx^\alpha dx^\beta, \tag{8.1}$$

where the Greek indices α, β run over the values 0 and 1. If we use the proper time τ to parametrize the observer's trajectory $x^\alpha(\tau)$, then the 2-velocity

$$u^\alpha(\tau) \equiv \frac{dx^\alpha(\tau)}{d\tau} = \left(\dot{t}(\tau), \dot{x}(\tau)\right) \tag{8.2}$$

satisfies the normalization condition

$$\eta_{\alpha\beta} u^\alpha u^\beta = \eta_{\alpha\beta} \dot{x}^\alpha(\tau) \dot{x}^\beta(\tau) = 1, \tag{8.3}$$

where the dot denotes the derivative with respect to the proper time τ. Taking the time derivative of (8.3), we find that the 2-acceleration $a^\alpha(\tau) \equiv \dot{u}^\alpha(\tau)$ is orthogonal to the velocity,

$$\eta_{\alpha\beta} a^\alpha u^\beta = 0. \tag{8.4}$$

To understand the physical meaning of constant acceleration, one can imagine a spaceship with a propulsion engine that exerts a constant force. In this case, we intuitively expect that the acceleration of the spaceship is constant. In an instantly comoving inertial frame the observer is at rest and $\dot{x}(\tau)=0$; hence $u^{\alpha}(\tau)=(1,0)$. It then follows from (8.4) that $a^{\alpha}(\tau)=(0,a)$, where $a=\text{const}$. Since this is valid at any moment of time τ, the condition of constant acceleration can be formulated in the following completely covariant form, applicable in *any* (not necessarily comoving) inertial frame:

$$\eta_{\alpha\beta}a^{\alpha}(\tau)\,a^{\beta}(\tau)=\eta_{\alpha\beta}\ddot{x}^{\alpha}(\tau)\,\ddot{x}^{\beta}(\tau)=-a^{2}. \tag{8.5}$$

We have seen in the previous chapters that the notion of a particle crucially depends on the definition of the positive-frequency modes. For an inertial observer, these modes should be defined with respect to the time t of some inertial frame (the notion of a particle is Lorentz-invariant). However, when we consider an accelerated observer it is natural to expect that the positive-frequency modes must be defined with respect to the proper time of this observer. Then a "particle" registered by an accelerated detector can be very different from a "particle" of an inertial observer. As a result an accelerated observer views the Minkowski vacuum as a state containing particles. To verify this statement we need to (a) determine the trajectory of the accelerated observer in an inertial frame, (b) construct an accelerated comoving coordinate frame, and finally (c) solve the wave equation and compare the definition of particles in both coordinate frames. All these steps are significantly simplified if we use the lightcone coordinates instead of t and x.

Lightcone coordinates The inertial lightcone coordinates are defined as

$$u\equiv t-x,\qquad v\equiv t+x \tag{8.6}$$

and metric (8.1) then becomes

$$ds^{2}=du\,dv=g^{(M)}_{\alpha\beta}\,dx^{\alpha}dx^{\beta}, \tag{8.7}$$

where $x^{0}\equiv u,\ x^{1}\equiv v$ and

$$g^{(M)}_{\alpha\beta}=\begin{pmatrix}0 & 1/2\\ 1/2 & 0\end{pmatrix} \tag{8.8}$$

is the Minkowski metric in the lightcone coordinates. One can easily see that the coordinate transformation

$$u\rightarrow\tilde{u}=\alpha u,\ v\rightarrow\tilde{v}=\frac{1}{\alpha}v, \tag{8.9}$$

with $\alpha=\text{const}$, preserves metric (8.7) and therefore corresponds to a Lorentz transformation.

Exercise 8.1
Find an expression for α in terms of the relative velocity of the two inertial frames.

The trajectory of an accelerated observer Let us now determine the trajectory of a uniformly accelerated observer in the inertial frame. In the lightcone coordinates this trajectory is described by

$$x^{\alpha}(\tau) = (u(\tau), v(\tau)). \tag{8.10}$$

Replacing $\eta_{\alpha\beta}$ by $g^{(M)}_{\alpha\beta}$ in (8.3), (8.5) and substituting (8.10) for $x^{\alpha}(\tau)$ we obtain

$$\dot{u}(\tau)\,\dot{v}(\tau) = 1, \tag{8.11}$$

$$\ddot{u}(\tau)\,\ddot{v}(\tau) = -a^2. \tag{8.12}$$

Since $\dot{u}(\tau) = 1/\dot{v}(\tau)$, we have

$$\ddot{u} = -\frac{\ddot{v}}{\dot{v}^2},$$

and (8.12) reduces to

$$\left(\frac{\ddot{v}}{\dot{v}}\right)^2 = a^2.$$

This equation can be easily integrated, yielding the result

$$v(\tau) = \frac{A}{a}e^{a\tau} + B,$$

where A and B are integration constants. It then follows from $\dot{u}(\tau) = 1/\dot{v}(\tau)$ that

$$u(\tau) = -\frac{1}{Aa}e^{-a\tau} + C,$$

where C is a further integration constant. Performing a Lorentz transformation (8.9), one can set $A = 1$. Furthermore, shifting the origin of the corresponding inertial frame we can make both the integration constants B and C vanish, and then the trajectory of the accelerated observer is described by

$$u(\tau) = -\frac{1}{a}e^{-a\tau}, \qquad v(\tau) = \frac{1}{a}e^{a\tau}. \tag{8.13}$$

Using definitions (8.6) and going back to the original Minkowski coordinates t and x we obtain

$$t(\tau) = \frac{v+u}{2} = \frac{1}{a}\sinh a\tau, \qquad x(\tau) = \frac{v-u}{2} = \frac{1}{a}\cosh a\tau. \tag{8.14}$$

Thus, the worldline of the accelerated observer is a branch of the hyperbola $x^2 - t^2 = a^{-2}$ in the (t, x) plane (see Fig. 8.1). For large $|t|$, the worldline approaches the lightcone. The observer arrives from $x = +\infty$, decelerates and stops at $x = a^{-1}$, then accelerates back towards infinity.

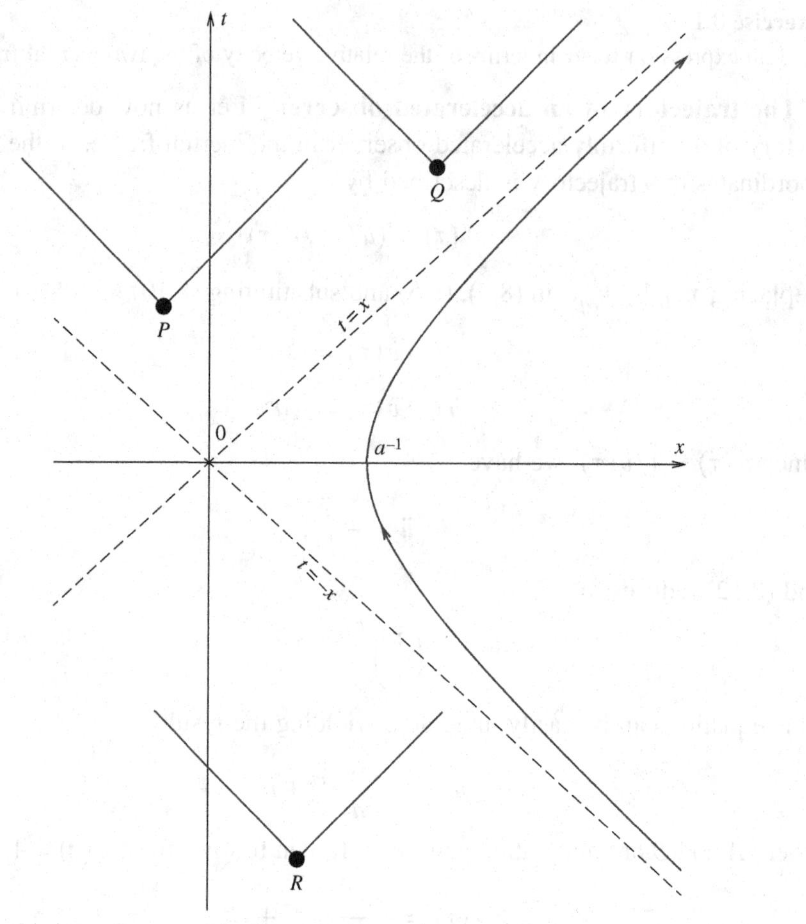

Fig. 8.1 The worldline of a uniformly accelerated observer (proper acceleration $a \equiv |\mathbf{a}|$) in Minkowski spacetime. The dashed lines show the lightcone. The observer cannot receive any signals from the events P, Q and cannot send signals to R.

8.2 Comoving frame of accelerated observer

As the next step, we will find an appropriate comoving frame $\left(\xi^0, \xi^1\right)$ for an accelerated observer. We are looking for a coordinate system in which the observer is at rest at $\xi^1 = 0$. In addition, the time coordinate ξ^0 must coincide with the proper time τ along the observer's worldline. Finally, we would like the metric in the comoving frame to be *conformally flat*,

$$ds^2 = \Omega^2 \left(\xi^0, \xi^1\right) \left[\left(d\xi^0\right)^2 - \left(d\xi^1\right)^2\right], \tag{8.15}$$

where $\Omega(\xi^0, \xi^1)$ is a function to be determined. The conformally flat form of the metric greatly simplifies quantization of fields. Our present task is to show that such a coordinate system can be found and to establish the relation between ξ^0, ξ^1 and the Minkowski coordinates t, x.

It is convenient to use the lightcone coordinates of the comoving frame,

$$\tilde{u} \equiv \xi^0 - \xi^1, \tilde{v} \equiv \xi^0 + \xi^1, \tag{8.16}$$

where metric (8.15) takes the form

$$ds^2 = \Omega^2 (\tilde{u}, \tilde{v}) \, d\tilde{u} \, d\tilde{v}. \tag{8.17}$$

In terms of the lightcone coordinates, the observer's worldline,

$$\xi^0 (\tau) = \tau, \quad \xi^1 (\tau) = 0, \tag{8.18}$$

takes the form

$$\tilde{v} (\tau) = \tilde{u} (\tau) = \tau. \tag{8.19}$$

Since ξ^0 is the proper time at the observer's location, the conformal factor $\Omega^2(\tilde{u}, \tilde{v})$ must satisfy

$$\Omega^2 (\tilde{u} = \tau, \tilde{v} = \tau) = 1. \tag{8.20}$$

Metrics (8.7) and (8.17) describe the same Minkowski spacetime in different coordinate systems and therefore

$$ds^2 = du \, dv = \Omega^2 (\tilde{u}, \tilde{v}) \, d\tilde{u} \, d\tilde{v}. \tag{8.21}$$

The functions $u (\tilde{u}, \tilde{v})$ and $v (\tilde{u}, \tilde{v})$ can depend only on one of two arguments, either $\tilde{u}$ or $\tilde{v}$; otherwise there will arise the terms $d\tilde{u}^2$ and $d\tilde{v}^2$ in the latter equality in (8.21). To be definite, let us choose

$$u = u (\tilde{u}), \qquad v = v (\tilde{v}). \tag{8.22}$$

We shall now determine the required functions $u(\tilde{u})$ and $v(\tilde{v})$.

Considering the observer's trajectory in two coordinate systems, we have

$$\frac{du (\tau)}{d\tau} = \frac{du (\tilde{u})}{d\tilde{u}} \frac{d\tilde{u} (\tau)}{d\tau}. \tag{8.23}$$

It follows from (8.13) and (8.19) that

$$\frac{du (\tau)}{d\tau} = e^{-a\tau} = -au (\tau), \quad \frac{d\tilde{u} (\tau)}{d\tau} = 1,$$

and (8.23) reduces to the following equation for $u (\tilde{u})$:

$$\frac{du}{d\tilde{u}} = -au; \tag{8.24}$$

hence

$$u = C_1 e^{-a\tilde{u}},$$

where C_1 is an integration constant. Similarly we find

$$v = C_2 e^{a\tilde{v}}.$$

Condition (8.20) restricts integration constants to satisfy $a^2 C_1 C_2 = -1$. Taking $C_2 = -C_1$ we obtain

$$u = -\frac{1}{a} e^{-a\tilde{u}}, \qquad v = \frac{1}{a} e^{a\tilde{v}}, \tag{8.25}$$

and

$$ds^2 = du\, dv = e^{a(\tilde{v}-\tilde{u})} d\tilde{u}\, d\tilde{v}. \tag{8.26}$$

With the help of the definitions in (8.6) and (8.16), one can rewrite relations (8.25) in terms of t, x and ξ^0, ξ^1:

$$t\left(\xi^0, \xi^1\right) = a^{-1} e^{a\xi^1} \sinh a\xi^0, \qquad x\left(\xi^0, \xi^1\right) = a^{-1} e^{a\xi^1} \cosh a\xi^0. \tag{8.27}$$

The metric in accelerated frame,

$$ds^2 = e^{2a\xi^1} \left[\left(d\xi^0\right)^2 - \left(d\xi^1\right)^2 \right], \tag{8.28}$$

describes the *Rindler spacetime*, which is locally equivalent to Minkowski spacetime and therefore has zero curvature. In Fig. 8.2 we show the hypersurfaces $\xi^0 = \text{const}$ and $\xi^1 = \text{const}$ in the (t, x) plane. The coordinates ξ^0, ξ^1 span the ranges

$$-\infty < \xi^0 < +\infty, \quad -\infty < \xi^1 < +\infty,$$

covering only one quarter of the Minkowski spacetime (the domain $x > |t|$). Hence, the coordinate system (ξ^0, ξ^1) is *incomplete*. The accelerated observer cannot measure distances larger than a^{-1} in the direction opposite to the acceleration. To see this, let us consider a hypersurface of constant time $\xi^0 = \text{const}$. An infinite range of the spacelike coordinate, $-\infty < \xi^1 < 0$, where $\xi^1 = 0$ is the observer's location, spans a *finite* physical distance,

$$d = \int_{-\infty}^{0} e^{a\xi^1} d\xi^1 = \frac{1}{a}.$$

Therefore, there is no comoving accelerated frame which could cover the entire Minkowski spacetime. The lightcone $t = x$ plays the role of the event horizon; for example, the events P and Q will never become visible to the accelerated observer.

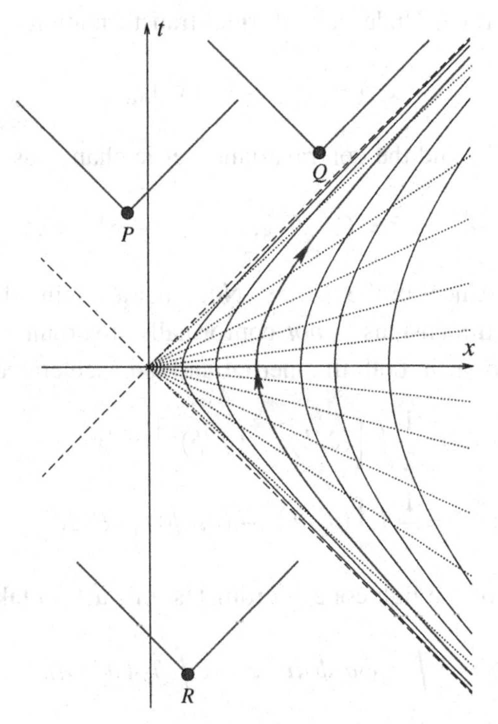

Fig. 8.2 The proper coordinate system of a uniformly accelerated observer in Minkowski spacetime. The solid hyperbolae are the lines of constant proper distance ξ^1; the hyperbola with arrows is $\xi^1 = 0$, or $x^2 - t^2 = a^{-2}$. The lines of constant ξ^0 are dotted. The dashed lines show the lightcone which corresponds to $\xi^1 = -a^{-1}$. The events P, Q, R are not covered by the proper coordinate system.

8.3 Quantum fields in inertial and accelerated frames

What a particular detector registers as a particle depends on the clocks used: Particles are determined by the positive frequency modes with respect to the proper time of the observer. An inertial observer defines these modes using the Minkowski time t, while the accelerated observer must use the proper time $\tau = \xi^0$. Since t and ξ^0 are related in a nontrivial way, one expects that the positive frequency mode with respect to t is a superposition of the positive and negative frequency modes with respect to ξ^0. As a result the Minkowski vacuum can appear as a state containing particles from the point of view of the accelerated observer.

The goal of this section is to show that this really takes place. The problem is greatly simplified for a massless scalar field in 1+1-dimensional spacetime because the action

$$S[\phi] = \frac{1}{2} \int g^{\alpha\beta} \phi_{,\alpha} \phi_{,\beta} \sqrt{-g} d^2x \tag{8.29}$$

is conformally invariant. Under a conformal transformation

$$g_{\alpha\beta} \to \tilde{g}_{\alpha\beta} = \Omega^2(x^\gamma) g_{\alpha\beta},$$

the determinant $\sqrt{-g}$ and the contravariant metric change as

$$\sqrt{-g} \to \sqrt{-\tilde{g}} = \Omega^2 \sqrt{-g}, \quad g^{\alpha\beta} \to \tilde{g}^{\alpha\beta} = \Omega^{-2} g^{\alpha\beta}, \tag{8.30}$$

and the factors Ω^2 cancel in the action. (Note that a minimally coupled massless scalar field in 3+1 dimensions is *not* conformally invariant.) This explains why the action looks similar in both the inertial and the accelerated frames:

$$\begin{aligned} S &= \frac{1}{2} \int \left[(\partial_t \phi)^2 - (\partial_x \phi)^2 \right] dt dx \\ &= \frac{1}{2} \int \left[(\partial_{\xi^0} \phi)^2 - (\partial_{\xi^1} \phi)^2 \right] d\xi^0 d\xi^1. \end{aligned} \tag{8.31}$$

Rewritten in terms of the lightcone coordinates, this action takes the form

$$S = 2 \int \partial_u \phi \partial_v \phi \, du dv = 2 \int \partial_{\tilde{u}} \phi \partial_{\tilde{v}} \phi \, d\tilde{u} d\tilde{v}.$$

The field equations

$$\partial_u \partial_v \phi = 0, \qquad \partial_{\tilde{u}} \partial_{\tilde{v}} \phi = 0,$$

have simple solutions,

$$\phi(u, v) = A(u) + B(v), \quad \phi(\tilde{u}, \tilde{v}) = \tilde{A}(\tilde{u}) + \tilde{B}(\tilde{v}), \tag{8.32}$$

where A, $\tilde{A}$ etc. are arbitrary smooth functions. In particular,

$$\phi \propto e^{-i\omega u} = e^{-i\omega(t-x)}$$

describes a right-moving, positive-frequency mode with respect to the Minkowski time t, while

$$\phi \propto e^{-i\Omega \tilde{u}} = e^{-i\Omega(\xi^0 - \xi^1)}$$

corresponds to a right-moving positive-frequency mode with respect to $\tau = \xi^0$. The solutions $\phi \propto e^{-i\omega v}$ and $\phi \propto e^{-i\Omega \tilde{v}}$ describe left-moving modes. Since $u = u(\tilde{u})$ and $v = v(\tilde{v})$, and the solutions of the wave equation have the form (8.32), the left- and right-moving modes do not affect each other and can be considered separately. In all formulae below we write explicitly only the right-moving modes. The reader can easily recover the contributions of the left-moving modes.

The actions in (8.31) have a canonical form. Therefore, in the domain $x > |t|$ of the spacetime where both coordinate frames overlap, we can immediately write the standard mode expansions for the field operator $\hat{\phi}$ as

$$\hat{\phi} = \int_0^\infty \frac{d\omega}{(2\pi)^{1/2}} \frac{1}{\sqrt{2\omega}} \left[e^{-i\omega u} \hat{a}_\omega^- + e^{i\omega u} \hat{a}_\omega^+ \right] + (\text{left-moving})$$

$$= \int_0^\infty \frac{d\Omega}{(2\pi)^{1/2}} \frac{1}{\sqrt{2\Omega}} \left[e^{-i\Omega \tilde{u}} \hat{b}_\Omega^- + e^{i\Omega \tilde{u}} \hat{b}_\Omega^+ \right] + (\text{left-moving}), \tag{8.33}$$

where both sets of operators $\hat{a}_\omega^\pm$ and $\hat{b}_\Omega^\pm$ satisfy the standard commutation relations:

$$\left[\hat{a}_\omega^-, \hat{a}_{\omega'}^+ \right] = \delta(\omega - \omega'), \quad \left[\hat{b}_\Omega^-, \hat{b}_{\Omega'}^+ \right] = \delta(\Omega - \Omega'), \text{ etc.} \tag{8.34}$$

Because we consider one spatial dimension, the normalization factor $(2\pi)^{1/2}$ in these formulae replaces the factor $(2\pi)^{3/2}$ used in the three-dimensional case.

The *Minkowski vacuum* $|0_M\rangle$ is the zero eigenvector of all annihilation operators $\hat{a}_\omega^-$, that is,

$$\hat{a}_\omega^- |0_M\rangle = 0.$$

In turn, the operators $\hat{b}_\Omega^-$ define the *Rindler vacuum* state $|0_R\rangle$,

$$\hat{b}_\Omega^- |0_R\rangle = 0.$$

The corresponding particle states are then built with the help of the creation operators in the standard way.

The states $|0_M\rangle$ and $|0_R\rangle$ are different and a natural question to ask is which of them is the "correct" vacuum. The answer to this question depends on the particular physical experiment considered. For example, for normalization of energy which contributes to the gravitational field we have to use the Minkowski vacuum $|0_M\rangle$. On the other hand, the detector of the accelerated observer reacts to particles associated with the Rindler vacuum $|0_R\rangle$. The detector remains unexcited only if the quantum field is in the state $|0_R\rangle$, while the Minkowski vacuum $|0_M\rangle$ is, from the point of view of the accelerated observer, a state containing particles. This is a manifestation of the Unruh effect.

In the rest of this chapter we find the relation between operators $\hat{a}_k^\pm$ and $\hat{b}_k^\pm$ and calculate the occupation numbers of the Rindler particles in the Minkowski vacuum state.

Remark: Rindler vacuum In contrast to the Minkowski vacuum, the Rindler vacuum is an unphysical state which is singular on the horizons $u = 0$ and $v = 0$. To get a rough idea why this is so, let us consider the appropriately regularized expectation values of the operators $(\partial_u \hat{\phi})^2$ and $(\partial_{\tilde{u}} \hat{\phi})^2$ for the Minkowski and Rindler vacuum states respectively. It follows from the expansions in (8.33) that

$$\langle 0_M | \left(\partial_u \hat{\phi}\right)^2 | 0_M \rangle = \langle 0_R | \left(\partial_{\tilde{u}} \hat{\phi}\right)^2 | 0_R \rangle , \tag{8.35}$$

and as a result we have

$$\langle 0_R | \left(\partial_u \hat{\phi}\right)^2 | 0_R \rangle = \left(\frac{\partial \tilde{u}}{\partial u}\right)^2 \langle 0_R | \left(\partial_{\tilde{u}} \hat{\phi}\right)^2 | 0_R \rangle = \frac{1}{a^2 u^2} \langle 0_M | \left(\partial_u \hat{\phi}\right)^2 | 0_M \rangle , \tag{8.36}$$

where (8.25) has been used. Thus, the expectation values of $(\partial_u \hat{\phi})^2$ taken for the Rindler and Minkowski vacua are related by a coordinate-dependent factor which becomes infinite on the future horizon $u = 0$. Because the Minkowski vacuum is a physically well-defined state, the Rindler vacuum is a singular state which requires an infinite energy to be prepared. Note that only the right-moving modes contribute to $(\partial_u \hat{\phi})^2$ and they are responsible for the singularity of the Rindler vacuum on the future horizon. Similarly, by considering $(\partial_v \hat{\phi})^2$ and $(\partial_{\tilde{v}} \hat{\phi})^2$, we can find that the left-moving modes lead to a singularity of the Rindler vacuum on the past horizon at $v = 0$.

8.4 Bogolyubov transformations

The operators $\hat{a}^{\pm}$ and $\hat{b}^{\pm}$ are related by the Bogolyubov transformations

$$\hat{b}_{\Omega}^{-} = \int_0^{\infty} d\omega \left[\alpha_{\Omega\omega} \hat{a}_{\omega}^{-} - \beta_{\Omega\omega} \hat{a}_{\omega}^{+}\right] . \tag{8.37}$$

Because the Rindler coordinates cover only a quarter of Minkowski spacetime, the inverse Bogolyubov transformation is not defined. The transformations (8.37) have a more general form than in (6.28) because *all* positive and negative frequency modes with respect to t contribute to the positive frequency mode with respect to τ, whereas the Bogolyubov transformations in (6.28) are "diagonal," with $\alpha_{\omega\Omega}$ and $\beta_{\omega\Omega}$ proportional to $\delta(\omega - \Omega)$. The normalization condition for the Bogolyubov coefficients,

$$\int_0^{\infty} d\omega \left(\alpha_{\Omega\omega} \alpha_{\Omega'\omega}^{*} - \beta_{\Omega\omega} \beta_{\Omega'\omega}^{*}\right) = \delta(\Omega - \Omega'), \tag{8.38}$$

follows from the compatibility of the commutation relations in (8.34). (This is the generalization of the condition $|\alpha_{\mathbf{k}}|^2 - |\beta_{\mathbf{k}}|^2 = 1$ we encountered before.)

Exercise 8.2

Derive (8.38).

Substituting (8.37) into (8.33) we infer that

$$\frac{1}{\sqrt{\omega}}e^{-i\omega u} = \int_0^\infty \frac{d\Omega'}{\sqrt{\Omega'}}\left(\alpha_{\Omega'\omega}e^{-i\Omega'\tilde{u}} - \beta^*_{\Omega'\omega}e^{i\Omega'\tilde{u}}\right). \tag{8.39}$$

Multiplying both sides of this relation by $\exp(\pm i\Omega\tilde{u})$, taking into account that

$$\int_{-\infty}^{+\infty} e^{i(\Omega-\Omega')\tilde{u}}d\tilde{u} = 2\pi\delta(\Omega-\Omega'),$$

and integrating over $\tilde{u}$ we obtain the result in terms of the Γ-function:

$$\left\{\begin{matrix}\alpha_{\Omega\omega}\\ \beta_{\Omega\omega}\end{matrix}\right. = \int_{-\infty}^{+\infty} e^{\mp i\omega u + i\Omega\tilde{u}}d\tilde{u} = \pm\frac{1}{2\pi}\sqrt{\frac{\Omega}{\omega}}\int_{-\infty}^{0}(-au)^{-\frac{i\Omega}{a}-1}e^{\mp i\omega u}du \tag{8.40}$$

$$= \pm\frac{1}{2\pi a}\sqrt{\frac{\Omega}{\omega}}e^{\pm\frac{\pi\Omega}{2a}}\exp\left(\frac{i\Omega}{a}\ln\frac{\omega}{a}\right)\Gamma\left(-\frac{i\Omega}{a}\right). \tag{8.41}$$

It follows that α and β obey the useful relation,

$$|\alpha_{\Omega\omega}|^2 = e^{\frac{2\pi\Omega}{a}}|\beta_{\Omega\omega}|^2. \tag{8.42}$$

Exercise 8.3
Derive (8.41).

Exercise 8.4
Derive (8.42) directly from (8.40) without using the Γ-function.

8.5 Occupation numbers and Unruh temperature

The vacua $|0_M\rangle$ and $|0_R\rangle$, annihilated by the operators $\hat{a}^-_\omega$ and $\hat{b}^-_\Omega$, are different. The Minkowski a-vacuum is a state with Rindler b-particles and vice versa. We now compute the number of b-particles in the a-vacuum state. The expectation value of the b-particle number operator $\hat{N}_\Omega \equiv \hat{b}^+_\Omega\hat{b}^-_\Omega$ in the Minkowski vacuum $|0_M\rangle$ is

$$\langle\hat{N}_\Omega\rangle \equiv \langle 0_M|\hat{b}^+_\Omega\hat{b}^-_\Omega|0_M\rangle = \langle 0_M|\left(\int d\omega\left[\alpha^*_{\omega\Omega}\hat{a}^+_\omega - \beta^*_{\omega\Omega}\hat{a}^-_\omega\right]\right)$$
$$\times\left(\int d\omega'\left[\alpha_{\omega'\Omega}\hat{a}^-_{\omega'} - \beta_{\omega'\Omega}\hat{a}^+_{\omega'}\right]\right)|0_M\rangle = \int d\omega\,|\beta_{\omega\Omega}|^2, \tag{8.43}$$

and this is interpreted as the mean number of particles with frequency Ω found by the accelerated observer.

For $\Omega' = \Omega$ the normalization condition (8.38) becomes

$$\int_0^\infty d\omega \left(|\alpha_{\Omega\omega}|^2 - |\beta_{\Omega\omega}|^2 \right) = \delta(0), \tag{8.44}$$

and taking into account (8.42) we find

$$\langle \hat{N}_\Omega \rangle = \int_0^{+\infty} d\omega \, |\beta_{\omega\Omega}|^2 = \left[\exp\left(\frac{2\pi\Omega}{a} \right) - 1 \right]^{-1} \delta(0). \tag{8.45}$$

The divergent factor $\delta(0)$ is due to the infinite volume of the entire space. If the field were quantized in a finite box of volume V, the momenta ω and Ω would be discrete and $\delta(0)$ would be replaced by the volume V, that is, $\delta(0) = V$. Thus the mean density of the particles with frequency Ω is

$$n_\Omega = \frac{\langle \hat{N}_\Omega \rangle}{V} = \left[\exp\left(\frac{2\pi\Omega}{a} \right) - 1 \right]^{-1}. \tag{8.46}$$

This is the main result of this chapter.

We have computed n_Ω only for right-moving modes (with positive momenta). The result for left-moving modes is obtained similarly. Massless particles detected by the accelerated detector in the Minkowski vacuum obey the Bose–Einstein distribution (8.46) with the *Unruh temperature*

$$T \equiv \frac{a}{2\pi}. \tag{8.47}$$

Thus an accelerated observer will see a thermal bath of particles. A physical interpretation of the Unruh effect is the following. The accelerated detector is coupled to the quantum vacuum fluctuations and these fluctuations act on the detector and excite it as if the detector were in a thermal bath with the temperature proportional to the acceleration. However, vacuum fluctuations in Minkowski spacetime cannot supply their own energy to excite the detector and they serve only as a "mediator" borrowing the energy from the agent responsible for acceleration. The acceleration required to produce a measurable temperature is enormous and therefore it is unlikely that the Unruh effect can be verified in the near future (see Exercise 1.6 on p. 12 for a numerical example). The energy spent by the accelerating agent is exponentially large compared with the energy in detected particles.

Finally, we note that the consideration above can be straightforwardly generalized to the case of four-dimensional spacetime, as well as to spinor- and vector-valued quantum fields.

9

Hawking effect. Thermodynamics of black holes

Summary Quantization of fields in a black hole background. Vacuum choice. Hawking radiation and black hole evaporation. Thermodynamics of black holes.

9.1 Hawking radiation

Black holes are massive objects which have such a strong gravitational field that even light cannot escape from them. According to the classical General Relativity, a black hole can only absorb matter and its size never decreases. In 1974 Hawking considered *quantum fields* in a *classical* black hole background and discovered that the black hole emits thermal particles and thus evaporates. This theoretical result came to a certain extent as a surprise. In fact, at that time one thought that particles can be produced only by a nonstatic gravitational field. For example, for a rotating black hole there exist negative-energy states outside its horizon, and therefore the gravitational field can convert a virtual particle–antiparticle pair into a pair of real particles with zero total energy. The positive-energy particle can then escape to infinity, while the negative-energy particle falls into the black hole. In this case the black hole can emit energy. This effect is known as *superradiance.* On the contrary, a nonrotating black hole has no negative energy states outside its horizon. Therefore, at first glance its mass cannot decrease and hence no particles can be produced from the quantum fluctuations outside the black hole horizon. To understand why particle production is nevertheless possible, we have to consider a virtual pair in the vicinity of the horizon. There are negative-energy states inside the horizon, and therefore one of the virtual particles (inside horizon) can have negative energy while the other one (outside horizon) has a positive energy. The first virtual particle can never escape from the black hole, but the second one can move away from the horizon to infinity thus becoming a real particle. As a result, the black hole can emit radiation and its mass decreases.

To get a rough estimate for the typical energy of the emitted quanta, we note that the de Broglie wavelength of particles with energies $E \leq \hbar c / r_g$, where $r_g = 2GM/c^2$ is the gravitational radius, is larger than the size of the black hole. Therefore, one could naively expect that the probability that such particles might escape falling into the black hole is non-negligible. If the produced particles were thermal, then their temperature must be of order

$$T \sim \frac{\hbar c}{r_g} \sim \frac{\hbar c^3}{GM}.$$

We would like to warn the reader that the arguments above should not be taken too seriously. They are simply a naive guess which presumably could help our intuitive understanding of the origin of the Hawking radiation. However, from these arguments it remains completely unclear whether and how the radiation is actually emitted.

In this chapter we derive the Hawking result for a massless scalar field in two-dimensional spacetime. We will see that the calculations can then be formally reduced to those we performed for an accelerated observer. After that we discuss how the results should be modified in the realistic case of the four-dimensional black hole.

9.1.1 Schwarzschild solution

A 4-dimensional nonrotating black hole without electric charge is described by the Schwarzschild metric,

$$ds^2 = \left(1 - \frac{2M}{r}\right) dt^2 - \frac{dr^2}{1 - \frac{2M}{r}} - r^2 \left(d\theta^2 + d\varphi^2 \sin^2 \theta\right), \tag{9.1}$$

where M is the mass of a black hole (from now on we use the natural units where $G = \hbar = c = 1$). To simplify the calculations, we first consider a two-dimensional black hole, assuming that its metric is the same as the time-radial part of the Schwarzschild metric:

$$ds^2 = g_{ab} dx^a dx^b = \left(1 - \frac{r_g}{r}\right) dt^2 - \frac{dr^2}{1 - \frac{r_g}{r}}, \tag{9.2}$$

where $r_g = 2M$. It is convenient to introduce the "tortoise coordinate" $r^*(r)$ according to

$$dr^* = \frac{dr}{1 - \frac{r_g}{r}},$$

so that

$$r^*(r) = r - r_g + r_g \ln\left(\frac{r}{r_g} - 1\right). \tag{9.3}$$

Metric (9.2) then becomes

$$ds^2 = \left(1 - \frac{r_g}{r(r^*)}\right)\left[dt^2 - dr^{*2}\right], \tag{9.4}$$

where r must be expressed through r^* using (9.3). The tortoise coordinate r^* is defined only for $r > r_g$ and varies in the range $-\infty < r^* < +\infty$. As r approaches r_g the coordinate r^* goes to $-\infty$ (this explains the origin of the name for this coordinate) and far away from the black hole $r^* \to r$ as $r \to \infty$.

Introducing the tortoise lightcone coordinates

$$\tilde{u} \equiv t - r^*, \quad \tilde{v} \equiv t + r^*, \tag{9.5}$$

we can rewrite metric (9.2) in the form

$$ds^2 = \left(1 - \frac{r_g}{r(\tilde{u}, \tilde{v})}\right) d\tilde{u}\, d\tilde{v}. \tag{9.6}$$

9.1.2 Kruskal–Szekeres coordinates

The Schwarzschild coordinates are singular on the horizon at $r = r_g$. The tortoise lightcone coordinates $\tilde{u}, \tilde{v}$ are also singular and they cover only the "exterior" of the black hole, $r > r_g$. To describe the entire spacetime, we need another coordinate system. It follows from (9.3) and (9.5) that

$$1 - \frac{r_g}{r} = \frac{r_g}{r} \exp\left(1 - \frac{r}{r_g}\right) \exp\left(\frac{\tilde{v} - \tilde{u}}{2r_g}\right),$$

and hence metric (9.6) can be rewritten as

$$ds^2 = \frac{r_g}{r} \exp\left(1 - \frac{r}{r_g}\right) e^{-\frac{\tilde{u}}{2r_g}} e^{\frac{\tilde{v}}{2r_g}} d\tilde{u}\, d\tilde{v}. \tag{9.7}$$

In the Kruskal–Szekeres lightcone coordinates, defined as

$$u = -2r_g \exp\left(-\frac{\tilde{u}}{2r_g}\right), \quad v = 2r_g \exp\left(\frac{\tilde{v}}{2r_g}\right), \tag{9.8}$$

this metric takes the form

$$ds^2 = \frac{r_g}{r(u, v)} \exp\left(1 - \frac{r(u, v)}{r_g}\right) du dv \tag{9.9}$$

and becomes regular at $r = r_g$. Thus the singularity that occurs in the Schwarzschild metric as $r \to r_g$ is merely a *coordinate singularity*, which can be removed by a coordinate transformation. A freely falling observer will see nothing peculiar while crossing the horizon. As defined in (9.8), the Kruskal–Szekeres coordinates vary in the intervals $-\infty < u < 0$ and $0 < v < +\infty$, covering the "exterior" of the black hole at $r > r_g$. However, they can be analytically extended to $u > 0$ and $v < 0$ where metric (9.9) is still well-defined. The Kruskal–Szekeres coordinates span the ranges $-\infty < u < \infty$ and $-\infty < v < \infty$, and thus cover the whole Schwarzschild spacetime.

The relation between the lightcone Kruskal coordinates u, v and the original Schwarzschild coordinates t, r can be easily found from (9.8), if we take into account (9.5) and (9.3):

$$uv = -4r_g^2 \exp\left(\frac{r^*}{r_g}\right) = -4r_g^2 \left(\frac{r}{r_g} - 1\right) \exp\left(\frac{r}{r_g} - 1\right), \tag{9.10}$$

$$\left(\frac{v}{u}\right)^2 = \exp\left(\frac{2t}{r_g}\right). \tag{9.11}$$

Note that these relations make sense (via analytic continuation) for an arbitrary u and v, that is they are valid also outside the range of applicability of (9.8). It follows from (9.10) that the black hole horizon, $r = r_g$, corresponds to $u = 0$ and $v = 0$. Thus, the Schwarzschild spacetime has in fact two horizons resolved only in the Kruskal–Szekeres coordinates. We see from (9.11) that $v = 0$ corresponds to $t = -\infty$ (the past horizon), while $u = 0$ corresponds to $t = +\infty$ (the future horizon).

Introducing the timelike and spacelike coordinates T and R according to

$$u = T - R, \quad v = T + R, \tag{9.12}$$

we can draw in the (T, R) plane the *Kruskal diagram* for the Schwarzschild spacetime, as shown in Fig. 9.1. Null geodesics $u = \text{const}$ and $v = \text{const}$ are straight lines at $\pm 45°$ angles in this plane. It follows from (9.10) that the hypersurfaces $r = \text{const}$ correspond to $uv = T^2 - R^2 = \text{const}$. For $r > r_g$, we have $uv < 0$ and the lines $r = \text{const}$ are timelike, while for $r < r_g$, where $uv > 0$, they are spacelike. Therefore, the Schwarzschild r coordinate has the usual interpretation as the spacelike radial coordinate only outside the horizon ($r > r_g$). We infer from (9.11) that the surfaces $t = \text{const}$ are straight lines in the (T, R) plane. The Schwarzschild coordinate t is interpreted as the time for $r > r_g$ but becomes a spatial coordinate for $r < r_g$. There is a physical singularity at $r = 0$ where the curvature invariants

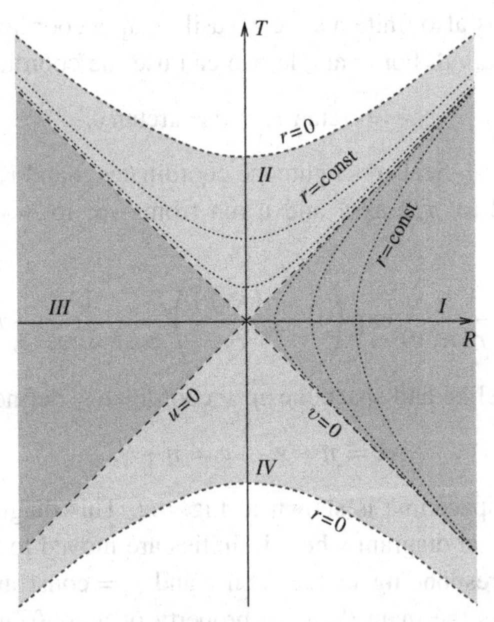

Fig. 9.1 A spacetime diagram in Kruskal–Szekeres coordinates (R, T). Shaded regions *I–IV* are the different asymptotic domains of the spacetime. Dashed lines represent the horizon $u = 0$ and $v = 0$. Dotted lines are surfaces of constant r. Thick dotted lines represent the singularity $r = 0$.

become infinite. This spacelike singularity (occurring at the given moment of time) corresponds in fact to two singularities resolved as

$$T = \pm\sqrt{R^2 + 4e^{-1} r_g^2}, \tag{9.13}$$

in T, R coordinates. The spacetime is not extendable beyond the singularities.

The spacetime shown in Fig. 9.1 describes four different physical objects. It has two singularities (one in past and the other in future), two horizons ($u = 0$ and $v = 0$) and two asymptotically flat regions (as $R \to \infty$ and $R \to -\infty$). Therefore, the fully extended Schwarzschild solution contains two white holes that share the same past singularity at $T = -\sqrt{R^2 + 4e^{-1} r_g^2}$ but have two different asymptotically flat regions as $R \to \pm\infty$, and two black holes sharing the same future singularity. What can be realized in nature is only one of these four possibilities. When we are interested in describing a black hole formed as a result of collapse, the relevant part of the diagram includes the regions *I* and *II* with omitted past horizon. It is clear that the physics of the formed black hole should not depend on the other domains of the fully extended Schwarzschild spacetime, shown in the Kruskal diagram.

Conformal diagram Sometimes it is convenient to represent the causal structure of the manifold using coordinates that vary within finite ranges. Then the spacetime diagram is also finite. One can easily map a coordinate with an infinite range to a finite interval. For example, we can use the coordinates

$$\varkappa = \arctan u, \quad \nu = \arctan v, \tag{9.14}$$

instead of the Kruskal–Szekeres lightcone coordinates u and v. These coordinates change from $-\pi/2$ to $\pi/2$ as u and v run from $-\infty$ to $+\infty$, and metric (9.9) takes the form

$$ds^2 = \frac{r_g}{r(\varkappa, \nu)} \exp\left(1 - \frac{r(\varkappa, \nu)}{r_g}\right) \frac{1}{\cos^2 \varkappa \cos^2 \nu} d\varkappa d\nu. \tag{9.15}$$

In the bounded timelike and spacelike η, χ coordinates, defined via

$$\varkappa = \eta - \chi, \quad \nu = \eta + \chi, \tag{9.16}$$

the Schwarzschild spacetime is shown in Fig. 9.2. This diagram can be viewed as a deformed Kruskal diagram where infinities are moved to a finite "distance." Null geodesics corresponding to $\varkappa = \text{const}$ and $\nu = \text{const}$ are straight lines at $\pm 45°$ angles. This is the main defining property of a conformal diagram, along with the fact that the diagram has a finite size. Although spacelike and timelike hypersurfaces can be arbitrarily deformed, null geodesics defining the causal structure of the manifold must always remain straight lines at $\pm 45°$ angles.

The black hole horizon $r = r_g$ corresponds to $\varkappa = 0$ and $\nu = 0$. In Fig. 9.2 one can easily recognize the different kind of infinities occurring in the Schwarzschild spacetime. In particular, there are two future timelike infinities i^+ where all timelike lines end ($T \to +\infty$, R is finite). The past timelike infinities ($T \to -\infty$, R is finite) are denoted by i^-. The spacelike infinities i^0 correspond to $|R| \to \infty$ at finite T. The straight lines connecting i^0 and i^+ are the future null infinities $\mathcal{J}^+$($T \to +\infty$ with $T+R$ or $T-R$ finite), the regions toward which null geodesics extend. The past null infinities $\mathcal{J}^-$ are the regions where null geodesics originate.

Exercise 9.1

Draw a conformal diagram for Minkowski spacetime.

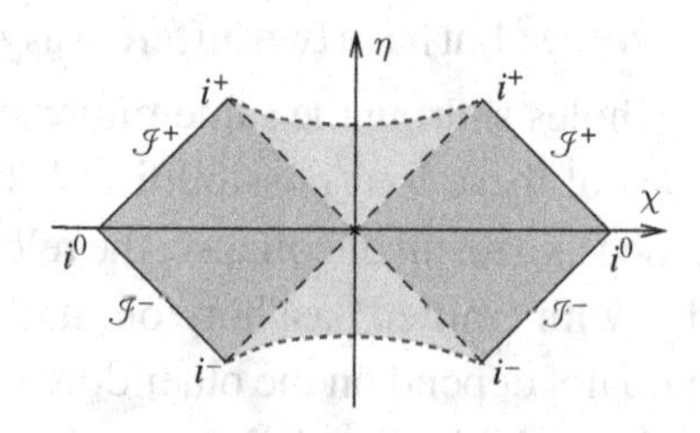

Fig. 9.2 A conformal diagram of the fully extended Schwarzschild spacetime, using the coordinates (9.15)–(9.16). Dashed lines represent the horizons. Thick dotted lines are the singularities. The shaded regions are the same as in Fig. 9.1.

9.1.3 Field quantization and Hawking radiation

Let us consider a scalar field with the action

$$S[\phi] = \frac{1}{2}\int g^{\alpha\beta}\phi_{,\alpha}\phi_{,\beta}\sqrt{-g}d^2x \tag{9.17}$$

in a two-dimensional spacetime. Since this action is conformally invariant, the solution of the scalar field equation can be written either in terms of the lightcone tortoise coordinates (9.5) as

$$\phi = \tilde{A}(\tilde{u}) + \tilde{B}(\tilde{v}), \tag{9.18}$$

or in the lightcone Kruskal–Szekeres coordinates as

$$\phi = A(u) + B(v), \tag{9.19}$$

where A, $\tilde{A}$ etc. are arbitrary smooth functions. The situation is similar to the case of the scalar field in the Rindler spacetime. In particular,

$$\phi \propto e^{-i\omega\tilde{u}} = e^{-i\omega(t-r^*)}$$

describes a right-moving positive-frequency mode with respect to time t, which propagates away from the black hole. The proper time of an observer at rest located far away from the black hole coincides with t since

$$ds^2 \to d\tilde{u}d\tilde{v} = dt^2 - dr^{*2}$$

as $r \to \infty$ (see (9.6)). Therefore this observer associates "particles" with the positive-frequency modes with respect to the time t. The expansion of the field operator,

$$\hat{\phi} = \int_0^\infty \frac{d\Omega}{(2\pi)^{1/2}} \frac{1}{\sqrt{2\Omega}} \left[e^{-i\Omega\tilde{u}}\hat{b}_\Omega^- + e^{i\Omega\tilde{u}}\hat{b}_\Omega^+ \right] + (\text{left-moving}), \tag{9.20}$$

determines the corresponding creation and annihilation operators $\hat{b}_\Omega^\pm$. As before, to simplify the formulae we do not write explicitly the contribution of the left-moving modes. The eigenstate $|0_B\rangle$ defined via

$$\hat{b}_\Omega^- |0_B\rangle = 0$$

is called the *Boulware vacuum.* It contains no particles from the point of view of the far-away observer. The tortoise coordinates cover only the part of the Schwarzschild spacetime outside the black hole horizon. In this sense they are similar to the Rindler coordinates of an accelerated observer, and the Boulware vacuum is similar to the Rindler vacuum. Therefore the Boulware vacuum $|0_B\rangle$ is singular on the black hole horizon and hence it is physically unacceptable (see remark at the end of Section 8.3). In particular, the regularized energy density diverges on the horizon, and hence the backreaction of the quantum fluctuations makes the picture of a quantum field in the background of the almost unperturbed classical black hole inconsistent.

The Kruskal–Szekeres coordinates are nonsingular on the horizon and cover the whole Schwarzschild spacetime. In this sense they are similar to the inertial coordinates in Minkowski spacetime. In the vicinity of the horizon,

$$ds^2 \to du dv = dT^2 - dR^2,$$

and particles registered by an observer crossing the horizon should be associated with positive-frequency modes with respect to the time T. Expanding the field operator in terms of the Kruskal–Szekeres lightcone coordinate,

$$\hat{\phi} = \int_0^\infty \frac{d\omega}{(2\pi)^{1/2}} \frac{1}{\sqrt{2\omega}} \left[e^{-i\omega u} \hat{a}_\omega^- + e^{i\omega u} \hat{a}_\omega^+ \right] + (\text{left-moving}), \tag{9.21}$$

we define the creation and annihilation operators $\hat{a}_\omega^\pm$ that determine the Kruskal "vacuum" state,

$$\hat{a}_\omega^- |0_\mathrm{K}\rangle = 0.$$

The state $|0_\mathrm{K}\rangle$ is obviously regular on the horizon. Moreover, it leads to a finite energy density (after a subtraction of the zero-point energy), which for a large black hole is small everywhere except at the singularities. As a result, the backreaction of quantum fluctuations is negligible and they do not destroy the classical background. Hence the Kruskal state $|0_\mathrm{K}\rangle$ is a natural candidate to be the true physical "vacuum" in the presence of the black hole.

From the point of view of a remote observer, the Kruskal vacuum $|0_\mathrm{K}\rangle$ contains particles. To determine their number density, we can simply exploit the formal similarity between the formulae describing the accelerated observer and the black hole in two dimensions. Comparing (9.8) and (8.25), we see that the Kruskal–Szekeres and tortoise lightcone coordinates are related exactly in the same way as the Minkowski and Rindler lightcone coordinates, if the acceleration a is replaced by the *surface gravity* $\kappa = (2r_g)^{-1}$. Then the mathematical similarity between the two cases becomes obvious:

Accelerated observer	Schwarzschild spacetime
Minkowski vacuum $\|0_\mathrm{M}\rangle$	Kruskal vacuum $\|0_\mathrm{K}\rangle$
Rindler vacuum $\|0_\mathrm{R}\rangle$	Boulware vacuum $\|0_\mathrm{B}\rangle$
Acceleration a	Surface gravity $\kappa = (2r_g)^{-1}$
$u = -a^{-1}\exp(-a\tilde{u})$	$u = -\kappa^{-1}\exp(-\kappa\tilde{u})$
$v = a^{-1}\exp(a\tilde{v})$	$v = \kappa^{-1}\exp(\kappa\tilde{v})$

Using this similarity and recalling (8.43) and (8.45), we find that the remote observer must see particles with the thermal spectrum,

$$\langle \hat{N}_\Omega \rangle \equiv \langle 0_K | \hat{b}^+_\Omega \hat{b}^-_\Omega | 0_K \rangle = \left[\exp\left(\frac{2\pi\Omega}{\kappa} \right) - 1 \right]^{-1} \delta(0), \tag{9.22}$$

corresponding to the temperature

$$T_H = \frac{\kappa}{2\pi} = \frac{1}{8\pi M}. \tag{9.23}$$

Note that the state $|0_K\rangle$ contains not only outgoing (right-moving) particles, but also the incoming (left-moving) particles with the same thermal spectrum. Only in this case the Kruskal vacuum is nonsingular on the past horizon ($v = 0$) of the eternal Schwarzschild spacetime (see the remark in Section 8.3). Therefore, in the presence of quantum fields the picture of an eternal black hole is consistent only if the black hole is placed in a thermal reservoir with the temperature T_H. Because the black hole absorbs particles, it should also emit them to maintain the equilibrium. It then follows that the black hole placed in an empty space must evaporate by emitting thermal radiation with the temperature given in (9.23). If we consider a non-eternal black hole formed as a result of gravitational collapse, the past horizon at $v = 0$ does not exist, and therefore there is no need to choose for the left-moving v-modes the Kruskal state annihilated simultaneously by the operators $\hat{a}^-$ entering (9.21) for both right- and left-moving modes. Instead, it is more natural to choose the state annihilated by the operators $\hat{a}^-$ for the right-moving modes and by the operators $\hat{b}^-$ for the left-moving modes. Because there is no past horizon, this state is also non-singular and does not contain incoming particles from the point of view of the remote observer. Hence this observer sees only thermal radiation coming from the black hole. The corresponding choice of the quantum state is justified by the consideration of the black hole formed as a result of a spherical shell collapse. In the remote past before the collapse takes place, the spacetime is almost flat and the initial "in"-state of quantum fields is the Minkowski vacuum containing no particles. One can show that in this case the final "out"-state of the quantum fields in the remote future contain the thermal flux of particles coming from the formed black hole. These particles are associated with the modes which spent a very long time near the horizon before they managed to escape (see Fig. 9.3).

One more way to derive the Hawking radiation is with the help of the renormalized energy-momentum tensor $T_{\mu\nu}$ of a quantum field in classical black hole background. It was explicitly calculated for conformal fields in two-dimensional spacetime,[1] and it was found that along with a local vacuum polarization contribution

[1] V. P. Frolov and G. A. Vilkovisky (1979), published in *Proceedings 2nd Marcel Grossmann Meeting, Trieste, Italy, Jul. 5–11, 1979* Grossmann Mts. 1979:0455; V. P. Frolov and G. A. Vilkovisky, *Phys. Lett. B* **106** (1981), 307.

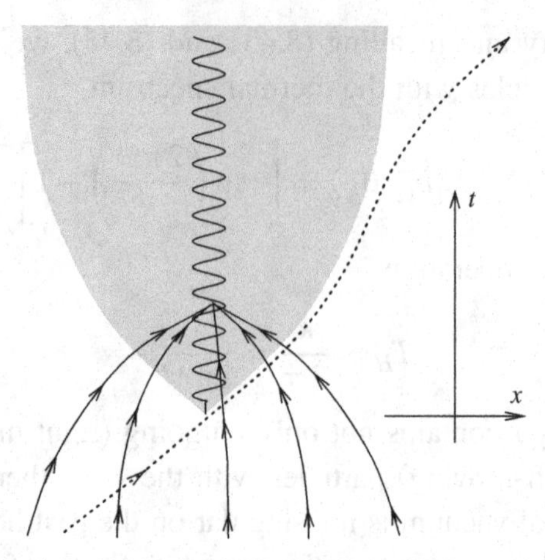

Fig. 9.3 Black hole (shaded region) formed by gravitational collapse of matter (lines with arrows). The wavy line marks the singularity at the BH center. A light-like trajectory (dotted line) may linger near the horizon (the boundary of the shaded region) for a long time before escaping to infinity.

the energy-momentum tensor also contains a non-local contribution from the Hawking thermal flux. The Hawking radiation has been derived in many different ways, and at present there is very little doubt that the Hawking effect is a valid and inevitable prediction of classical general relativity combined with quantum field theory.

Exercise 9.2

Rewrite (9.23) in SI units and compute the Hawking temperature for black holes of masses $M_1 = M_\odot = 2 \cdot 10^{30}$kg (one solar mass), $M_2 = 10^{15}$g, and $M_3 = 10^{-5}$g (of order of the Planck mass).

Exercise 9.3

(a) Estimate the typical wavelength of photons radiated by a black hole of mass M and compare it with the size of the black hole (the Schwarzschild radius $R = 2M$).

(b) The temperature of a sufficiently small black hole can be high enough to efficiently produce baryons (e.g. protons) as components of the Hawking radiation. Estimate the required mass M of such black holes and compare their Schwarzschild radius with the size of the proton (its Compton wavelength).

9.1.4 Hawking effect in 3+1 dimensions

So far we have considered the two-dimensional spacetime. It is natural to ask how the derivations and results above are modified for the realistic $3+1$-dimensional case. Let us first consider a minimally coupled massless scalar field and decompose it into spherical harmonics,

$$\phi(t, r, \theta, \varphi) = \sum_{l,m} \phi_{lm}(t, r) Y_{lm}(\theta, \varphi). \tag{9.24}$$

Substituting this expansion into the wave equation

$$^{(4)}\Box\phi = 0 \tag{9.25}$$

we find that (9.25) in the black hole background (9.1) reduces to

$$\left[{}^{(2)}\Box + \left(1 - \frac{r_g}{r}\right)\left(\frac{r_g}{r^3} + \frac{l(l+1)}{r^2}\right)\right] \phi_{lm}(t, r) = 0 \tag{9.26}$$

for each mode $\phi_{lm}(t, r)$. Formally, this equation describes a wave propagating in the two-dimensional spacetime with metric (9.2) and an extra potential

$$V_l(r) = \left(1 - \frac{r_g}{r}\right)\left(\frac{r_g}{r^3} + \frac{l(l+1)}{r^2}\right). \tag{9.27}$$

Even for the spherically symmetric mode $\phi_{00}(t, r)$, the potential V does not vanish, as shown in Fig. 9.4.

A wave escaping the black hole needs to propagate through the potential $V_l(r)$, and this decreases the intensity of the wave and changes the resulting spectrum by a *greybody factor* $\Gamma_l(\Omega) < 1$,

$$n_\Omega = \Gamma_l(\Omega)\left[\exp\left(\frac{\Omega}{T_H}\right) - 1\right]^{-1}. \tag{9.28}$$

Because the greybody factor is entirely due to the potential outside the black hole horizon it is clear that this factor is not directly related to the quantum origin of the Hawking radiation. Therefore, the basic features of the derivation above "survive" without significant alteration for the four-dimensional case. This result

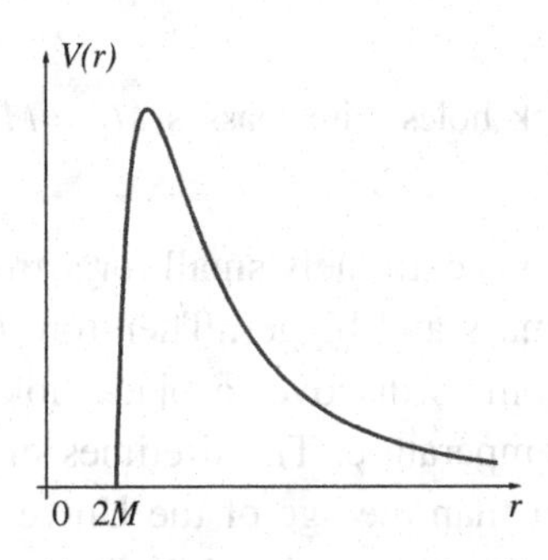

Fig. 9.4 The potential $V(r)$ for the propagation of the spherically symmetric mode $\phi_{00}(t, r)$ in $3+1$ dimensions.

can be generalized for the case of a massive scalar field, and also for vector and spinor fields. The main conclusion is that the black hole must emit all possible species of particles, each having the Hawking thermal spectrum corrected by the corresponding greybody factor.

9.2 Thermodynamics of black holes

Let us consider an isolated black hole placed in empty space. The black hole emits radiation and thus loses its mass. To calculate the flux L of the radiated energy, we can treat the black hole as a spherical body with the surface area

$$A = 4\pi r_{\rm g}^2 = 16\pi M^2$$

and surface temperature $T_H = (8\pi M)^{-1}$. Using the Stefan–Boltzmann law, we obtain

$$L = \Gamma\gamma\sigma T_{\rm H}^4 A = \frac{\Gamma\gamma}{15360\pi M^2},$$

where Γ is the coefficient correcting for the greybody factors, γ is the total number of the "massless" degrees of freedom, and $\sigma = \pi^2/60$ is the Stefan–Boltzmann constant in Planck units. Then the mass of the black hole decreases with time according to

$$\frac{dM}{dt} = -L = -\frac{\Gamma\gamma}{15360\pi M^2}. \tag{9.29}$$

The solution of this equation with the initial condition $M|_{t=0} = M_0$ is

$$M(t) = M_0\left(1 - \frac{t}{t_{\rm L}}\right)^{1/3}, \quad t_{\rm L} \equiv 5120\pi\frac{M_0^3}{\Gamma\gamma}.$$

Thus an isolated black hole has a finite lifetime $t_{\rm L} \sim M_0^3$. During this time the black hole evaporates, and its temperature becomes very high at the last stages of evaporation, increasing in inverse proportion to its mass.

Exercise 9.4

Estimate the lifetimes of black holes with masses $M_1 = M_\odot = 2\cdot 10^{30}$kg, $M_2 = 10^{15}$g, $M_3 = 10^{-5}$g.

The Hawking temperature is extremely small for astrophysical black holes with masses of order the solar mass and bigger. Therefore the Hawking effect could be observed only if astronomers discover a black hole near the end of its life, with a very high surface temperature. The lifetimes of astrophysically plausible black holes are much larger than the age of the Universe which is of order 10^{10} years. To evaporate within this time, a black hole must be lighter than $\sim 10^{15}$g (see Exercise 9.4). Such black holes could not have been formed as a result of stellar collapse and must be *primordial,* i.e. created at very early times when the universe was extremely dense and hot. There is currently no direct observational evidence for the existence of such primordial black holes.

It is almost certain that the final stage of the black hole evaporation cannot be described using classical general relativity. When the mass of the black hole reaches 10^{-5}g, its size is of order the Planck scale (10^{-33}cm), and so one expects that nonperturbative quantum gravity effects must become dominant. It is possible that these effects stabilize a "remnant," and a microscopic black hole with mass 10^{-5}g does not radiate, similarly to electrons in atoms that do not radiate on the lowest orbit.

9.2.1 Laws of black hole thermodynamics

Prior to the discovery of the black hole radiation, J. Bekenstein conjectured in 1971 that black holes must have a nonvanishing intrinsic entropy. Let us consider a black hole that absorbs matter with nonzero entropy. If the black hole entropy were zero, then the total entropy of the system including the black hole would decrease, which would violate the second law of thermodynamics. Based on such *gedanken* experiments, J. Bekenstein concluded that the second law of thermodynamics can be preserved only if one attributes to a black hole an intrinsic entropy S_{BH} proportional to its surface area A. However, the coefficient of proportionality between S_{BH} and A could not be fixed until the discovery of the Hawking radiation.

Differentiating the expression for the surface area $A = 16\pi M^2$, we can rewrite it in the form

$$dM = \frac{1}{8\pi M} d\left(\frac{A}{4}\right), \tag{9.30}$$

which reminds us of the first law of thermodynamics,

$$dE = TdS. \tag{9.31}$$

Taking into account that $dM = dE$ and $(8\pi M)^{-1} = T_{\mathrm{H}}$, we conclude from (9.30) that the black hole entropy must be equal to

$$S_{\mathrm{BH}} = \frac{1}{4}A = 4\pi M^2, \tag{9.32}$$

and the first law of black hole thermodynamics becomes

$$dM = T_{\mathrm{H}} dS_{\mathrm{BH}}. \tag{9.33}$$

The entropy of a typical astrophysically plausible black hole is extremely large. For instance, a black hole of one solar mass has the entropy $S_{\odot} \sim 10^{76}$. This large entropy implies a large number of possible microstates hidden behind the horizon of the black hole. The origin of these microstates can in principle be related with the number of possible matter configurations from which a black hole of a given mass could be formed.

Remark A static black hole without charge is completely characterized by its mass M. The first law of black hole thermodynamics can be generalized also for the case of rotating and charged black holes.

Taking into account the entropy of black holes, the *generalized second law* of thermodynamics states that the total entropy of all the black holes and of ordinary matter never decreases,

$$\delta S_{\text{total}} = \delta S_{\text{matter}} + \delta S_{\text{BH}} \geq 0.$$

In classical general relativity it has been established that the combined area of all black horizons cannot decrease. This statement applies not only to adiabatic processes but also to strongly out-of-equilibrium situations, such as a collision of black holes with the resulting merger.

Black holes in heat reservoirs Ordinary thermodynamical systems can be in a stable equilibrium with an infinite heat reservoir. However, this is not true for black holes because they have a *negative* heat capacity C_{BH}. In other words, black holes become *colder* when they absorb energy. Taking into account that $E(T) = M = (8\pi T)^{-1}$, we obtain

$$C_{\text{BH}} = \frac{\partial E}{\partial T} = -\frac{1}{8\pi T^2} < 0.$$

A black hole surrounded by an infinite thermal bath with the temperature $T < T_{\text{BH}}$ would emit radiation and become even hotter. The process of evaporation is not halted in an infinite thermal reservoir whose temperature T remains constant. Similarly, a black hole placed inside an infinite reservoir with a higher temperature $T > T_{\text{BH}}$ will be absorbing radiation and becoming colder. In either case, no stable equilibrium is possible. A stable equilibrium can be achieved only if the black hole is placed in a reservoir with a finite energy. In this case the radiation emitted by the black hole changes the temperature in the reservoir. The following exercise shows under which conditions a black hole can be stabilized with respect to absorption or emission of radiation.

Exercise 9.5

(a) Find the range of heat capacities C_r of a heat reservoir for which a black hole of mass M is in a stable equilibrium with the reservoir.

(b) Assume that the reservoir is a completely reflecting cavity of volume V filled with thermal radiation (massless fields). The energy of the radiation is $E_r = \sigma\gamma VT^4$, where the constant γ characterizes the number of degrees of freedom in the radiation fields. Determine the largest volume V for which a black hole of mass M can remain in a stable equilibrium with the surrounding radiation.
Hint: A stable equilibrium is the state with the largest total entropy.

10

The Casimir effect

Summary Zero-point energy in the presence of boundaries. Regularization and renormalization. Casimir force.

The Casimir effect is an experimentally verified prediction of quantum field theory. It is manifested by a force of attraction between two *uncharged* conducting plates in a vacuum. This force cannot be explained except by considering the vacuum fluctuations of the quantized electromagnetic field. The presence of the conducting plates makes the electromagnetic field vanish on the surfaces of the plates, causing a finite shift ΔE of the zero-point energy. This shift depends on the distance L between the plates, and as a result there arises the *Casimir force*:

$$F(L) = -\frac{d}{dL}\Delta E(L).$$

This theoretical prediction has been verified experimentally.[1]

10.1 Vacuum energy between plates

A realistic description of the Casimir effect requires quantization of the electromagnetic field in the presence of conducting plates. To simplify the calculations, we consider a two-dimensional massless scalar field $\phi(t, x)$ between two plates at $x = 0$ and $x = L$, imposing the boundary conditions

$$\phi(t, x)|_{x=0} = \phi(t, x)|_{x=L} = 0,$$

which are supposed to be due to the presence of the plates. With these boundary conditions the general solution of the classical equation of motion,

$$\partial_t^2\phi - \partial_x^2\phi = 0,$$

[1] For example, a recent measurement of the Casimir force to 1% precision is described in: U. Mohideen and A. Roy, *Phys. Rev. Lett.* **81** (1998), 4549.

becomes

$$\phi(t,x) = \sum_{n=-\infty}^{\infty} \left(A_n e^{-i\omega_n t} + B_n e^{i\omega_n t}\right) \sin \omega_n x, \quad \omega_n \equiv \frac{|n|\pi}{L}. \tag{10.1}$$

In quantum theory only the modes present in (10.1) "survive" in the expansion of the field operator $\hat{\phi}$, which becomes

$$\hat{\phi}(t,x) = \sqrt{\frac{1}{L}} \sum_{n=1}^{\infty} \frac{\sin \omega_n x}{\sqrt{\omega_n}} \left[\hat{a}_n^- e^{-i\omega_n t} + \hat{a}_n^+ e^{i\omega_n t}\right]. \tag{10.2}$$

The resulting zero-point energy per unit length between the plates is then

$$\varepsilon_0 \equiv \frac{1}{L} \langle 0| \hat{H} |0\rangle = \frac{1}{2L} \sum_k \omega_k = \frac{\pi}{2L^2} \sum_{n=1}^{\infty} n. \tag{10.3}$$

Exercise 10.1

(a) Show that the mode expansion (10.2) yields the standard commutation relations $[\hat{a}_m^-, \hat{a}_n^+] = \delta_{mn}$.

(b) Derive (10.3).

Hint: Use the identities which hold for integers m, n:

$$\int_0^L dx \sin \frac{m\pi x}{L} \sin \frac{n\pi x}{L} = \int_0^L dx \cos \frac{m\pi x}{L} \cos \frac{n\pi x}{L} = \frac{L}{2} \delta_{mn}. \tag{10.4}$$

10.2 Regularization and renormalization

The zero-point energy density ε_0 is divergent and must be first regularized and then renormalized. A *regularization* means introducing an extra parameter (cutoff scale) into the theory to make the divergent quantity finite unless that parameter is set either to zero or infinity depending on the concrete regularization procedure used. Usually there exist different possible ways to regularize the divergent quantities. However, different regularization procedures (fortunately) lead to the same final physical results. After a regularization, one obtains an asymptotic expansion of the regularized divergent quantity at small (or large) values of the cutoff. This asymptotic expansion may contain divergent powers and logarithms of the cutoff scale, as well as finite terms. *Renormalization* gives a physical justification for removing the divergent terms and leaves us with finite contributions responsible for physical effects.

We shall now apply this procedure to (10.3). As a first step, we replace ε_0 by the regularized quantity

$$\varepsilon_0(L;\alpha) = \frac{\pi}{2L^2} \sum_{n=1}^{\infty} n \exp\left[-\frac{n\alpha}{L}\right], \tag{10.5}$$

where α is the cutoff parameter. It is easy to see that the series in (10.5) converges for $\alpha > 0$, while the original divergent expression is recovered in the limit $\alpha \to 0$.

Remark We regularize the series by $\exp(-n\alpha/L)$ and not by $\exp(-n\alpha)$ or $\exp(-nL\alpha)$. The motivation is that the physically significant quantity is $\omega_n = \pi n/L$, therefore the cutoff factor should be a function of ω_n.

A straightforward calculation gives

$$\varepsilon_0(L;\alpha) = -\frac{\pi}{2L}\frac{\partial}{\partial\alpha}\sum_{n=1}^{\infty}\exp\left[-\frac{n\alpha}{L}\right] = \frac{\pi}{2L^2}\frac{\exp\left(-\frac{\alpha}{L}\right)}{\left[1-\exp\left(-\frac{\alpha}{L}\right)\right]^2}.$$

For small α this expression can be expanded in a Laurent series,

$$\varepsilon_0(L;\alpha) = \frac{\pi}{8L^2}\frac{1}{\sinh^2\frac{\alpha}{2L}} = \frac{\pi}{2\alpha^2} - \frac{\pi}{24L^2} + \frac{1}{L^2}O\left(\frac{\alpha^2}{L^2}\right). \tag{10.6}$$

As $\alpha \to 0$, the first term here diverges as α^{-2}, the second term remains finite and further terms vanish. The crucial fact is that the singular term does not depend on the distance L between the plates and can be interpreted as the energy density of the zero-point fluctuations in Minkowski spacetime without boundaries. This zero-point energy density can be thought of as the limit of $\varepsilon_0(L)$ as $L \to \infty$ and, as it is clear from (10.6), must be exactly equal to the first divergent term in (10.6). On the other hand, we have agreed to ignore an infinite energy of zero-point fluctuations in Minkowski space, assuming that this energy does not contribute to any physically relevant quantities. This is the physical justification for omitting the divergent contribution to (10.6).

Subtracting from (10.6) the vacuum energy density and removing the cutoff (taking the limit $\alpha \to 0$), we obtain

$$\Delta\varepsilon_{\text{ren}}(L) = \lim_{\alpha\to 0}\left[\varepsilon_0(L;\alpha) - \lim_{L\to\infty}\varepsilon_0(L;\alpha)\right] = -\frac{\pi}{24L^2}. \tag{10.7}$$

After we have decided to fix the normalization point and to attribute a "zero" energy to vacuum fluctuations in Minkowski spacetime, there remains no more freedom to renormalize the finite shift of the energy density due to the presence of the plates. Therefore this energy shift is physical. The corresponding Casimir force between the plates is

$$F = -\frac{d}{dL}\Delta E = -\frac{d}{dL}(L\Delta\varepsilon_{\text{ren}}) = -\frac{\pi}{24L^2},$$

where the negative sign tells us that the plates are pulled toward each other.

A similar calculation in four-dimensional spacetime gives the Casimir force per unit area between two uncharged parallel plates as

$$f = -\frac{\pi^2}{240} L^{-4}.$$

Remark: negative energy Note that the shift of the energy density in (10.7) is negative. Quantum field theory generally admits quantum states with a negative expectation value of energy.

Riemann's zeta function regularization An elegant and quick way to calculate the finite energy shift due to the plates is with the help of Riemann's zeta (ζ) function defined by the series

$$\zeta(x) = \sum_{n=1}^{\infty} \frac{1}{n^x}, \tag{10.8}$$

which converges for real $x > 1$. An analytic continuation extends this function to all (complex) x, except $x = 1$ where $\zeta(x)$ has a pole.

The divergent sum $\sum_{n=1}^{\infty} n$ appearing in (10.3) is *formally* equivalent to the series for $\zeta(x)$ with $x = -1$. The ζ function obtained via analytic continuation is, however, finite at $x = -1$ and is equal to[2]

$$\zeta(-1) = -\frac{1}{12}.$$

This motivates us to replace the divergent sum $\sum_{n=1}^{\infty} n$ in (10.3) by the number $-\frac{1}{12}$. After this substitution, we immediately obtain the result (10.7).

At a first glance, this procedure may appear miraculous and lacking of physical explanation of neglecting divergences, unlike the straightforward renormalization approach. However, it has been verified in many cases that the results obtained using the ζ function method are in agreement with more direct renormalization procedures.

[2] This result requires a complicated proof. See e.g. H. Bateman and A. Erdelyi, *Higher Transcendental Functions*, vol. 1 (McGraw-Hill, New York, 1953).

Part II

Path integrals and vacuum polarization

11

Path integrals

Summary A path integral in the Hamiltonian and Lagrangian formalism.

In the first part of this book, we used *canonical quantization* to introduce the concept of "particle" and to calculate the particle production in a few interesting cases. It turns out that it is more convenient to study the vacuum polarization effects using the path integral method, which provides an alternative description of a quantum system. In this chapter we introduce path integrals by considering the evolution of a simple quantum-mechanical system.

11.1 Evolution operator. Propagator

We recall that in the Schrödinger picture, the state vector $|\psi(t)\rangle$ obeys the Schrödinger equation

$$i\hbar\frac{\partial|\psi(t)\rangle}{\partial t}=\hat{H}\,|\psi(t)\rangle\,. \tag{11.1}$$

In the case of a time-independent Hamiltonian $\hat{H}$, the *formal solution* of this equation is

$$|\psi(t)\rangle=\hat{U}\,(t,t_0)\,|\psi(t_0)\rangle\,, \tag{11.2}$$

where

$$\hat{U}\,(t,t_0)\equiv\exp\left[-\frac{i}{\hbar}\,(t-t_0)\,\hat{H}\right] \tag{11.3}$$

is called the *evolution operator*. This operator is unitary if $\hat{H}$ is Hermitian. It is also clear that the composition of two subsequent evolution operators is equal to the evolution operator for the combined timespan, that is

$$\hat{U}\,(t_1,t_2)\,\hat{U}\,(t_2,t_3)=\hat{U}\,(t_1,t_3)\,.$$

Because $\hat{U}(t, t_0)$ depends only on $t - t_0$, we have

$$\hat{U}(\Delta t_1)\,\hat{U}(\Delta t_2) = \hat{U}(\Delta t_1 + \Delta t_2) = \hat{U}(\Delta t_2)\,\hat{U}(\Delta t_1),$$

and all the operators $\hat{U}(\Delta t)$ for any Δt commute.[1]

In the coordinate representation, a quantum state $|\psi(t)\rangle$ is described by a wave function (or functional)

$$\psi(q, t) = \langle q|\,\psi(t)\rangle,$$

where the variable q is a "symbolic combined notation" which includes all the degrees of freedom of the system; we recall that for N degrees of freedom, $q \equiv \{q_1, q_2, \ldots q_N\}$, while in the case of a real scalar field, $q(t) \equiv \{\phi_{\mathbf{x}}(t)\}$ and the spatial coordinate $\mathbf{x}$ is to be considered as a continuous "index." Given a wave function $\psi(q, t_0)$ at some initial moment of time $t = t_0$, we can express the wave function $\psi(q, t)$ for $t > t_0$ as

$$\begin{aligned}\psi(q, t) = \langle q|\,\psi(t)\rangle &= \langle q|\,\hat{U}(t, t_0)\left(\int Dq_0\,|q_0\rangle\,\langle q_0|\right)|\psi(t_0)\rangle \\ &\equiv \int Dq_0 K(q, q_0; t, t_0)\,\psi(q_0, t_0),\end{aligned} \tag{11.4}$$

where the matrix element

$$K(q, q_0; t, t_0) \equiv \langle q|\,\hat{U}(t, t_0)\,|q_0\rangle,$$

called the *propagator*, gives the quantum-mechanical amplitude for the transition between an initial state $|q_0\rangle$ at time t_0 and a final state $|q\rangle$ at time t. In deriving (11.4), we used the decomposition of the unit operator

$$\hat{1} = \int Dq\,|q\rangle\,\langle q|\,.$$

For the case of N degrees of freedom, the measure is $Dq = dq_1 \ldots dq_N$, while in field theory Dq is a functional measure which will be specified later.

11.2 Propagator as a path integral

The propagator can be expressed as an integral over all trajectories connecting the initial and the final states (a *path integral*). To derive the path integral representation for the propagator, we first consider a system with one degree of freedom and calculate the matrix element

$$K(q_f, q_0; t_f, t_0) \equiv \langle q_f|\,\hat{U}(t_f, t_0)\,|q_0\rangle\,.$$

[1] If the Hamiltonian is explicitly time-dependent, then the evolution operators do not generally commute, but the composition rule is obviously still valid.

In this section we will use a subscript index to distinguish different quantum states $|q_0\rangle$, $|q_1\rangle$, etc., of the same system with one degree of freedom q. One should not confuse this subscript with an index enumerating the different degrees of freedom.

Let us divide the time interval (t_0, t_f) into $N+1$ small intervals $(t_0, t_1), \ldots,$ $(t_k, t_{k+1}), \ldots, (t_N, t_f)$. Eventually we shall pass to the limit $N \to \infty$ and $\Delta t_k \equiv t_{k+1} - t_k \to 0$. For convenience, we sometimes use the alternative notation t_{N+1} for t_f and t_i for t_0. The evolution operator $\hat{U}(t_f, t_0)$ is equal to the product of the evolution operators for all the intermediate ranges (t_k, t_{k+1}),

$$\hat{U}(t_f, t_0) = \hat{U}(t_f, t_N) \ldots \hat{U}(t_1, t_0) = \prod_{k=0}^{N} \hat{U}(t_{k+1}, t_k),$$

and therefore the propagator

$$K(q_f, q_0; t_f, t_0) = \langle q_f | \hat{U}(t_f, t_0) | q_0 \rangle = \langle q_f | \prod_{k=0}^{N} \hat{U}(t_{k+1}, t_k) | q_0 \rangle \tag{11.5}$$

can be expressed in terms of the intermediate propagators. Inserting N decompositions of unity

$$\hat{1} = \int dq\, |q\rangle \langle q|$$

into (11.5), we have

$$\begin{aligned}
&\langle q_f | \prod_{k=0}^{n} \hat{U}(t_{k+1}, t_k) | q_0 \rangle \\
&= \langle q_f | \hat{U}(t_{N+1}, t_N) \left[\textstyle\int dq_N\, |q_N\rangle \langle q_N| \right] \hat{U}(t_N, t_{N-1}) \ldots \\
&\quad \ldots \left[\textstyle\int dq_{k+1} |q_{k+1}\rangle \langle q_{k+1}| \right] \hat{U}(t_{k+1}, t_k) \\
&\quad \times \left[\textstyle\int dq_k\, |q_k\rangle \langle q_k| \right] \ldots \left[\textstyle\int dq_1\, |q_1\rangle \langle q_1| \right] \hat{U}(t_1, t_0) |q_0\rangle ,
\end{aligned}$$

where $q_0, q_1, q_2, \ldots$ label the different quantum states at the moments of time $t_0, t_1, t_2, \ldots$ respectively. Thus, it follows that the resulting propagator is the N-fold integrated product of the propagators of all intermediate time intervals:

$$K(q_f, q_0; t_f, t_0) = \int dq_N dq_{N-1} \ldots dq_1 \left(\prod_{k=0}^{N} K(q_{k+1}, q_k; t_{k+1}, t_k) \right). \tag{11.6}$$

The product of $(N+1)$ intermediate propagators,

$$K(q_{k+1}, q_k; t_{k+1}, t_k) = \langle q_{k+1} | \hat{U}(t_{k+1}, t_k) | q_k \rangle ,$$

is equal to the quantum-mechanical amplitude for a chain of transitions $|q_0\rangle \to |q_1\rangle \to \ldots \to |q_f\rangle$. This amplitude describes a certain class of "constrained transitions," for which the particle passes from q_0 to q_f while visiting the intermediate points q_k at the times t_k (see Fig. 11.1). So the formula (11.6) shows that the

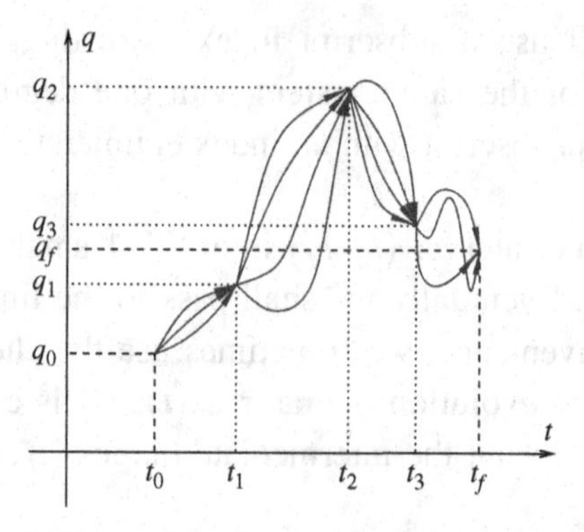

Fig. 11.1 A "constrained transition" with fixed intermediate points $q_1, \ldots, q_n$. The multiple lines connecting the points q_k indicate that the motion of the quantum particle between the specified points is not described by a single classical path.

total amplitude for the transition from the initial state $|q_0\rangle$ to the final state $|q_f\rangle$ is found by integrating the constrained transition amplitude over all the possible intermediate values $q_1, \ldots, q_N$.

To calculate the resulting propagator $K(q_f, q_0; t_f, t_0)$, we thus need first to find an explicit expression for intermediate propagators in the limit of small Δt_k. Expanding the evolution operator as

$$\hat{U}(t_{k+1}, t_k) = \exp\left[-\frac{i}{\hbar}\Delta t_k \hat{H}\right] = 1 - \frac{i}{\hbar}\Delta t_k \hat{H} + O(\Delta t_k^2), \tag{11.7}$$

we find

$$K(q_{k+1}, q_k; t_{k+1}, t_k) = \langle q_{k+1}| \left(1 - \frac{i}{\hbar}\Delta t_k \hat{H}\right) |q_k\rangle + O(\Delta t_k^2). \tag{11.8}$$

The matrix element of the Hamiltonian can be calculated by inserting the decomposition of unity in the momentum representation,[2]

$$\hat{1} = \int dp_k \, |p_k\rangle \langle p_k|,$$

into $\langle q_{k+1}| \hat{H} |q_k\rangle$; as a result we obtain

$$\langle q_{k+1}| \hat{H} |q_k\rangle = \int dp_k \langle q_{k+1}|p_k\rangle \langle p_k| \hat{H} |q_k\rangle. \tag{11.9}$$

The matrix element $\langle q_{k+1}|p_k\rangle$ was calculated before (see (2.35)) and is equal to

$$\langle q_{k+1}|p_k\rangle = \frac{1}{\sqrt{2\pi\hbar}} \exp\left(\frac{ip_k q_{k+1}}{\hbar}\right). \tag{11.10}$$

[2] Again we use p_k simply to distinguish the momentum eigenstates at time t_k.

To find $\langle p_k | \hat{H} | q_k \rangle$, we must first reorder all the operators $\hat{p}$ in the Hamiltonian and move them to the left, so that $\hat{H}(\hat{p}, \hat{q})$ is rewritten as

$$\hat{H} = \sum_J f_J(\hat{p}) g_J(\hat{q}).$$

This can be done using the fundamental commutation relation $[\hat{q}, \hat{p}] = i\hbar$. As a result, we find

$$\begin{aligned}\langle p_k | \hat{H} | q_k \rangle &= \left(\sum_J f_J(p_k) g_J(q_k) \right) \times \langle p_k | q_k \rangle \equiv H(p_k, q_k) \langle p_k | q_k \rangle \\ &= \frac{1}{\sqrt{2\pi\hbar}} H(p_k, q_k) \exp\left(-\frac{i p_k q_k}{\hbar} \right). \end{aligned} \tag{11.11}$$

For example, if

$$\hat{H} = \hat{q}\hat{p}^2\hat{q} = \hat{p}^2\hat{q}^2 + 2i\hbar\hat{p}\hat{q},$$

we have

$$H(p_k, q_k) = p_k^2 q_k^2 + 2i\hbar p_k q_k.$$

Taking into account (11.10) and (11.11), it follows from (11.9) that

$$\langle q_{k+1} | \hat{H} | q_k \rangle = \int \frac{dp_k}{2\pi\hbar} H(p_k, q_k) \exp \frac{i p_k (q_{k+1} - q_k)}{\hbar}, \tag{11.12}$$

and the propagator in (11.8) becomes

$$\begin{aligned} &K(q_{k+1}, q_k; t_{k+1}, t_k) \\ &= \int \frac{dp_k}{2\pi\hbar} \left[1 - \frac{i\Delta t_k}{\hbar} H(p_k, q_k) + O\left(\Delta t_k^2\right) \right] \exp \frac{i p_k (q_{k+1} - q_k)}{\hbar} \\ &= \int \frac{dp_k}{2\pi\hbar} \exp\left[\frac{i\Delta t_k}{\hbar} \left(p_k \frac{q_{k+1} - q_k}{\Delta t_k} - H(p_k, q_k) \right) + O\left(\Delta t_k^2\right) \right]. \end{aligned} \tag{11.13}$$

Substituting this result into (11.6), we obtain

$$\begin{aligned} K(q_f, q_0; t_f, t_0) = &\int \left[\prod_{k=1}^{N} \frac{dq_k dp_k}{2\pi\hbar} \right] \frac{dp_0}{2\pi\hbar} \\ &\times \exp\left[\sum_{k=0}^{N} \frac{i\Delta t_k}{\hbar} \left(p_k \frac{q_{k+1} - q_k}{\Delta t_k} - H(p_k, q_k) \right) + O\left(\Delta t_k^2\right) \right]. \end{aligned} \tag{11.14}$$

We would like to point out that (11.14) involves $(N+1)$ integrations over p_k but only N integrations over q_k.

Now let us consider the limit $N \to \infty$ and $\Delta t_k \to 0$. When the number of intermediate points t_k becomes infinitely large, one can introduce the functions $q(t), p(t)$ such that $q_k = q(t_k)$ and $p_k = p(t_k)$, and replace the sum in (11.14) by an integral over t,

$$\lim_{N\to\infty} \sum_{k=0}^{N} \left[\frac{i}{\hbar} \Delta t_k \left(p_k \frac{q_{k+1} - q_k}{\Delta t_k} - H(p_k, q_k) \right) + O\left(\Delta t_k^2\right) \right]$$
$$= \int_{t_0}^{t_f} \frac{i}{\hbar} dt \left[p(t) \frac{dq(t)}{dt} - H(p(t), q(t)) \right]. \qquad (11.15)$$

Note that the contribution of quadratic corrections $O\left(\Delta t_k^2\right)$, which we neglected to calculate, vanishes in the limit $N \to \infty$. Considering equal time intervals $\Delta t = (t_f - t_0)/(N+1)$, we have

$$\sum_{k=0}^{N} O\left(\Delta t_k^2\right) \sim N \Delta t^2 = N \frac{(t_f - t_0)^2}{(N+1)^2} \to 0$$

as $N \to \infty$.

The integration over infinitely many intermediate values q_k, p_k in (11.14) is naturally interpreted as the integration over *all the functions* $q(t)$, $p(t)$ such that $q(t_0) = q_0$, $q(t_f) = q_f$. An integral of this kind is called a *functional integral* or a *path integral.* In the limit $N \to \infty$, the $(2N+1)$-fold integration over dp_k and dq_k becomes an infinite-dimensional integration measure. Introducing a symbolic notation for this measure,

$$\mathcal{D}p\mathcal{D}q \equiv \lim_{N\to\infty} \left[\prod_{k=1}^{N} \frac{dq_k dp_k}{2\pi\hbar} \right] \frac{dp_0}{2\pi\hbar}, \qquad (11.16)$$

one can rewrite (11.14) as

$$K\left(q_f, q_0; t_f, t_0\right) = \int_{q(t_0)=q_0}^{q(t_f)=q_f} \mathcal{D}p\mathcal{D}q \exp\left[\frac{i}{\hbar} \int_{t_0}^{t_f} (p\dot{q} - H(p,q))\, dt \right]. \qquad (11.17)$$

We would like to stress once more that this formal expression must be understood as the limit of the finite-dimensional integral in (11.14) as $N \to \infty$. The expression in the exponent in (11.17) is the classical Hamiltonian action (2.19), and the boundary conditions are exactly those needed for the Hamiltonian action principle (see Section 2.2).

The generalization to the case of an arbitrary number of degrees of freedom, including an infinite number of degrees of freedom needed in field theory, is largely straightforward.

Exercise 11.1

Derive the path integral: (a) for a system with an arbitrary finite number of degrees of freedom; (b) for a scalar field with Hamiltonian (4.11).

So far we considered only systems with time-independent Hamiltonians, but the path integral formalism also applies to time-dependent Hamiltonians.

Exercise 11.2

Derive the path integral expression for the propagator $\langle q_f | \hat{U}(t_f, t_0) | q_0 \rangle$ in the case of an explicitly time-dependent Hamiltonian $\hat{H}(\hat{p}, \hat{q}, t)$.

Hint: Operators $\hat{H}(\hat{p}, \hat{q}, t)$ at different times t may not commute and should be manipulated more carefully. The explicit form (11.3) of the evolution operator which holds only for time-independent Hamiltonians is not actually needed for the derivation of the path integral; only the approximation (11.7) is important.

11.3 Lagrangian path integrals

If the Hamiltonian is a quadratic function of the momenta, then the integration over the momenta can be performed explicitly and the path integral is simplified. Let us consider, for example, a system with the Hamiltonian

$$\hat{H}(\hat{p}, \hat{q}) = \frac{\hat{p}^2}{2m} + V(\hat{q}). \tag{11.18}$$

In this case, the integral in (11.13) becomes

$$K(q_{k+1}, q_k; t_{k+1}, t_k)$$
$$= \int \frac{dp_k}{2\pi\hbar} \exp\left[\frac{i\Delta t_k}{\hbar}\left(\frac{q_{k+1}-q_k}{\Delta t_k}p_k - \frac{p_k^2}{2m} - V(q_k)\right) + O(\Delta t^2)\right] \tag{11.19}$$

and can be calculated with the help of the following formula for the Gaussian integral,

$$\int_{-\infty}^{+\infty} dx \exp\left[-\frac{ax^2}{2} + ibx\right] = \sqrt{\frac{2\pi}{a}} \exp\left[-\frac{b^2}{2a}\right],$$

which holds also for complex a, b as long as the integral converges. The result is

$$K(q_{k+1}, q_k; t_{k+1}, t_k)$$
$$= \sqrt{\frac{m}{2\pi i\hbar\Delta t_k}} \exp\left[\frac{i\Delta t_k}{\hbar}\left(m\frac{(q_{k+1}-q_k)^2}{2\Delta t_k^2} - V(q_k)\right) + O(\Delta t^2)\right]. \tag{11.20}$$

Substituting this result in (11.6) and taking the limit $N \to \infty$, $\Delta t_k \to 0$, we finally obtain

$$K\left(q_f, q_0; t_f, t_0\right) \equiv \left\langle q_f \right| \hat{U}\left(t_f, t_0\right) \left| q_0 \right\rangle = \int_{q(t_0)=q_0}^{q(t_f)=q_f} \mathcal{D}q \exp\left(\frac{iS}{\hbar}\right), \quad (11.21)$$

where

$$S = \lim_{N\to\infty} \sum_{k=0}^{N} \left(m \frac{\left(q_{k+1} - q_k\right)^2}{2\Delta t_k^2} - V\left(q_k\right) \right) \Delta t_k = \int_{t_0}^{t_f} \left(\frac{m\dot{q}^2}{2} - V(q) \right) dt \quad (11.22)$$

is the Lagrangian action and

$$\mathcal{D}q = \lim_{N\to\infty} \frac{1}{\sqrt{2\pi i\hbar m^{-1}\Delta t_0}} \prod_{k=1}^{N} \frac{dq_k}{\sqrt{2\pi i\hbar m^{-1}\Delta t_k}} \quad (11.23)$$

is the measure for the Lagrangian path integral. Notice that this measure contains $N+1$ factors proportional to $\Delta t^{-1/2}$, but only N integrations.

The Lagrangian path integral (11.21) is the original form of the path integral introduced by R. Feynman. However, we would like to stress that although the propagator for any system can be always written as the Hamiltonian path integral (11.17), the Lagrangian formulation of the path integral exists only for the systems with Hamiltonians quadratic in momenta.

Exercise 11.3
Derive the Lagrangian path integral for a system with the Hamiltonian

$$\hat{H} = f\left(\hat{q}\right)\hat{p}^2 f\left(\hat{q}\right) + V\left(\hat{q}\right).$$

Exercise 11.4
Derive the measure in the Lagrangian path integral for a scalar field with the action (4.5).

11.4 Propagators for free particle and harmonic oscillator

Now we will explicitly calculate the propagators in two simple cases, namely, for a "free particle" and for a "particle"[3] in a quadratic potential $V(q,t) \propto q^2$ that can also explicitly depend on time. This last case covers both the usual harmonic oscillator with a constant frequency, as well as a harmonic oscillator interacting with a time-dependent external field. We will see that the calculation of the path integral for a quadratic potential can be reduced to the calculation for a free particle.

[3] Of course, the results will be applicable not only to a quantum-mechanical particle but also to any system with one degree of freedom $q(t)$ which satisfies the equation of motion $\ddot{q} = -\partial V/\partial q$ with an appropriate potential $V(q,t)$.

11.4.1 Free particle

To compute the propagator for a free particle, we use the formula (11.20) with $V(q) = 0$. Neglecting terms $O(\Delta t^2)$ in the limit of small Δt, we have

$$\begin{aligned}
&\int dq_k K\left(q_{k+1}, q_k; t_{k+1}, t_k\right) K\left(q_k, q_{k-1}; t_k, t_{k-1}\right) \\
&= \frac{m}{2\pi i\hbar\sqrt{\Delta t_k \Delta t_{k-1}}} \int dq_k \exp\left[\frac{im}{\hbar}\left(\frac{\left(q_{k+1}-q_k\right)^2}{2\Delta t_k} + \frac{\left(q_k - q_{k-1}\right)^2}{2\Delta t_{k-1}}\right)\right] \\
&= \sqrt{\frac{m}{2\pi i\hbar\left(\Delta t_k + \Delta t_{k-1}\right)}} \exp\left[\frac{im}{\hbar}\frac{\left(q_{k+1}-q_{k-1}\right)^2}{2\left(\Delta t_k + \Delta t_{k-1}\right)}\right] \\
&= K\left(q_{k+1}, q_{k-1}; t_{k+1}, t_{k-1}\right).
\end{aligned} \tag{11.24}$$

In this manner we can integrate over all the intermediate points q_k to obtain the following result for the propagator of a free particle,

$$K\left(q_f, q_i; t_f, t_i\right) = \sqrt{\frac{m}{2\pi i\hbar\left(t_f - t_i\right)}} \exp\left[\frac{im\left(q_f - q_i\right)^2}{2\hbar\left(t_f - t_i\right)}\right]. \tag{11.25}$$

This result can be rewritten in a more elegant form if we note that the action for a free particle calculated along the classical trajectory,

$$q(t) = \frac{1}{t_f - t_i}\left[\left(t_f - t\right) q_i + \left(t - t_i\right) q_f\right],$$

satisfying the boundary conditions $q(t_i) = q_i$ and $q\left(t_f\right) = q_f$, is equal to

$$S_{\mathrm{cl}} = \frac{m}{2}\int_{t_i}^{t_f} \dot{q}^2 dt = \frac{m}{2}\int_{t_i}^{t_f}\left(\frac{d(q\dot{q})}{dt} - q\ddot{q}\right) dt = \frac{m}{2}\, q\dot{q}\Big|_{t_i}^{t_f} = \frac{m}{2}\frac{\left(q_f - q_i\right)^2}{t_f - t_i}. \tag{11.26}$$

Then (11.25) becomes

$$K = \left(-\frac{1}{2\pi i\hbar}\frac{\partial^2 S_{\mathrm{cl}}}{\partial q_f \partial q_i}\right)^{1/2} \exp\left(\frac{i}{\hbar} S_{\mathrm{cl}}\right). \tag{11.27}$$

11.4.2 Quadratic potential

Next we consider a system with the action

$$S = \frac{1}{2}\int\left(\dot{q}^2 + f(t)\, q^2\right) dt, \tag{11.28}$$

assuming that $f(t)$ is an arbitrary function of time. In particular, for a harmonic oscillator $f(t) = -\omega^2 = \text{const}$. The variable $q(t)$ satisfies the classical equation

$$\ddot{q} - f(t)\, q = 0. \tag{11.29}$$

Let us assume that $u(t)$ is a solution of this equation, and introduce a new variable

$$Q \equiv u^{-1} q \tag{11.30}$$

instead of q. Then the action (11.28) can be rewritten in terms of the new variable,

$$S = \frac{1}{2}\int_i^f \left[u^2 \dot{Q}^2 + \frac{d}{dt}\left(u\dot{u}Q^2\right) + \left(\dot{u}^2 - \frac{d}{dt}(u\dot{u}) + f(t)\, u^2\right) Q^2 \right] dt. \tag{11.31}$$

Taking into account that u satisfies (11.29), we obtain

$$S = \frac{1}{2}\left(\frac{\dot{u}_f}{u_f} q_f^2 - \frac{\dot{u}_i}{u_i} q_i^2\right) + \frac{1}{2}\int_i^f u^2 \dot{Q}^2 dt, \tag{11.32}$$

where $u_i \equiv u(t_i)$, $q_i \equiv q(t_i)$, ... If we introduce a new time variable

$$\eta \equiv \int \frac{dt}{u^2} \tag{11.33}$$

instead of t, the action (11.32) becomes

$$S = \frac{1}{2}\left(\frac{\dot{u}_f}{u_f} q_f^2 - \frac{\dot{u}_i}{u_i} q_i^2\right) + \frac{1}{2}\int_{\eta_i}^{\eta_f} Q'^2(\eta)\, d\eta, \tag{11.34}$$

where the prime denotes the derivative with respect to η. Thus, knowing a solution of the classical equation of motion, we can reduce the (off-shell) action for a quadratic potential to the action for a free particle with coordinate Q. Rigorously speaking, the transformations above make sense only if the solution $u(t)$ does not vanish and the integral in (11.33) converges. This is obviously not true for the usual harmonic oscillator with a positive ω^2. However, for the moment we neglect this problem and proceed assuming that the solution u has the required properties.

To calculate the path integral in (11.21) with the action (11.34), we first have to express the measure (11.23) in terms of the new variables Q, η instead of q, t. Substituting

$$\Delta t_k = \int_{\eta_k}^{\eta_{k+1}} u^2 d\eta = u^2(\eta_k)\, \Delta\eta_k + \frac{1}{2} \left.\frac{du^2}{d\eta}\right|_{\eta_k} \Delta\eta_k^2 + O\left(\Delta\eta_k^3\right)$$

$$= u_k^2 \Delta\eta_k \exp\left[\frac{1}{2u^2}\left.\frac{du^2}{d\eta}\right|_k \Delta\eta_k\right] + O\left(\Delta\eta_k^3\right) \tag{11.35}$$

into (11.23), we obtain

$$\mathcal{D}q = \lim_{N\to\infty} \frac{e^{-\frac{1}{2}\left.\frac{d\ln u}{d\tau}\right|_0 \Delta\eta_0}}{u_i\sqrt{2\pi i\hbar\Delta\eta_0}} \prod_{k=1}^{N} \frac{e^{-\frac{1}{2}\left.\frac{d\ln u}{d\tau}\right|_k \Delta\eta_k} u_k^{-1} dq_k}{\sqrt{2\pi i\hbar\Delta\eta_k}}$$

$$= \frac{1}{u_i}\exp\left[-\frac{1}{2}\int_{\eta_i}^{\eta_f} \frac{d\ln u}{d\eta} d\eta\right] \mathcal{D}Q = \frac{1}{\sqrt{u_i u_f}} \mathcal{D}Q, \tag{11.36}$$

where $\mathcal{D}Q$ is the standard measure

$$\mathcal{D}Q = \lim_{N\to\infty} \frac{1}{\sqrt{2\pi i\hbar\Delta\eta_0}} \prod_{k=1}^{N} \frac{dQ_k}{\sqrt{2\pi i\hbar\Delta\eta_k}}.$$

Hence, taking into account (11.34), the original path integral reduces to

$$K = \int \mathcal{D}q \exp\left(\frac{i}{2\hbar}\int \left(\dot{q}^2 + f(t)\, q^2\right) dt\right)$$

$$= \frac{1}{\sqrt{u_i u_f}} \exp\left[\frac{i}{2\hbar}\left(\frac{\dot{u}_f}{u_f} q_f^2 - \frac{\dot{u}_i}{u_i} q_i^2\right)\right] \times \int \mathcal{D}Q \exp\left(\frac{i}{2\hbar}\int Q'^2 d\eta\right). \tag{11.37}$$

The obtained integral over Q was calculated above (see (11.25)); the result is

$$\int \mathcal{D}Q \exp\left(\frac{i}{2\hbar}\int Q'^2 d\eta\right) = \sqrt{\frac{1}{2\pi i\hbar\left(\eta_f - \eta_i\right)}} \exp\left[\frac{i\left(Q_f - Q_i\right)^2}{2\hbar\left(\eta_f - \eta_i\right)}\right]. \tag{11.38}$$

To find an explicit expression for $\eta_f - \eta_i$ in terms of the original time t, it is convenient to use two independent solutions $u(t)$, $v(t)$ of (11.29). Since the Wronskian

$$W \equiv u\dot{v} - \dot{u}v \tag{11.39}$$

is constant in time, we have

$$\eta_f - \eta_i = \frac{1}{W}\int_{t_i}^{t_f} \left(\frac{u\dot{v} - \dot{u}v}{u^2}\right) dt = \frac{1}{W}\int_{t_i}^{t_f} \frac{d}{dt}\left(\frac{v}{u}\right) dt = \frac{u_i v_f - u_f v_i}{W u_i u_f}. \tag{11.40}$$

The solutions $u(t)$, $v(t)$ play an auxiliary role in the calculation and can be chosen arbitrarily as long as $W \neq 0$. To simplify the final expressions, we normalize the solutions u, v in such a way that

$$u_i v_f - u_f v_i = 1. \tag{11.41}$$

Substituting (11.38) in (11.37) and taking into account (11.40) and (11.30), we obtain

$$K=\sqrt{\frac{W}{2\pi i\hbar}}\exp\left[\frac{i}{2\hbar}\left(\frac{Wu_i+\dot{u}_f}{u_f}q_f^2+\frac{Wu_f-\dot{u}_i}{u_i}q_i^2-2Wq_iq_f\right)\right]. \tag{11.42}$$

This expression can be further simplified if we note that

$$W=u_i\dot{v}_i-\dot{u}_iv_i=u_f\dot{v}_f-\dot{u}_fv_f \tag{11.43}$$

and use (11.41). The result is

$$K=\sqrt{\frac{W}{2\pi i\hbar}}\exp\left[\frac{i}{2\hbar}\left(\left(u_i\dot{v}_f-v_i\dot{u}_f\right)q_f^2+\left(u_f\dot{v}_i-v_f\dot{u}_i\right)q_i^2-2Wq_iq_f\right)\right]. \tag{11.44}$$

Exercise 11.5
Verify that (11.44) can be written as

$$K=\left(-\frac{1}{2\pi i\hbar}\frac{\partial^2 S_{\rm cl}}{\partial q_f\partial q_i}\right)^{1/2}\exp\left(\frac{i}{\hbar}S_{\rm cl}\right), \tag{11.45}$$

where $S_{\rm cl}$ is the action calculated for the classical trajectory satisfying the corresponding boundary conditions.

Harmonic oscillator In the case of a harmonic oscillator with constant frequency, we set $f(t)=-\omega^2=\text{const}$ and take as two independent fundamental solutions

$$u(t)=\frac{\cos\omega t}{\sqrt{\sin\omega T}},\quad v(t)=\frac{\sin\omega t}{\sqrt{\sin\omega T}},$$

which are normalized in agreement with (11.41); here $T=t_f-t_i$. Substituting these expressions in (11.44), we obtain

$$K\left(q_f,q_i;t_f,t_i\right)=\sqrt{\frac{\omega}{2\pi i\hbar\sin\omega T}}\exp\left[\frac{i\omega}{2\hbar\sin\omega T}\left(\left(q_f^2+q_i^2\right)\cos\omega T-2q_iq_f\right)\right]. \tag{11.46}$$

11.4.3 Euclidean path integral

The Euclidean oscillator is obtained from the usual oscillator by an analytic continuation in the time variable to pure imaginary times $t=-i\tau$, where τ is a real parameter. This procedure is called the *Wick rotation*, and it simplifies many calculations in quantum field theory. It can be thought as "rotating" the real axis in the complex t plane by 90 degrees to transform it into the imaginary axis. Because under the Wick rotation $(t=-i\tau)$ the Lorentzian metric

$$ds^2=dt^2-d\mathbf{x}^2$$

becomes (apart from an irrelevant overall sign) the Euclidean metric

$$ds^2 = -d\tau^2 - d\mathbf{x}^2,$$

the variable τ is called the *Euclidean time.* Obviously the transition to complex time is motivated primarily by mathematical convenience and has no physical meaning (for example the complex values such as $t = -(4i)$ s cannot be interpreted as moments of time). Having obtained a result using the Euclidean time τ, one can then perform an analytic continuation back to real (*Lorentzian*) time t.

Under the Wick rotation the action for harmonic oscillator with constant frequency ω becomes

$$S = iS_{\mathrm{E}},$$

where

$$S_{\mathrm{E}} \equiv \frac{1}{2}\int \left[\left(\frac{dq}{d\tau}\right)^2 + \omega^2 q^2\right] d\tau \tag{11.47}$$

is the Euclidean action. Then the path integral in (11.22) takes the following form,

$$K_{\mathrm{E}}\left(q_f, q_i; \tau_f, \tau_i\right) = \int_{q(\tau_i)=q_i}^{q(\tau_f)=q_f} \mathcal{D}q \exp\left(-\frac{S_{\mathrm{E}}}{\hbar}\right), \tag{11.48}$$

with the measure

$$\mathcal{D}q = \lim_{N\to\infty} \frac{1}{\sqrt{2\pi\hbar\Delta\tau_0}} \prod_{k=1}^{N} \frac{dq_k}{\sqrt{2\pi\hbar\Delta\tau_k}}.$$

In distinction from (11.22), this Euclidean path integral is mathematically well-defined. The factors

$$\exp\left[-\frac{\left(q_{k+1}-q_k\right)^2}{2\hbar\Delta\tau_k}\right] \tag{11.49}$$

coming from the "kinetic term" in Euclidean action are sometimes included in the definition of the measure, which is then called the *Wiener measure*. Given $\Delta\tau_k$, we see that for a "typical path" $q_{k+1} - q_k \sim \sqrt{\hbar\Delta\tau_k}$; hence

$$\frac{q_{k+1}-q_k}{\Delta\tau_k} \sim \sqrt{\frac{\hbar}{\Delta\tau_k}} \to \infty \quad \text{as } \Delta\tau_k \to 0.$$

Therefore one says sometimes that the main contribution to the path integral comes from trajectories which are non-differentiable at every point.

The "propagator" for the Euclidean oscillator, given by the path integral (11.48), is equal to

$$K_{\mathrm{E}}\left(q_f, q_i; \tau_f, \tau_i\right)$$
$$=\sqrt{\frac{\omega}{2\pi\hbar\sinh\omega T}}\exp\left[-\frac{\omega}{2\hbar\sinh\omega T}\left(\left(q_f^2+q_i^2\right)\cosh\omega T-2q_iq_f\right)\right], \quad (11.50)$$

where $T=\tau_f-\tau_i$.

Exercise 11.6
Following the strategy of this section, derive (11.50) and verify the legitimacy of every step in this derivation.

It is easy to see that (11.50) follows from (11.46) by considering pure imaginary times $t=-i\tau$, where τ is a real parameter, and making an analytic continuation. Inversely, the propagator for a harmonic oscillator can be obtained from the Euclidean propagator by making the inverse Wick rotation, that is by replacing $\tau\to it$ where t is real. Therefore in general it is very convenient to calculate first the mathematically well-defined Euclidean path integral and only then make an analytic continuation of the result to the Lorentzian spacetime. As we pointed out, the Lorentzian path integral is strictly speaking not well-defined; certain steps performed to calculate it are not completely legitimate and are mainly justified by the final result for the propagator which must satisfy the corresponding Schrödinger equation.

11.4.4 Ground state as a path integral

The wave function of the ground (vacuum) state of a harmonic oscillator in the coordinate basis can be written as a path integral with the appropriate boundary conditions. We will see later that the corresponding expressions are very useful in quantum field theory and therefore we derive them here. The ground state $|0\rangle$ is defined by the equation

$$\hat{a}^-|0\rangle=\sqrt{\frac{\omega}{2}}\left[\hat{q}+\frac{i}{\omega}\hat{p}\right]|0\rangle=0,$$

which in the coordinate basis takes the form

$$\left(\frac{\hbar}{\omega}\frac{\partial}{\partial q}+q\right)\psi_0(q)=0,$$

where $\psi_0(q)\equiv\langle q|0\rangle$. The normalized solution of this equation is

$$\psi_0(q)=\left(\frac{\omega}{\pi\hbar}\right)^{1/4}\exp\left(-\frac{\omega q^2}{2\hbar}\right). \quad (11.51)$$

It follows from (11.50) that

$$\psi_0(q) \equiv \langle q|0\rangle = \lim_{\tau_f\to\infty}\left(\frac{\pi\hbar}{\omega}\right)^{1/4} e^{\frac{\omega T}{2}} K_{\mathrm{E}}\left(q_f=0, q_i=q; \tau_f, \tau_i\right)$$

$$= \lim_{\tau_i\to-\infty}\left(\frac{\pi\hbar}{\omega}\right)^{1/4} e^{\frac{\omega T}{2}} K_{\mathrm{E}}\left(q_f=q, q_i=0; \tau_f, \tau_i\right). \tag{11.52}$$

In turn, the expression for K_{E} can be written as a path integral for the Euclidean harmonic oscillator, hence

$$\langle q|0\rangle = \lim_{\tau_f\to\infty}\left(\frac{\pi\hbar}{\omega}\right)^{1/4} e^{\frac{\omega T}{2}} \int_{q(\tau_i)=q}^{q(\tau_f)=0} \mathcal{D}q \exp\left(-\frac{S_{\mathrm{E}}}{\hbar}\right)$$

$$= \lim_{\tau_i\to-\infty}\left(\frac{\pi\hbar}{\omega}\right)^{1/4} e^{\frac{\omega T}{2}} \int_{q(\tau_i)=0}^{q(\tau_f)=q} \mathcal{D}q \exp\left(-\frac{S_{\mathrm{E}}}{\hbar}\right). \tag{11.53}$$

Note that after taking the limits $\tau\to\pm\infty$ the Wick rotation to the Lorentzian spacetime does not change the result because $\langle q|0\rangle$ does not depend on time.

12

Effective action

Summary Effective action for a driven harmonic oscillator and in general. Backreaction and vacuum polarization. Semiclassical gravity. Euclidean effective action as a functional determinant. Zeta (ζ) functions and renormalization of determinants. Computation of ζ functions using heat kernels.

The path integral technique allows us to introduce a concept of effective action in a natural way. The effective action turns out to be very useful for calculating expectation values and matrix elements of various operators and, in particular, for the energy-momentum tensor of a quantum field. To simplify the presentation we first consider the driven harmonic oscillator using the path integral method, define the effective action in this case, and then generalize the results to an arbitrary quantum system.

12.1 Driven harmonic oscillator (continuation)

12.1.1 Green's functions and matrix elements

In Chapter 3 we have considered a driven quantum oscillator and expressed the results for the expectation values and matrix elements in terms of the Green's functions (see Section 3.3). These Green's functions are solutions of the inhomogeneous equation,

$$\frac{\partial^2}{\partial t^2}G(t,t') + \omega^2 G(t,t') = \delta(t-t'), \tag{12.1}$$

with appropriate initial and boundary conditions. It is straightforward to verify that the equation for the driven harmonic oscillator,

$$\frac{d^2}{dt^2}q(t) + \omega^2 q(t) = J(t), \tag{12.2}$$

is solved for an arbitrary $J(t)$ by the following expression,

$$q(t) = \int_{-\infty}^{+\infty} J(t')\, G(t, t')\, dt', \tag{12.3}$$

where $G(t, t')$ is the corresponding *Green's function.* The Green's function can be interpreted as the oscillator's response to a sudden jolt, that is, to a force $J(t) = \delta(t - t')$ acting only at time $t = t'$ and conferring a unit of momentum to the oscillator.

Since equation (12.2) is of second order, its solution is specified uniquely if two conditions are imposed on the function $q(t)$. For example, a typical problem to be considered is the computation of the response of an oscillator initially at rest to a force $J(t)$ that is absent for $t < t_0$. In this case, the initial conditions, $q(t) = \dot{q}(t) = 0$ for $t < t_0$, are satisfied if we use in (12.3) the *retarded Green's function* $G_{\text{ret}}(t, t')$, defined as the solution of (12.1) for which $G_{\text{ret}}(t, t') = 0$ for all $t \le t'$. Indeed if the driving force $J(t)$ is absent until $t = t_0$, then it follows from (12.3) with $G = G_{\text{ret}}$ that $q(t) = 0$ for all $t \le t_0$, i.e. the oscillator remains at rest until the force is switched on.

Exercise 12.1

Verify that the retarded Green's function for a harmonic oscillator is

$$G_{\text{ret}}(t, t') = \theta(t - t') \frac{\sin \omega (t - t')}{\omega}. \tag{12.4}$$

The *Feynman Green's function* $G_{\text{F}}(t, t')$ is defined as the solution of (12.1) which satisfies the "in-out" conditions

$$G_{\text{F}}(t, t') \to e^{-i\omega t}, \quad t \to +\infty,$$
$$G_{\text{F}}(t, t') \to e^{+i\omega t}, \quad t \to -\infty.$$

As we have seen before, the Feynman Green's function for a harmonic oscillator is

$$G_{\text{F}}(t, t') = \frac{i}{2\omega} e^{-i\omega |t - t'|}. \tag{12.5}$$

Using the Green's functions above, we can rewrite the results in (3.23) and (3.28) as

$$\langle 0_{\text{in}} | \hat{q}(t) | 0_{\text{in}} \rangle = \int_{-\infty}^{+\infty} G_{\text{ret}}(t, t') J(t') dt', \tag{12.6}$$

$$\frac{\langle 0_{\text{out}} | \hat{q}(t) | 0_{\text{in}} \rangle}{\langle 0_{\text{out}} | 0_{\text{in}} \rangle} = \int_{-\infty}^{+\infty} G_{\text{F}}(t, t') J(t') dt'. \tag{12.7}$$

Note that these relations hold for all t. Other matrix elements can also be expressed through the Green's functions (see Exercise 3.4 on page 41).

The retarded Green's function describes the *causal* influence of an external force on the future evolution of the system. The Feynman Green's function, however, corresponds to a time-symmetric effect, namely the source $\delta(t-t')$ at time $t=t'$ affects both the future and the past evolution of the system. This *acausal* Green's function appears, however, in quantum-mechanical matrix elements.

12.1.2 Euclidean Green's function

As we have mentioned above, one can use the Wick rotation to define the Euclidean action and Euclidean path integrals. A driven harmonic oscillator (in Lorentzian time) satisfies the equation

$$\frac{d^2q}{dt^2}+\omega^2 q = J(t). \tag{12.8}$$

Assuming that the function $J(t)$ is analytic in a sufficiently large domain of the complex t plane, we can treat equation (12.8) as a differential equation in complex time. Then $q(t)$ and $J(t)$ become complex-valued functions that satisfy (12.8) for all complex t within the mentioned domain. Considering pure imaginary values of "time" $t=-i\tau$ (with real τ), we thus obtain the equation of the *Euclidean driven oscillator*,

$$-\frac{d^2q(\tau)}{d\tau^2}+\omega^2 q(\tau) = J(\tau). \tag{12.9}$$

Since this equation does not explicitly involve complex numbers, one may consider only *real-valued* $J(\tau)$ and $q(\tau)$. A real function $q(\tau)$ describes a *Euclidean trajectory* or *Euclidean path*.

Remark A real-valued function $q(\tau)$ may become complex-valued after an analytic continuation back to the Lorentzian time t. Similarly, a real analytic function of Lorentzian time $q(t)$ is in general not real-valued at $t=-i\tau$. We shall see below that a Euclidean path $q(\tau)$ *cannot* be also interpreted as an analytic continuation of the physically relevant solution $q(t)$. However, formally introduced real-valued Euclidean trajectories $q(\tau)$ will not explicitly enter the final results we are interested in.

As before, we assume that the driving force $J(\tau)$ vanishes outside of a finite period of Euclidean time τ. In that case, it is natural to require that the response $q(\tau)$ to that force does not grow for large $|\tau|$, or in other words, that there exists a number C such that

$$|q(\tau\to\pm\infty)| < C < \infty.$$

For sufficiently large $|\tau|$ there is no force and the solutions of (12.9) are $\exp(\pm\omega\tau)$. Hence, $q(\tau)$ remains bounded only if $q(\tau) \propto \exp(+\omega\tau)$ for $\tau \to -\infty$ and $q(\tau) \propto \exp(-\omega\tau)$ for $\tau \to +\infty$. This leads to the conditions

$$\lim_{\tau\to\pm\infty} q(\tau) = 0, \tag{12.10}$$

which are the natural boundary conditions for Euclidean trajectories selecting the "ground state" of the oscillator.

The general solution of equation (12.9) can be expressed as

$$q(\tau) = \int_{-\infty}^{+\infty} d\tau' G_{\mathrm{E}}(\tau, \tau') J(\tau'), \tag{12.11}$$

where $G_{\mathrm{E}}(\tau, \tau')$ is the *Euclidean Green's function* which satisfies the conditions

$$\lim_{\tau\to\pm\infty} G_{\mathrm{E}}(\tau, \tau') = 0.$$

These conditions specify $G_{\mathrm{E}}(\tau, \tau')$ uniquely (see Exercise 12.2),

$$G_{\mathrm{E}}(\tau, \tau') = \frac{1}{2\omega} e^{-\omega|\tau-\tau'|}, \tag{12.12}$$

and with the above Green's function, the solution (12.11) satisfies the boundary conditions (12.10) for any force $J(\tau)$ acting during a finite period of Euclidean time.

Exercise 12.2

Derive the result (12.12) by solving the equation

$$\left[-\frac{\partial^2}{\partial\tau^2} + \omega^2\right] G_{\mathrm{E}}(\tau, \tau') = \delta(\tau - \tau') \tag{12.13}$$

with the boundary conditions $|G_{\mathrm{E}}(\tau, \tau')| \to 0$ for $\tau \to \pm\infty$.

Relation between G_{E} and G_{F}

The similarity between the Euclidean and the Feynman Green's functions is apparent from a comparison of (12.5) and (12.12). Substituting $\tau = it$ in equation (12.11), one can verify that the analytic continuation of the solution $q(\tau)$ back to real times t yields the unphysical solution $q_{\mathrm{F}}(t)$,

$$q(\tau) = \int_{-\infty}^{+\infty} d\tau' G_{\mathrm{E}}(\tau, \tau') J(\tau') \quad \xrightarrow{\tau = it} \quad q_{\mathrm{F}}(t) = \int_{-\infty}^{+\infty} dt' G_{\mathrm{F}}(t, t') J(t').$$

Both the Feynman and the Euclidean Green's functions are symmetric in their two arguments. One might be tempted to say that they are analytic continuations of each other, except for the fact that neither of the two Green's functions $G_{\mathrm{E}}(\tau, \tau')$

and $G_F(t,t')$ are analytic in t or t'. Strictly speaking, only the restrictions of $G_F(t,t')$ to $t > t'$ or to $t < t'$ are analytic functions such that $G_F(t,t')$ for $t > t'$ is the analytic continuation of $iG_E(\tau,\tau')$ for $\tau < \tau'$ and vice versa.

Note that the retarded Green's function $G_{ret}(t,t')$ is also non-analytic. Generally, a Green's function cannot be an analytic function of t or t' in the entire complex plane. This can be explained by considering the requirements imposed on the Green's functions. On physical grounds, we expect that if the force $J(t)$ is active only during a finite time interval $0 < t < T$, then the influence of $J(t)$ should not grow as $|t| \to \infty$. However, it is a standard result of complex variable theory that there exist no nonconstant analytic functions that are uniformly bounded in the entire complex plane.

12.1.3 Introducing effective action

The results quoted above for a driven harmonic oscillator can be also derived using the path integral methods and introducing the useful concept of effective action. Effective action is a powerful method of calculations and its extensive presentation is far beyond the scope of this textbook. Here we introduce the effective action in a simple case of a driven harmonic oscillator and then employ it to describe the interaction of quantum systems with a classical external gravitational field (background).

To begin, let us express the matrix element $\langle 0_{out}|0_{in}\rangle$ in terms of the path integral. Assuming that the force $J(t)$ is acting only during finite time interval $0 < t < T$, the vacuum state $|0_{in}\rangle$ is defined as the state which is annihilated by the operator $\hat{a}^-(t_1)$ taken at some moment of time $t_1 < 0$,

$$\hat{a}^-(t_1)\,|0_{in}\rangle = 0. \tag{12.14}$$

On the other hand, the state $|0_{out}\rangle$ is annihilated by the operator $\hat{a}^-(t_2)$, where $t_2 > T$. These two annihilation operators are related by

$$\hat{a}^-(t_2) = \hat{U}^{-1}(t_2,t_1)\hat{a}^-(t_1)\,\hat{U}(t_2,t_1), \tag{12.15}$$

where $\hat{U}(t_2,t_1)$ is the evolution operator. Comparing

$$\hat{U}^{-1}(t_2,t_1)\hat{a}^-(t_1)\,\hat{U}(t_2,t_1)\,|0_{out}\rangle = 0 \tag{12.16}$$

with (12.14), we conclude that

$$|0_{in}\rangle = \hat{U}(t_2,t_1)\,|0_{out}\rangle \tag{12.17}$$

and hence

$$\begin{aligned}\langle 0_{out}|0_{in}\rangle &= \langle 0_{in}|\,\hat{U}(t_2,t_1)\,|0_{in}\rangle \\ &= \int dq_2 dq_1\,\langle 0_{in}|q_2\rangle\,\langle q_2|\,\hat{U}(t_2,t_1)\,|q_1\rangle\,\langle q_1|0_{in}\rangle\end{aligned} \tag{12.18}$$

Taking into account (11.52), we can rewrite this equation as

$$\langle 0_{\text{out}}|0_{\text{in}}\rangle = N_\omega \int dq_1 dq_2 K_{\text{E}}\left(q_f = 0, q_2; \tau_f \to +\infty\right) \times K\left(q_2, q_1; t_2, t_1\right) K_{\text{E}}\left(q_1, q_i = 0; \tau_i \to -\infty\right),$$

where N_ω is a J-independent factor. After performing a Wick rotation in $K(q_2, q_1; t_2, t_1)$, the expression in the right hand of this equation can be written as the Euclidean path integral

$$\int_{q(\tau_i=-\infty)=0}^{q(\tau_f=+\infty)=0} \mathcal{D}q\, e^{-S_{\text{E}}[J(\tau), q(\tau)]}, \tag{12.19}$$

where the Euclidean action for a driven harmonic oscillator is

$$S_{\text{E}}[q(\tau)] = -iS[q(t)]_{t=-i\tau} = \int_{\tau_0}^{\tau_f} d\tau \left[\frac{1}{2}\dot{q}^2 + \frac{\omega^2}{2}q^2 - J(\tau)q\right],$$

and $\dot{q} \equiv dq/d\tau$. In a sense, the functional $S_{\text{E}}[q(\tau)]$ is the *analytic continuation* of the functional $-iS[q(t)]$ to pure imaginary values of t; the factor $(-i)$ is introduced for convenience.

In the *Euclidean path integral* (12.19) the integration is performed over all *real-valued* Euclidean trajectories $q(\tau)$ constrained by boundary conditions $q(\tau_i = -\infty) = 0$ and $q\left(\tau_f = +\infty\right) = 0$. It can be viewed as the analytic continuation of the Lorentzian path integral

$$\int \mathcal{D}q\, e^{iS[J(\tau), q(\tau)]}, \tag{12.20}$$

which yields the matrix element $\langle 0_{\text{out}}|0_{\text{in}}\rangle$. Therefore one can expect to obtain useful results by computing first the well-defined Euclidean path integral and then performing an analytic continuation back to the real time t. This motivates us to define the *Euclidean effective action* as the functional $\Gamma_{\text{E}}[J(\tau)]$ determined by the relation

$$e^{-\Gamma_{\text{E}}[J(\tau)]} = \int_{q(-\infty)=0}^{q(+\infty)=0} \mathcal{D}q\, e^{-S_{\text{E}}[J(\tau), q(\tau)]}. \tag{12.21}$$

Note that $\Gamma_{\text{E}}[J]$ is a functional of J but not of q. After calculating the Euclidean effective action, we obtain the *Lorentzian effective action* $\Gamma_{\text{L}}[J(t)]$ by an analytic continuation of $\Gamma_{\text{E}}[J(\tau)]$ with an extra factor i:

$$\Gamma_{\text{L}}[J(t)] \equiv i\Gamma_{\text{E}}[J(\tau)]_{\tau=it}. \tag{12.22}$$

The matrix element $\langle 0_{\text{out}}|0_{\text{in}}\rangle$ is proportional to $\exp(i\Gamma_{\text{L}})$.

The correspondence between the Lorentzian and the Euclidean path integrals is not, however, very straightforward. Normally the path integral involves trajectories that are not necessarily analytic functions. So it is not obvious that an analytic continuation back to the Lorentzian time yields physically meaningful results. Below we shall see how the expressions obtained from Euclidean calculations can lead to the correct answers found in Chapter 3.

Remark While the path integral (12.20) involves a rapidly oscillating exponential, its Euclidean analog (12.21) contains a rapidly decaying expression and is expected to converge better. A mathematically rigorous definition of functional integration is currently available only for Euclidean path integrals. It is also easier in practice to perform calculations with the Euclidean action. These are the reasons for introducing the Wick rotation.

12.1.4 Calculating effective action for a driven oscillator

Unlike the Lorentzian action, the Euclidean action S_E is often bounded from below[1] and the minimum of the action is achieved at the classical Euclidean trajectory $q_{cl}(\tau)$. For instance, for the driven Euclidean oscillator,

$$\frac{1}{2}\dot{q}^2 + \frac{1}{2}\omega^2 q^2 - Jq = \frac{1}{2}\dot{q}^2 + \frac{1}{2}\left(\omega q - \frac{J}{\omega}\right)^2 - \frac{1}{2}\frac{J^2}{\omega^2} \geq -\frac{1}{2}\frac{J^2}{\omega^2}.$$

The dominant contribution to the path integral in (12.21) comes from paths $q(\tau)$ with the smallest value of the action. These are the paths near a solution $q_{cl}(\tau)$ of the classical Euclidean equation of motion,

$$\frac{\delta S_E[q]}{\delta q(\tau)} = 0,$$

which for a driven Euclidean oscillator takes the form (see (12.9))

$$-\frac{d^2}{d\tau^2} q_{cl} + \omega^2 q_{cl} = J(\tau). \tag{12.23}$$

The solution of this equation with the boundary conditions

$$\lim_{\tau \to \pm\infty} q_{cl}(\tau) = 0 \tag{12.24}$$

is given by

$$q_{cl}(\tau) = \int_{-\infty}^{+\infty} d\tau' G_E(\tau, \tau') J(\tau'). \tag{12.25}$$

[1] This is not always the case. For instance, the Euclidean action for general relativity is bounded neither from below nor from above.

The path integral (12.21) contains contributions not only from $q_{\mathrm{cl}}(\tau)$ but also from neighbor paths whose action S_{E} is only slightly larger than the minimum value $S_{\mathrm{E}}[q_{\mathrm{cl}}]$. To calculate the path integral, it is convenient to split $q(\tau)$ as

$$q(\tau) \equiv q_{\mathrm{cl}}(\tau) + \tilde{q}(\tau).$$

It is clear that the deviation $\tilde{q}(\tau)$ from the classical path should also satisfy the boundary conditions $\tilde{q}(\pm\infty) = 0$.

The path integral can now be rewritten as an integral over $\tilde{q}(\tau)$ with

$$\mathcal{D}q = \mathcal{D}[q_{\mathrm{cl}}(\tau) + \tilde{q}(\tau)] = \mathcal{D}\tilde{q}.$$

The measure $\mathcal{D}q$ is the limit of a product of the form $dq(\tau_1)\ldots dq(\tau_n)$. Each integration variable $q(\tau_k)$ can be shifted by a *constant* amount $q_{\mathrm{cl}}(\tau_k)$, and then

$$dq(\tau_k) = d[q_{\mathrm{cl}}(\tau_k) + \tilde{q}(\tau_k)] = d\tilde{q}(\tau_k)$$

because $q_{\mathrm{cl}}(\tau_k)$ is a fixed number. Thus we have

$$\int_{q(-\infty)=0}^{q(+\infty)=0} \mathcal{D}q\, e^{-S_{\mathrm{E}}[q(\tau)]} = \int_{\tilde{q}(-\infty)=0}^{\tilde{q}(+\infty)=0} \mathcal{D}\tilde{q}\, e^{-S_{\mathrm{E}}[q_{\mathrm{cl}}(\tau)+\tilde{q}(\tau)]}. \tag{12.26}$$

Integrating by parts, we can rewrite the action $S_{\mathrm{E}}[q_{\mathrm{cl}}(\tau) + \tilde{q}(\tau)]$ as

$$\begin{aligned} S_{\mathrm{E}}[q_{\mathrm{cl}} + \tilde{q}] &= \int \left[\frac{1}{2}\left(\dot{q}_{\mathrm{cl}} + \dot{\tilde{q}}\right)^2 + \frac{\omega^2}{2}(q_{\mathrm{cl}} + \tilde{q})^2 - (q_{\mathrm{cl}} + \tilde{q})\, J\right] d\tau \\ &= \left[\frac{1}{2}\dot{q}_{\mathrm{cl}} q_{\mathrm{cl}} + \dot{q}_{\mathrm{cl}}\tilde{q}\right]_{-\infty}^{+\infty} + \frac{1}{2}\int\left(\dot{\tilde{q}}^2 + \omega^2\tilde{q}^2\right) d\tau - \int q_{\mathrm{cl}} J d\tau \\ &\quad + \int\left(-\ddot{q}_{\mathrm{cl}} + \omega^2 q_{\mathrm{cl}} - J\right)\tilde{q}\, d\tau + \int\left[-\frac{1}{2}\ddot{q}_{\mathrm{cl}} q_{\mathrm{cl}} + \frac{\omega^2}{2} q_{\mathrm{cl}}^2\right] d\tau \\ &= \frac{1}{2}\int\left(\dot{\tilde{q}}^2 + \omega^2\tilde{q}^2\right) d\tau - \frac{1}{2}\int q_{\mathrm{cl}} J d\tau, \end{aligned}$$

where we have used the boundary conditions for q and $\tilde{q}$ as well as the equation of motion (12.23) to eliminate $\ddot{q}_{\mathrm{cl}}$. Substituting the resulting expression into (12.26), we obtain

$$\int_{q(-\infty)=0}^{q(+\infty)=0} \mathcal{D}q\, e^{-S_{\mathrm{E}}[q(\tau)]} = \exp\left(\frac{1}{2}\int q_{\mathrm{cl}} J d\tau\right) \int_{\tilde{q}(-\infty)=0}^{\tilde{q}(+\infty)=0} \mathcal{D}\tilde{q}\, e^{-\frac{1}{2}\int\left(\dot{\tilde{q}}^2 + \omega^2\tilde{q}^2\right)d\tau}. \tag{12.27}$$

Note that the remaining path integral in (12.27) does not depend on $J(\tau)$. Let us denote that integral by N_ω. We do not need an explicit expression for N_ω because we are interested only in the J-dependent part of the effective action. Hence the final result can be written as

$$\int_{q(-\infty)=0}^{q(+\infty)=0} \mathcal{D}q\, e^{-S_E[q(\tau)]} = N_\omega \exp\left[\frac{1}{2}\int_{-\infty}^{+\infty} q_{cl}(\tau)J(\tau)d\tau\right]$$

$$= N_\omega \exp\left[\frac{1}{2}\int J(\tau)J(\tau')G_E(\tau,\tau')d\tau d\tau'\right], \qquad (12.28)$$

where we have used (12.25).

Recalling the definition in (12.21), we infer from (12.28) the effective action for the driven oscillator:

$$\Gamma_E\left[J(\tau)\right] = -\frac{1}{2}\int J(\tau)J(\tau')G_E(\tau,\tau')d\tau d\tau' - \ln N_\omega. \qquad (12.29)$$

The Lorentzian effective action is obtained by analytic continuation. We set $d\tau d\tau' = -dtdt'$ and replace the Euclidean Green's function G_E in (12.29) by $-iG_F$. The result is

$$\Gamma_L\left[J(t)\right] = \frac{1}{2}\int J(t)J(t')G_F(t,t')dtdt' - i\ln N_\omega$$

$$= \frac{i}{2}\left|J_0\right|^2 + \int J(t)J(t')\frac{\sin\omega\left|t-t'\right|}{4\omega}dt\,dt' - i\ln N_\omega, \qquad (12.30)$$

where J_0 is defined in (3.13).

12.1.5 Matrix elements

We note that the expression $\exp\left(i\Gamma_L\left[J\right]\right)$ indeed coincides with the matrix element $\langle 0_{out}|0_{in}\rangle$,

$$\langle 0_{out}|0_{in}\rangle = \exp\left(-\frac{1}{2}\left|J_0\right|^2\right),$$

up to a phase factor that can be absorbed into the definition of $|0_{out}\rangle$, and a J-independent normalization factor N_ω.

Other matrix elements can also be expressed via path integrals and ultimately through the effective action in a similar way. Let us consider, for example, the matrix element $\langle 0_{out}|\hat{q}(t_1)|0_{in}\rangle$. The operator $\hat{q}(t_1)$ can be expressed in terms of the position operator $\hat{q}$ at some earlier moment of time $t_0 < 0$ (at which the vacuum state $|0_{in}\rangle$ is defined) as

$$\hat{q}(t_1) = \hat{U}_{10}^{-1}\hat{q}\hat{U}_{10}, \qquad (12.31)$$

where we have introduced the shortcut notation $\hat{U}_{10} \equiv \hat{U}(t_1, t_0)$. It follows from (12.17) that the vacuum state $\langle 0_{\text{out}}|$, defined at $t_2 > T$, is related to the vacuum state $\langle 0_{\text{in}}|$ as

$$\langle 0_{\text{out}}| = \langle 0_{\text{in}}| \hat{U}_{20}. \tag{12.32}$$

Taking into account the composition rule

$$\hat{U}_{20}\hat{U}_{10}^{-1} \equiv \hat{U}(t_2, t_0)\hat{U}(t_2, t_0) = \hat{U}(t_2, t_1) \equiv \hat{U}_{21},$$

and using (12.31) and (12.32), we obtain

$$\langle 0_{\text{out}}| \hat{q}(t_1) |0_{\text{in}}\rangle = \langle 0_{\text{in}}| \hat{U}_{21}\hat{q}\hat{U}_{10} |0_{\text{in}}\rangle. \tag{12.33}$$

With the help of the decomposition of the unit operator, this result can be rewritten as

$$\langle 0_{\text{out}}| \hat{q}(t_1) |0_{\text{in}}\rangle = \int dq_2 dq_1 dq_0 \langle 0_{\text{out}}|q_2\rangle K_{21} q_1 K_{10} \langle q_0|0_{\text{in}}\rangle, \tag{12.34}$$

where $K_{21} \equiv K(q_2, q_1; t_2, t_1)$, $K_{10} \equiv K(q_1, q_0; t_1, t_0)$ are the propagators. After performing an analytic continuation, the expression on the right-hand side of this equation can be interpreted as the well-defined path integral similar to the integral defining the Euclidean effective action in (12.21).

Formally, we may treat the expression on the right-hand side in (12.34) as the Lorentzian path integral

$$\int \mathcal{D}q\, q(t_1)\, e^{iS[J,q]}$$

and relate it to the formally defined Lorentzian effective action

$$e^{i\Gamma_L[J(t)]} = \int \mathcal{D}q\, e^{iS[J,q]}. \tag{12.35}$$

We will manipulate these integrals as if they were well-defined, for instance, when computing the functional derivatives of Γ_L. These operations should, however, be understood as the analogous manipulations on the Euclidean path integral, followed by an analytic continuation to the Lorentzian time.

Since the external field J enters linearly into the action,

$$S[q, J] = S_0[q] + \int J(t)q(t)dt, \tag{12.36}$$

the functional derivative of S with respect to $J(t_1)$ is

$$\frac{\delta S[q, J]}{\delta J(t_1)} = q(t_1),$$

and thus we may write

$$\int \mathcal{D}q\, q(t_1)\, e^{iS[J,q]} = \frac{1}{i}\frac{\delta}{\delta J(t_1)} \int \mathcal{D}q\, e^{iS[J,q]} = \frac{1}{i}\frac{\delta}{\delta J(t_1)} \exp(i\Gamma_L[J]).$$

Finally, for the normalized matrix element one obtains

$$\frac{\langle 0_{\rm out}|\,\hat{q}\,(t_1)\,|0_{\rm in}\rangle}{\langle 0_{\rm out}|0_{\rm in}\rangle} = \frac{\int \mathcal{D}q q\,(t_1)\, e^{iS[q,J]}}{\int \mathcal{D}q e^{iS[q,J]}}$$

$$= \exp\left(-i\Gamma_L\,[J]\right) \frac{1}{i}\frac{\delta}{\delta J\,(t_1)} \exp\left(i\Gamma_L\,[J]\right) = \frac{\delta\Gamma_L\,[J]}{\delta J\,(t_1)}. \qquad (12.37)$$

Substituting (12.30) into (12.37) yields

$$\frac{\langle 0_{\rm out}|\,\hat{q}\,(t_1)\,|0_{\rm in}\rangle}{\langle 0_{\rm out}|0_{\rm in}\rangle} = \frac{\delta\Gamma_{\rm L}\,[J]}{\delta J\,(t_1)} = \int J(t) G_{\rm F}\,(t_1, t)\,dt, \qquad (12.38)$$

where we have used the symmetry of the Feynman Green's function, $G_{\rm F}(t, t') = G_{\rm F}(t', t)$. This result coincides with (12.7).

The matrix element $\langle 0_{\rm out}|\,\hat{q}\,(t_2)\,\hat{q}\,(t_1)\,|0_{\rm in}\rangle$ can also be calculated with the help of the effective action. Assuming that $t_0 < 0$ and $t_3 > T$ and taking into account that

$$\hat{q}\,(t_1) = \hat{U}_{10}^{-1}\hat{q}\hat{U}_{10}, \quad \hat{q}\,(t_2) = \hat{U}_{20}^{-1}\hat{q}\hat{U}_{20}, \quad \langle 0_{\rm out}| = \langle 0_{\rm in}|\,\hat{U}_{30},$$

we obtain

$$\langle 0_{\rm out}|\,\hat{q}\,(t_2)\,\hat{q}\,(t_1)\,|0_{\rm in}\rangle = \langle 0_{\rm in}|\,\hat{U}_{30}\hat{U}_{20}^{-1}\hat{q}\hat{U}_{20}\hat{U}_{10}^{-1}\hat{q}\hat{U}_{10}\,|0_{\rm in}\rangle\,. \qquad (12.39)$$

If $t_0 < t_1 < t_2 < t_3$ then the following composition rules are valid,

$$\hat{U}_{30}\hat{U}_{20}^{-1} = \hat{U}_{32}, \quad \hat{U}_{20}\hat{U}_{10}^{-1} = \hat{U}_{21}, \qquad (12.40)$$

and (12.39) becomes

$$\langle 0_{\rm out}|\,\hat{q}\,(t_2)\,\hat{q}\,(t_1)\,|0_{\rm in}\rangle = \langle 0_{\rm in}|\,\hat{U}_{32}\hat{q}\hat{U}_{21}\hat{q}\hat{U}_{10}\,|0_{\rm in}\rangle\,. \qquad (12.41)$$

In turn, the expression on the right-hand side is proportional to the path integral

$$\int \mathcal{D}q q\,(t_2)\, q\,(t_1)\, e^{iS[J,q]},$$

and hence

$$\frac{\langle 0_{\rm out}|\,\hat{q}\,(t_2)\,\hat{q}\,(t_1)\,|0_{\rm in}\rangle}{\langle 0_{\rm out}|0_{\rm in}\rangle} = \frac{\int \mathcal{D}q q\,(t_2)\, q\,(t_1)\, e^{iS[J,q]}}{\int \mathcal{D}q e^{iS[J,q]}}.$$

For the action (12.36) we have

$$\int \mathcal{D}q q\,(t_2)\, q\,(t_1)\, e^{iS[J,q]} = \frac{1}{i}\frac{\delta}{\delta J(t_1)}\frac{1}{i}\frac{\delta}{\delta J(t_2)} \int \mathcal{D}q e^{iS[J,q]},$$

and therefore

$$\frac{\langle 0_{\rm out}|\,\hat{q}\,(t_2)\,\hat{q}\,(t_1)\,|0_{\rm in}\rangle}{\langle 0_{\rm out}|0_{\rm in}\rangle} = \exp\left(-i\Gamma_{\rm L}\,[J]\right) \frac{1}{i}\frac{\delta}{\delta J\,(t_2)}\frac{1}{i}\frac{\delta}{\delta J\,(t_1)} \exp\left(i\Gamma_{\rm L}\,[J]\right)$$

$$= \frac{\delta \Gamma_{\mathrm{L}}}{\delta J(t_2)} \frac{\delta \Gamma_L}{\delta J(t_1)} - i \frac{\delta^2 \Gamma_{\mathrm{L}}}{\delta J(t_2)\, \delta J(t_1)}. \quad (12.42)$$

The functional derivatives are evaluated as in (12.38),

$$\left[\int G_{\mathrm{F}}(t, t_2)\, J(t) dt\right]\left[\int G_{\mathrm{F}}(t', t_1)\, J(t') dt'\right] - i G_{\mathrm{F}}(t_2, t_1). \quad (12.43)$$

(The derivation above is applicable only for $t_2 > t_1$; otherwise, the composition rules (12.40) would be invalid.) Then the J-independent term is equal to

$$-i G_{\mathrm{F}}(t_2, t_1) = \frac{1}{2\omega} \exp(-i\omega\, |t_2 - t_1|) = \frac{1}{2\omega} \exp(-i\omega\,(t_2 - t_1)). \quad (12.44)$$

The result in (12.43) coincides with the answer in Exercise 3.4b (see p. 41).

Note that the operators $\hat{q}(t_1)$ and $\hat{q}(t_2)$ do not commute in general, and the path integral always gives the matrix elements for the *time-ordered product* of these operators:

$$T\hat{q}(t_2)\, \hat{q}(t_1) \equiv \begin{cases} \hat{q}(t_1)\, \hat{q}(t_2), & t_1 > t_2; \\ \hat{q}(t_2)\, \hat{q}(t_1), & t_1 < t_2. \end{cases}$$

The reader can easily check that the composition rules needed for reducing the matrix element to the path integral form can be applied only if the operators are time-ordered.

12.1.6 The effective action "recipe"

Comparing (12.6) and (12.7), we find that for the case of a driven harmonic oscillator the only difference between the "in-out" matrix element and the "in-in" expectation value is the presence of the retarded Green's function G_{ret} instead of G_{F}. Replacing G_{F} by G_{ret} in the final expression for the matrix element, we have

$$\langle 0_{\mathrm{in}} |\, \hat{q}(t)\, | 0_{\mathrm{in}} \rangle = \left. \frac{\delta \Gamma_L[J]}{\delta J(t)} \right|_{G_{\mathrm{F}} \to G_{\mathrm{ret}}}. \quad (12.45)$$

Note that the replacement $G_{\mathrm{F}} \to G_{\mathrm{ret}}$ is to be performed *after* computing the functional derivative. The expression (12.45) is again a functional of $J(t)$ as it should be since the expectation value of $\hat{q}$ depends on the force J.

A rigorous justification of the result (12.45) is beyond the scope of this book. Therefore we simply formulate our findings as the following recipe for computing the "in-out" matrix elements and the "in-in" expectation values for a quantum system coupled to a classical background.

(i) After performing the Wick rotation $t = -i\tau$, compute the Euclidean effective action $\Gamma_{\mathrm{E}}[J(\tau)]$ defined in (12.21). As we have seen, the result for the Euclidean effective action involves the Euclidean Green's function G_{E}.

(ii) By an analytic continuation back to the Lorentzian time t according to (12.22), obtain the Lorentzian effective action

$$\Gamma_L[J(t)] = i\,\Gamma_E[J(\tau)]|_{\tau=it},$$

where the Euclidean Green's function G_E is replaced by the Feynman Green's function, namely, $G_E \to -iG_F$.

(iii) Using formal manipulations with the Lorentzian path integral, express the required matrix element via functional derivatives of $\Gamma_L[J]$ with respect to J.

(iv) Compute the required functional derivatives, keeping the Feynman Green's function G_F. The result is the "in-out" matrix element for the corresponding operator.

(v) The replacement of G_F by G_{ret} in the final result, performed after computing the functional derivatives, yields the "in-in" expectation value in an appropriate vacuum state $|0_{in}\rangle$.

Exercise 12.3*

Considering the ratio

$$\frac{\int a^+(t)a^-(t)e^{iS[q,J]}\mathcal{D}q}{\int e^{iS[q,J]}\mathcal{D}q}$$

compute the expectation value $\langle 0_{in}|\,\hat{a}^+(t)\hat{a}^-(t)\,|0_{in}\rangle$ by following the recipe described above. Compare the result obtained to (3.20).

Hint: First consider the (Lorentzian) action with two auxiliary external forces $J^{\pm}(t)$,

$$S\left[q, J^+, J^-\right] = \int \left(\frac{1}{2}\dot{q}^2 - \frac{\omega^2}{2}q^2 + J^+a^+ + J^-a^-\right) dt.$$

Here $a^{\pm}(t)$ are the variables introduced in Section 3.1 and $J^- = (J^+)^*$. An integration by parts converts this action into the standard form.

12.1.7 Backreaction

As we have seen, the effective action allows us to find how an external classical "background" J influences the behavior of a quantum system. On the other hand, the effective action can also be used to determine the backreaction of the vacuum fluctuations on the classical background.

In realistic situations the background $J(t)$ is itself a dynamical field described by a classical action $S_B[J]$. In the absence of interactions between q and J, the equation of motion for the classical background would be

$$\frac{\delta S_B[J]}{\delta J(t)} = 0.$$

In the presence of interactions between classical background J and a quantum system q the total (classical) action is given by

$$S_{total} = S_{int}[q, J] + S_B[J],$$

and it is obvious that the quantum system influences the behavior of the classical background. Assuming that at some initial moment of time the quantum system was in the vacuum state, we can describe the modified dynamics of J by a "total" effective action $S_{\text{eff}}[J]$, which accounts for the backreaction of the quantum system on the classical background. The action $S_{\text{eff}}[J]$ is the functional of J only and is naturally defined as

$$\exp(iS_{\text{eff}}[J]) \equiv \int \mathcal{D}q \exp(iS_{\text{int}}[q, J] + iS_{\text{B}}[J])$$
$$= \exp(i\Gamma_{\text{L}}[J] + iS_{\text{B}}[J]), \quad (12.46)$$

where $\Gamma_L[J]$ is the effective action (12.35). Then the modified equation of motion for the background,

$$\frac{\delta S_{\text{eff}}[J]}{\delta J(t)} = \left.\frac{\delta \Gamma_{\text{L}}[J]}{\delta J(t)}\right|_{G_{\text{F}} \to G_{\text{ret}}} + \frac{\delta S_{\text{B}}[J]}{\delta J(t)} = 0, \quad (12.47)$$

takes into account the averaged influence of the quantum fluctuations on the background. As explained in the previous section, the replacement $G_{\text{F}} \to G_{\text{ret}}$ is needed to obtain physically meaningful results. Equation (12.47) is valid only when the quantum system was initially in the "in" vacuum state; it needs to be modified if the initial state is different from the vacuum state. This is apparent from the derivation of the effective action where the "in" vacuum state $|0_{\text{in}}\rangle$ enters explicitly.

Remark: a more rigorous derivation Note that we have not integrated over $J(t)$ in the path integral (12.46); in other words, $J(t)$ remains a classical variable while $q(t)$ is quantized. However, there is no consistent way of formulating a physical theory where some degrees of freedom are quantized while others remain classical. For instance, the equation of motion for the classical variable J will contain a quantum operator,

$$\frac{\delta S_{\text{B}}[J]}{\delta J(t)} = -\hat{q},$$

which will force the operator $\hat{q}$ to be proportional to $\hat{1}$. Thus there will be no solutions satisfying the Heisenberg commutation relations. A consistent derivation of equation (12.47) can be performed only by starting with a fully quantized system $(\hat{q}, \hat{J})$ and subsequently making a suitable approximation appropriate for a nearly classical degree of freedom J. A brief derivation is presented in Appendix 2.

12.2 Effective action in external gravitational field

In this book we concentrate mostly on the behavior of quantum fields in the external classical gravitational field. An important problem to be solved is the computation of the backreaction of quantum fluctuations of matter fields on gravitation. We can consider a quantum scalar field in a classical curved background similarly to a driven harmonic oscillator. In this case the field $\hat{\phi}$ is analogous to the quantum oscillator from the preceding consideration, the only difference being that $\hat{\phi}$ has infinitely many degrees of freedom. The metric $g_{\alpha\beta}$ plays the role of the classical background J. Our main task is to calculate the expectation value of the energy-momentum tensor

$$\langle 0_{\text{in}} | \hat{T}_{\alpha\beta} \left(\hat{\phi}, g_{\gamma\delta} \right) | 0_{\text{in}} \rangle \tag{12.48}$$

assuming an initial vacuum state $|0_{\text{in}}\rangle$ for the quantum scalar field $\hat{\phi}$. In general this energy-momentum tensor (EMT) describes both the particle production and the vacuum polarization effects. The number density of the particles produced by the external gravitational field depends on the whole preceding history of the evolution of the background field. Therefore the contribution of the produced particles to the EMT is described by the non-local expressions. On the other hand, the vacuum polarization is related to the "deformation of the vacuum fluctuations" by the external gravitational field at a given moment of time and hence it is described by local terms that depend only on the local curvature characterizing the gravitational field at a given location. Because the notion of a particle in an external field is not well-defined, one cannot unambiguously split the local and nonlocal contributions to the induced EMT. Nevertheless, this is possible in the leading order. We will see that the leading local contributions to the induced EMT can be calculated for an arbitrary curved background, while the determination of the nonlocal contributions is a much more difficult problem that has not been solved in the general case. To determine the expectation value of the induced EMT (12.48), we can use the effective action, which in this case is defined as

$$\exp\left(i\Gamma_{\text{L}}\left[g_{\mu\nu}\right]\right) \equiv \int \exp\left(iS^{(m)}\left[g_{\mu\nu}, \phi\right]\right) \mathcal{D}\phi,$$

where $S^{(m)}[g_{\mu\nu}, \phi]$ is the action for the matter field ϕ in the presence of gravity. Taking into account that the classical EMT is given by

$$T_{\alpha\beta}(x) = \frac{2}{\sqrt{-g}} \frac{\delta S^{(m)}\left[g_{\mu\nu}, \phi\right]}{\delta g^{\alpha\beta}(x)},$$

we can formally express the vacuum expectation value (or matrix element) of the quantum EMT as

$$\langle \hat{T}_{\alpha\beta}(x) \rangle = \frac{\int T_{\alpha\beta}(x) \exp\left(iS^{(m)}[g, \phi]\right) \mathcal{D}\phi}{\int \exp\left(iS^{(m)}[g, \phi]\right) \mathcal{D}\phi}$$

$$= \exp(-i\Gamma_{\rm L}) \frac{2}{i\sqrt{-g}} \frac{\delta}{\delta g^{\alpha\beta}(x)} \exp(i\Gamma_{\rm L}) = \frac{2}{\sqrt{-g}} \frac{\delta\Gamma_L\left[g_{\mu\nu}\right]}{\delta g^{\alpha\beta}(x)}.$$

In the same way as for a driven harmonic oscillator, one can define the Lorentzian effective action via the analytic continuation of the corresponding Euclidean effective action. Then, one expects to get $\langle 0_{\rm in}|\hat{T}_{\alpha\beta}|0_{\rm in}\rangle$ by substituting the retarded Green's function instead of the Feynman Green's function in the above result for $\langle \hat{T}_{\alpha\beta}\rangle$. The justification of this prescription, which we verified for a driven oscillator, is beyond the scope of this book. One can show that this procedure is also valid for the one-loop effective action in the case of external gravitational field, when the quantum matter fields are initially in the ground state.[2]

Taking into account the backreaction of quantum fields, the Einstein equations become

$$R_{\alpha\beta} - \frac{1}{2} g_{\alpha\beta} R = 8\pi G \langle 0_{\rm in}|\hat{T}_{\alpha\beta}|0_{\rm in}\rangle, \tag{12.49}$$

where the expectation value of the induced EMT simultaneously accounts for the produced particles and for the shift of zero point energy of the "field harmonic oscillators" in the classical curved background or, in other words, for the vacuum polarization effects.

12.2.1 Euclidean action for scalar field

Let us consider the scalar field ϕ, described by the action

$$S\left[\phi, g_{\mu\nu}\right] = \frac{1}{2}\int d^{2\omega}x\sqrt{-g}\left(g^{\mu\nu}\phi_{,\mu}\phi_{,\nu} - V(x)\phi^2\right), \tag{12.50}$$

where $g_{\mu\nu}(x)$ is the spacetime metric and the external potential $V(x)$ plays the role of the effective mass of the field ϕ. The (Greek) indices run from 0 to $2\omega - 1$, where 2ω is the number of spacetime dimensions. Depending on the choice of $V(x)$, this action can represent both minimally coupled and conformally coupled fields. We assume that the metric $g_{\mu\nu}(x)$ and the potential $V(x)$ are given functions.

We now perform an analytic continuation of action (12.50) to the Euclidean time. With this purpose in mind, we first consider a purely real change of coordinates $x \to \tilde{x}$. In the new coordinates action (12.50) takes the form

$$S = \frac{1}{2}\int d^{2\omega}\tilde{x}\sqrt{-\tilde{g}}\left(\tilde{g}^{\mu\nu}\phi_{,\mu}\phi_{,\nu} - V\phi^2\right). \tag{12.51}$$

[2] See the paper A. Barvinsky and G. Vilkovisky, *Nucl. Phys. B* **282** (1987), 163.

One can formally consider the particular complex coordinate transformation

$$x \equiv (t, \mathbf{x}) \to \tilde{x} \equiv (\tau, \mathbf{x}) = (it, \mathbf{x}), \tag{12.52}$$

which corresponds to the Wick rotation and brings the metric to the Euclidean form with the signature $(----)$ in the four-dimensional case. It is convenient to change the overall sign of the metric and use the signature convention $(++++)$. Introducing the Euclidean metric $g^{(\mathrm{E})}_{\mu\nu} = -\tilde{g}_{\mu\nu}$ and using the notation $x^{(\mathrm{E})}$ instead of $\tilde{x}$ we can rewrite (12.51) as

$$S = -\frac{1}{2}\int d^{2\omega}x^{(\mathrm{E})}\sqrt{-g^{(\mathrm{E})}}\left(g^{\mu\nu}_{(\mathrm{E})}\phi_{,\mu}\phi_{,\nu} + V\phi^2\right).$$

It follows that the Euclidean action is given by

$$S_{\mathrm{E}}\left[\phi, g_{\mu\nu}\right] \equiv \frac{1}{i}S\left[\phi, g_{\mu\nu}\right]_{t=-i\tau} = \frac{1}{2}\int d^{2\omega}x\sqrt{g}\left(g^{\mu\nu}\phi_{,\mu}\phi_{,\nu} + V\phi^2\right), \tag{12.53}$$

where $g_{\mu\nu}$ is the Euclidean metric. In the next two chapters, we shall perform all calculations exclusively with the Euclidean metric. Therefore, for brevity we skip the index E everywhere. Assuming that the field ϕ decays quickly enough at infinity, we can integrate by parts in (12.53) and, omitting the boundary terms, rewrite the action (12.53) in the following convenient form,

$$\begin{aligned} S_{\mathrm{E}}\left[\phi, g_{\mu\nu}\right] &= \frac{1}{2}\int d^{2\omega}x\left[-\phi\left(\sqrt{g}\,g^{\mu\nu}\phi_{,\mu}\right)_{,\nu} + \sqrt{g}V\phi^2\right] \\ &= \frac{1}{2}\int d^{2\omega}x\sqrt{g}\left[\phi(x)\hat{F}\phi(x)\right], \end{aligned} \tag{12.54}$$

where

$$\hat{F} = -\Box + V$$

and

$$\Box\phi \equiv \frac{1}{\sqrt{g}}\partial_\mu\left[\sqrt{g}\,g^{\mu\nu}\partial_\nu\phi\right]$$

is the covariant Laplace operator.

The Euclidean field ϕ and the Euclidean metric $g_{\mu\nu}$ now must be chosen to be *real-valued* functions despite the fact that the "coordinate transformation" of a metric with the Lorentzian signature leads in general to a complex-valued metric under the Wick rotation.

The Euclidean field $\phi(x)$ satisfies the equation of motion

$$\hat{F}\phi = \left[-\Box + V(x)\right]\phi = 0, \tag{12.55}$$

which immediately follows from action (12.54).

12.3 Effective action as a functional determinant

The Euclidean effective action $\Gamma_E[g_{\mu\nu}]$ is defined as

$$\exp\left(-\Gamma_E\left[g_{\mu\nu}\right]\right)=\int\mathcal{D}\phi\exp\left(-S_E\left[\phi,g_{\mu\nu}\right]\right),\tag{12.56}$$

where a suitable generally covariant measure $\mathcal{D}\phi$ in the space of functions $\phi(x)$ should still be determined. The direct derivation of this measure beginning with the canonical quantization is too cumbersome and we will give here only heuristic arguments to justify our choice for this measure.

To this end, let us consider the eigenvalue problem

$$\hat{F}\phi_n(x)=\left[-\Box+V(x)\right]\phi_n(x)=\lambda_n\phi_n(x),\tag{12.57}$$

in the space of functions with the following naturally defined covariant scalar product,

$$(f,g)\equiv\int d^{2\omega}x\sqrt{g}\,f(x)g(x).\tag{12.58}$$

For mathematical convenience, we can consider a finite box and impose the corresponding boundary conditions on the field $\phi(x)$, such that the spectrum of eigenvalues λ_n is discrete ($n=0,1,\ldots$). In this case the operator $\hat{F}$ is self-adjoint and the set of all its eigenfunctions forms a complete orthonormal basis, that is,

$$\int d^{2\omega}x\sqrt{g}\,\phi_m(x)\phi_n(x)=\delta_{mn}.\tag{12.59}$$

An arbitrary function $\phi(x)$ can then be expanded as

$$\phi(x)=\sum_{n=0}^{\infty}c_n\phi_n(x),\tag{12.60}$$

where the coefficients

$$c_n=\int d^{2\omega}x\sqrt{g}\,\phi(x)\phi_n(x)\tag{12.61}$$

are the coordinates of the function $\phi(x)$ with respect to the basis $\{\phi_n\}$ in the infinite-dimensional functional space.

Substituting (12.60) into (12.54), we find

$$S_E\left[\phi,g_{\mu\nu}\right]=\frac{1}{2}\int d^{2\omega}x\sqrt{g}\sum_{m,n}c_mc_n\lambda_m\phi_m\phi_n=\frac{1}{2}\sum_n c_n^2\lambda_n.$$

In other words, the quadratic action (12.54) is diagonalized in the basis of the eigenfunctions $\{\phi_n\}$.

Once an orthonormal system of eigenfunctions $\{\phi_n(x)\}$ is chosen, the coefficients c_n characterize the space of all functions $\phi(x)$ over which we have to perform the functional integration in (12.56). Therefore the measure $\mathcal{D}\phi$ can be expressed entirely in terms of c_n. This measure must be generally covariant. Because both c_n and the eigenvalues λ_n are coordinate-independent, we are motivated to define the generally covariant functional measure as $\mathcal{D}\phi = \prod f(c_n) dc_n$. The simplest choice for $f(c)$ is to take this function to be a constant, and a comparison with the usual path integral measure in flat space suggests

$$\mathcal{D}\phi = \prod_n \frac{dc_n}{\sqrt{2\pi}}. \tag{12.62}$$

Then the path integral (12.56) is evaluated as

$$\int \exp\left(-S_{\mathrm{E}}\left[\phi, g_{\mu\nu}\right]\right) \mathcal{D}\phi = \int \prod_{n=0}^{\infty} \frac{dc_n}{\sqrt{2\pi}} \exp\left(-\frac{1}{2}\lambda_n c_n^2\right) = \left[\prod_{n=0}^{\infty} \lambda_n\right]^{-1/2}.$$

Remark: boundary conditions In the Euclidean space, it is natural to impose zero boundary conditions on the basis functions $\phi_n(x)$. After an analytic continuation to Lorentzian time, these boundary conditions will become the "in-out" boundary conditions (see Section 12.1.1 where such boundary conditions were used to define the Feynman Green's function). These boundary conditions depend on the choice of the "in" and "out" vacua in the spacetime. Therefore, this choice is implicit in the definition of the functional determinant.

It is well known that the product of all eigenvalues of a *finite*-dimensional operator is equal to its determinant. Assuming that there exists a suitable generalization of the determinant for infinite-dimensional operators, we can formally rewrite the Euclidean effective action as

$$\Gamma_{\mathrm{E}}\left[g_{\mu\nu}\right] = \frac{1}{2} \ln \prod_{n=0}^{\infty} \lambda_n \equiv \frac{1}{2} \ln \det \hat{F}. \tag{12.63}$$

Thus, the computation of effective action is now reduced to the problem of calculating the determinant of a differential operator (a *functional determinant*). However, it is clear that a functional determinant is not a well-defined quantity. For example, the eigenvalues λ_n of the differential operator $-\Box$ grow with n and their product diverges. A finite result can be obtained only after an appropriate regularization and renormalization of the determinant.

12.3.1 Reformulation of the eigenvalue problem

To compute the functional determinant of the operator $\hat{F} \equiv -\Box + V$, we will first reformulate the eigenvalue problem (12.57), (12.59) in terms of some auxiliary

Hilbert space, which has a complete basis of "generalized vectors" $|x\rangle$ normalized as

$$\langle x|x'\rangle = \delta(x - x'),$$

and where the decomposition of the unit operator takes the form

$$\hat{1} = \int d^{2\omega}x\, |x\rangle\, \langle x|\,. \tag{12.64}$$

Our task is to determine a Hermitian operator $\hat{O}$ that has precisely the same eigenvalues λ_n as in (12.57), that is,

$$\hat{O}\,|\psi_n\rangle = \lambda_n\,|\psi_n\rangle\,, \tag{12.65}$$

when its eigenvectors $|\psi_n\rangle$ are normalized as

$$\langle\psi_n|\psi_m\rangle = \delta_{nm}. \tag{12.66}$$

Remark: Hilbert space $\neq$ quantum mechanics The appearance of a Hilbert space and of the Dirac notation does not mean that the vectors $|\psi\rangle$ are states of some quantum system. We use the Hilbert space formalism because it simplifies the manipulations when we calculate the renormalized functional determinants. It is possible but more cumbersome to work directly with the partial differential equations.

Using the decomposition of the unit operator (12.64), we can rewrite the eigenvalue problem (12.65), (12.66) in the coordinate basis in the following way,

$$\int d^{2\omega}x'\, \langle x|\,\hat{O}\,|x'\rangle\, \psi_n(x') = \lambda_n \psi_n(x), \tag{12.67}$$

$$\int d^{2\omega}x\, \psi_n(x)\psi_m(x) = \delta_{nm}, \tag{12.68}$$

where $\psi(x) \equiv \langle x|\psi\rangle$. Comparing (12.67), (12.68) to (12.57), (12.59), we see that these two eigenvalue problems are equivalent if

$$\psi_n(x) = g^{1/4}(x)\phi_n(x) \tag{12.69}$$

and

$$\langle x|\,\hat{O}\,|x'\rangle = g^{1/4}(x)\,(-\Box_x + V)\left[g^{-1/4}(x)\delta(x - x')\right]. \tag{12.70}$$

Thus we reduced the eigenvalue problem for the operator $\hat{F}$ to the equivalent problem for a different operator $\hat{O}$. Notice that the eigenvectors of $\hat{O}$ are normalized without involving $\sqrt{g}$ and hence the normalization of the eigenvectors plays the crucial role for the eigenvalue problem.

12.3.2 Zeta function

To calculate the product of the eigenvalues of some matrix, one does not need at all to solve the eigenvalue problem. In case of a finite-dimensional matrix, this product is simply equal to the determinant of the matrix and can be easily calculated. In case of a differential operator, which corresponds to an infinite-dimensional matrix, the product of the eigenvalues is in general infinite. Therefore, to obtain physically meaningful results we must first regularize it and give a physical interpretation to the removal of infinities. The regularization procedure will be justified later; here we mainly concentrate on the calculation of the finite part of functional determinants using the method of zeta (ζ) function.

For an operator $\hat{O}$ with eigenvalues λ_n, we define the *zeta function* $\zeta_{\hat{O}}(s)$ by

$$\zeta_{\hat{O}}(s) \equiv \mathrm{Tr}\,(\hat{O}^{-s}) \equiv \sum_{n=0}^{\infty} \left(\frac{1}{\lambda_n}\right)^s . \tag{12.71}$$

This function is similar to the Riemann's ζ function (10.8) except for the summation over the eigenvalues λ_n instead of the natural numbers. The sum in (12.71) converges for large enough real s, and for those s for which this sum diverges we define $\zeta_{\hat{O}}(s)$ by an analytic continuation. As a result the function $\zeta_{\hat{O}}(s)$ is finite and well-defined almost everywhere except the poles where it is infinite. Notice that the procedure of analytic continuation automatically removes the divergences (for the physical interpretation of this mathematical trick, see Chapter 14).

It follows from (12.71) that

$$\frac{d\zeta_{\hat{O}}(s)}{ds} = \frac{d}{ds}\sum_n e^{-s\ln\lambda_n} = -\sum_n e^{-s\ln\lambda_n}\ln\lambda_n,$$

and therefore

$$\ln\det\hat{O} = \ln\prod_n \lambda_n = \sum_n \ln\lambda_n = -\left.\frac{d\zeta_{\hat{O}}(s)}{ds}\right|_{s=0} . \tag{12.72}$$

The function $\zeta_{\hat{O}}(s)$ is usually regular at $s = 0$, so the derivative $d\zeta_M/ds$ is finite. Then equation (12.72) can be treated as the *definition* of the regularized determinant of $\hat{O}$. Of course, this definition coincides with the standard one for finite-dimensional operators.

Equation (12.72) is the main result of the ζ function method. We stress that the derivations of equations (12.63) and (12.72) are *formal* (i.e. mathematically not well-defined) because we manipulated sums such as $\sum_n \ln\lambda_n$ as if they were finite. Lacking a rigorous justification, one has to treat such formal manipulations with caution. In many cases, the answers obtained using the ζ function method have been verified by other, more direct regularization and renormalization procedures. For this reason the use of ζ function is considered as a valuable method of dealing with divergences in quantum field theory.

To demonstrate how this method works, let us compute the determinant of the Laplace operator $\hat{O} = -\partial_x^2$ in a one-dimensional box of size L. The operator $-\partial_x^2$ is self-adjoint in the space of square-integrable functions $f(x)$ satisfying the boundary conditions $f(0) = f(L) = 0$. The eigenvalues and the eigenfunctions are

$$-\frac{\partial^2}{\partial x^2} f_n = \lambda_n f_n, \quad f_n(x) = \sin\frac{\pi n x}{L}, \quad \lambda_n = \frac{\pi^2 n^2}{L^2}, \; n = 1, 2, \ldots,$$

and hence

$$\zeta_{\hat{O}}(s) = \sum_{n=1}^{\infty} \frac{1}{\lambda_n^s} = \frac{L^{2s}}{\pi^{2s}} \sum_{n=1}^{\infty} \frac{1}{n^{2s}} = \frac{L^{2s}}{\pi^{2s}} \zeta(2s),$$

where $\zeta(s)$ is the Riemann's zeta function. Therefore

$$\det(-\partial_x^2) = -\left.\frac{d}{ds}\right|_{s=0} \zeta_M(s) = -\left.\frac{d}{ds}\right|_{s=0} \left[\frac{L^{2s}}{\pi^{2s}} \zeta(2s)\right] = \ln(2L),$$

where we have used the following formulae,

$$\zeta(0) = -\frac{1}{2}, \quad \zeta'(0) = -\frac{1}{2}\ln(2\pi),$$

proved in the theory of the Riemann's ζ function.

12.3.3 Heat kernel

The calculation of $\zeta_{\hat{O}}(s)$ can be reduced to the problem of solving of the partial differential equation for the "heat kernel." Given a Hermitian operator $\hat{O}$ with positive eigenvalues λ_n and a complete set of eigenvectors $|\psi_n\rangle$, the *heat kernel* operator is defined as

$$\hat{K}(\tau) \equiv \exp(-\hat{O}\tau) = \sum_n e^{-\lambda_n \tau} |\psi_n\rangle \langle\psi_n| . \tag{12.73}$$

It is obvious that $\hat{K}(\tau)|_{\tau=0} = \hat{1}$ and that $\hat{K}(\tau)$ is well-defined for $\tau > 0$. The real parameter τ is sometimes called the "proper time," although of course it has nothing to do with the physical time. The variable τ is auxiliary and it will eventually disappear from the physical results.

Now we shall show that $\zeta_{\hat{O}}(s)$ can be expressed through the *trace* of the heat kernel operator. The trace of the operator does not depend on the choice of the orthonormal basis and hence

$$\mathrm{Tr}\,\hat{K}(\tau) \equiv \sum_n \langle\psi_n|\hat{K}(\tau)|\psi_n\rangle = \sum_n e^{-\lambda_n \tau}.$$

Recalling the definition of Euler's Γ function (see Appendix A1.3),

$$\Gamma(s) \equiv \int_0^\infty e^{-\tau}\tau^{s-1}d\tau = \lambda^s \int_0^\infty e^{-\lambda\tau}\tau^{s-1}d\tau, \quad \operatorname{Re} s > 0,$$

we obtain

$$\zeta_{\hat{O}}(s) = \sum_n (\lambda_n)^{-s} = \frac{1}{\Gamma(s)}\int_0^\infty \left(\sum_n e^{-\lambda_n\tau}\right)\tau^{s-1}d\tau$$

$$= \frac{1}{\Gamma(s)}\int_0^{+\infty}\left[\operatorname{Tr}\hat{K}(\tau)\right]\tau^{s-1}d\tau. \tag{12.74}$$

The integral converges for the same range of s for which the sum in (12.71) is well-defined.

At first glance, it appears more difficult to calculate $\hat{K}(\tau)$ than $\zeta_{\hat{O}}(s)$ because the definition (12.73) involves not only the eigenvalues λ_n but also the eigenvectors $|\psi_n\rangle$. However, as we will see, the matrix elements of $\hat{K}(\tau)$ in the coordinate basis $|x\rangle$ satisfy a known differential equation. Therefore, one can determine $\operatorname{Tr}\hat{K}$ by simply solving that equation.

It follows from the definition of the heat kernel that

$$\frac{d\hat{K}(\tau)}{d\tau} = -\hat{O}\hat{K}(\tau). \tag{12.75}$$

Inserting the decomposition of the unit operator (12.64) into (12.75), we find

$$\frac{dK(x,x',\tau)}{d\tau} = -\int d^{2\omega}x'' \left\langle x\right|\hat{O}\left|x''\right\rangle K(x'',x',\tau), \tag{12.76}$$

where $K(x,x',\tau) \equiv \langle x|\hat{K}(\tau)|x'\rangle$. Taking into account that $\hat{K}(\tau=0) = \hat{1}$, we obtain the "initial conditions" for this equation,

$$K(x,x',\tau=0) = \langle x|\hat{K}(\tau=0)\left|x'\right\rangle = \delta(x-x'). \tag{12.77}$$

As we have noted above, the trace of the operator does not depend on the choice of the orthonormal basis, therefore

$$\operatorname{Tr}\hat{K}(\tau) = \int d^{2\omega}x\, K(x,x,\tau). \tag{12.78}$$

Thus, to calculate the functional determinant of an operator $\hat{O}$ with the matrix elements $\langle x|\hat{O}|x'\rangle$ we must: (1) solve the heat kernel equation (12.76) with the initial conditions (12.77) and determine $K(x,x',\tau)$; (2) substitute (12.78) into (12.74) and integrate over the proper time τ for those s for which the integral converges; (3) analytically continue the obtained function $\zeta_{\hat{O}}(s)$ and calculate $\ln\det\hat{O}$ according to (12.72). The corresponding Euclidean effective action is

$\frac{1}{2}\ln\det\hat{O}$; the Lorentzian effective action is obtained via analytic continuation with the replacement of the Green's functions.

We conclude this section by solving the heat kernel equation for the operator (12.70) in a flat one-dimensional Euclidean space, assuming that $V = 0$. In this case,

$$\langle x|\,\hat{O}\,|x''\rangle = -\frac{\partial^2}{\partial x^2}\delta(x - x''),$$

and the heat kernel equation (12.76) becomes

$$\frac{dK(x, x', \tau)}{d\tau} = \frac{\partial^2 K(x, x', \tau)}{\partial x^2}. \tag{12.79}$$

Equation (12.79) resembles the equation describing the propagation of heat in a homogeneous medium; this explains the origin of the names "heat kernel" and "proper time." Substituting the Fourier transform

$$K(x, x', \tau) = \int \frac{dk}{2\pi} e^{ikx} \tilde{K}(k, x', \tau)$$

into (12.79) and (12.77), we have

$$\frac{d\tilde{K}(k, x', \tau)}{d\tau} = -k^2 \tilde{K}(k, x', \tau), \quad \tilde{K}\,(k, x', \tau)\big|_{\tau=0} = e^{-ikx'}.$$

This equation can be easily solved,

$$\tilde{K}(k, x', \tau) = e^{-\tau k^2 - ikx'},$$

and performing the inverse Fourier transform we finally obtain

$$K(x, x', \tau) = \frac{1}{\sqrt{4\pi\tau}} \exp\left[-\frac{(x - x')^2}{4\tau}\right].$$

13

Calculation of heat kernel

Summary Perturbative solution for the heat kernel in curved space. The Seeley–DeWitt expansion.

In this chapter we calculate the heat kernel for the operator describing a scalar field in a gravitational background. This is the first step towards determining the effective action in the curved spacetime. The calculations are rather cumbersome and will be presented in full detail.

The matrix elements of the operator $\hat{O}$ we are interested in are (see (12.70)):

$$\begin{aligned}\langle x|\hat{O}|x'\rangle &= g^{1/4}\left(-\Box_x^{(g)}+V\right)\left[g^{-1/4}\delta(x-x')\right]\\ &= -g^{1/4}\frac{1}{\sqrt{g}}\frac{\partial}{\partial x^\nu}\left[g^{\mu\nu}\sqrt{g}\frac{\partial}{\partial x^\mu}\left(g^{-1/4}\delta(x-x')\right)\right]\\ &\quad + V(x)\delta(x-x').\end{aligned}$$

It is obvious that in this case the heat kernel equation cannot be solved exactly for an arbitrary metric $g_{\mu\nu}$ and a potential V. Therefore we will develop perturbation theory assuming that the potential V is small and that the (Euclidean) metric is a small deviation from the flat metric,

$$g_{\mu\nu}(x)=\delta_{\mu\nu}+h_{\mu\nu}(x),\quad g^{\mu\nu}(x)=\delta^{\mu\nu}+h^{\mu\nu}(x), \tag{13.1}$$

where $|h_{\alpha\beta}|\ll 1$. (In fact, even if the spacetime is strongly curved, one can always choose a locally inertial coordinate frame where the metric can be written in this form in the vicinity of any point.) In this case, the operator $\hat{O}$ can be written as

$$\hat{O}=-\Box-\hat{s}\left[h_{\mu\nu},V\right], \tag{13.2}$$

and one can develop the perturbation theory in $h_{\mu\nu}$ and V.

Exercise 13.1
Verify that the operator $\hat{O}$ can be rewritten in the form (13.2), where the matrix elements of $\Box$ are

$$\langle x|\Box|x'\rangle \equiv \delta^{\mu\nu}\partial_\mu\partial_\nu\delta(x-x'), \tag{13.3}$$

and

$$\hat{s}\left[h_{\mu\nu}, V\right] = \hat{h}+\hat{\Gamma}+\hat{P}, \tag{13.4}$$

where the operators $\hat{h}$, $\hat{\Gamma}$, $\hat{P}$ are defined by specifying their matrix elements as follows,

$$\langle x|\hat{h}|x'\rangle \equiv h^{\mu\nu}\partial_\mu\partial_\nu\delta(x-x'), \tag{13.5}$$

$$\langle x|\hat{\Gamma}|x'\rangle \equiv h^{\mu\nu}_{,\nu}\partial_\mu\delta(x-x'), \tag{13.6}$$

$$\langle x|\hat{P}|x'\rangle \equiv P(x)\delta(x-x'). \tag{13.7}$$

Here the partial derivatives are taken with respect to x ($\partial_\mu \equiv \partial/\partial x^\mu$) and

$$P(x) \equiv -\frac{1}{4}g^{\mu\nu}g^{\alpha\beta}h_{\alpha\beta,\mu\nu} - \frac{1}{4}g^{\mu\nu}h^{\alpha\beta}_{,\mu}h_{\alpha\beta,\nu}$$
$$-\frac{1}{4}h^{\mu\nu}_{,\nu}g^{\alpha\beta}h_{\alpha\beta,\mu} - \frac{1}{16}g^{\mu\nu}g^{\alpha\beta}g^{\kappa\lambda}h_{\alpha\beta,\mu}h_{\kappa\lambda,\nu} - V.$$

Hint: Use the identity $(\ln g)_{,\mu} = g^{\alpha\beta}g_{\alpha\beta,\mu}$.

13.1 Perturbative expansion for the heat kernel

If $\hat{s}$ can be treated as a "small correction" to the operator $\Box$, one can develop a perturbative expansion for the heat kernel,

$$\hat{K}(\tau) = \hat{K}_0(\tau)+\hat{K}_1(\tau)+\hat{K}_2(\tau)+\ldots, \tag{13.8}$$

where $\hat{K}_n(\tau)$ are operators of n-th order in $\hat{s}$. Substituting this expansion into the heat kernel equation

$$\frac{d\hat{K}}{d\tau} = (\Box+\hat{s})\,\hat{K},$$

we obtain in the leading order

$$\frac{d\hat{K}_0}{d\tau} = \Box\hat{K}_0, \tag{13.9}$$

while the first-order term $\hat{K}_1(\tau)$ satisfies

$$\frac{d}{d\tau}\hat{K}_1 = \Box\hat{K}_1+\hat{s}\hat{K}_0. \tag{13.10}$$

We will calculate here only the first-order terms. The higher-order corrections $\hat{K}_2, \ldots$ can be calculated following the general strategy explained below. The "initial conditions" for equations (13.9) and (13.10) can obviously be taken as

$$\hat{K}_0(0) = \hat{1}, \quad \hat{K}_1(0) = 0. \tag{13.11}$$

Then the formal solution of (13.9) is

$$\hat{K}_0(\tau) = \exp(\tau\Box). \tag{13.12}$$

Equation (13.10) can now be solved using the standard method of variation of constants and keeping in mind that the operators $\Box$, $\hat{K}_0(\tau)$, and $\hat{s}$ do not commute. Let us set

$$\hat{K}_1(\tau) = \hat{K}_0(\tau)\hat{C}(\tau),$$

where $\hat{C}(\tau)$ is an operator to be determined. Substituting this expansion into (13.10) and taking into account that $\hat{K}_0(\tau)$ satisfies (13.9), we obtain

$$\hat{K}_0(\tau)\frac{d}{d\tau}\hat{C}(\tau) = \hat{s}\hat{K}_0(\tau) \tag{13.13}$$

and hence

$$\hat{C}(\tau) = \int_0^\tau d\tau' \, \hat{K}_0^{-1}(\tau')\hat{s}\hat{K}_0(\tau'). \tag{13.14}$$

The integration starts at $\tau' = 0$ to satisfy the initial condition $\hat{C}(0) = 0$. It follows from (13.12) that

$$\hat{K}_0(\tau)\hat{K}_0(\tau') = \hat{K}_0(\tau+\tau'), \quad \tau > 0, \ \tau' > 0,$$

therefore $\hat{K}_0^{-1}(\tau) = \hat{K}_0(-\tau)$ and the final result for $\hat{K}_1(\tau)$ is

$$\hat{K}_1(\tau) = \int_0^\tau d\tau' \, \hat{K}_0(\tau - \tau')\hat{s}\hat{K}_0(\tau'). \tag{13.15}$$

Remark: inverting the heat kernel Note that (13.14) involves the inverse heat kernel $\hat{K}_0^{-1}(\tau) = \hat{K}_0(-\tau)$ which is undefined for most functions. Indeed, the operator $\hat{K}_0(\tau)$ with $\tau < 0$ can be applied only to those functions which decay extremely quickly for large $|x|$. However, the potentially problematic operator $\hat{C}(\tau)$ does not enter the final formula (13.15), which contains only $\hat{K}_0(\tau - \tau')$ and $\hat{K}_0(\tau')$ with $\tau - \tau' \geq 0$ and $\tau \geq 0$.

13.1.1 Matrix elements

To find the trace of the heat kernel, we first need to calculate the matrix elements of $\hat{K}_0$ and $\hat{K}_1$. Taking into account (13.3), we obtain

$$\langle x| \hat{K}_0(\tau) |y\rangle = \langle x| e^{\tau\Box} |y\rangle = e^{\tau\Box_x}\delta(x-y),$$

where $\Box_x$ indicates that the Laplace operator contains derivatives with respect to x rather than y. Substituting the Fourier transform of the δ function in 2ω dimensions,

$$\delta(x-y) = \int \frac{d^{2\omega}k}{(2\pi)^{2\omega}} e^{ik\cdot(x-y)},$$

we make use of the fact that the operator $e^{\tau\Box_x}$ acting on $e^{ik\cdot x}$ brings a factor of $e^{-\tau k^2}$. In other words,

$$e^{\tau\Box_x} \int \frac{d^{2\omega}k}{(2\pi)^{2\omega}} e^{ik\cdot(x-y)} = \int \frac{d^{2\omega}k}{(2\pi)^{2\omega}} e^{-\tau k^2 + ik\cdot(x-y)}.$$

This can be seen more formally by expanding $e^{\tau\Box_x}$ in the power series,

$$\begin{aligned}
\langle x|\hat{K}_0(\tau)|y\rangle = e^{\tau\Box_x}\delta(x-y) &= \int \frac{d^{2\omega}k}{(2\pi)^{2\omega}} \left[\sum_{n=0}^{\infty} \frac{(\tau\Box_x)^n}{n!}\right] e^{ik\cdot(x-y)} \\
&= \int \frac{d^{2\omega}k}{(2\pi)^{2\omega}} \left[\sum_{n=0}^{\infty} \frac{(-\tau k^2)^n}{n!}\right] e^{ik\cdot(x-y)} \\
&= \int \frac{d^{2\omega}k}{(2\pi)^{2\omega}} e^{-\tau k^2 + ik\cdot(x-y)}.
\end{aligned}$$

The resulting Gaussian integral can be easily calculated; the result is

$$\langle x|\hat{K}_0(\tau)|y\rangle = \frac{1}{(4\pi\tau)^{\omega}} \exp\left[-\frac{(x-y)^2}{4\tau}\right]. \tag{13.16}$$

This expression coincides with the Green's function of the heat equation in 2ω spatial dimensions.

The calculation of the matrix elements of $\hat{K}_1$ is more cumbersome. First we note that $\hat{K}_1$ is linear in $\hat{s}$, while $\hat{s} = \hat{h} + \hat{\Gamma} + \hat{P}$. Therefore the result for $\hat{K}_1$ can be written as

$$\hat{K}_1 = \hat{K}_1^h + \hat{K}_1^\Gamma + \hat{K}_1^P,$$

and the diagonal matrix elements (the only ones needed for the calculation of the trace) are

$$\langle x|\hat{K}_1|x\rangle = \langle x|\hat{K}_1^h|x\rangle + \langle x|\hat{K}_1^\Gamma|x\rangle + \langle x|\hat{K}_1^P|x\rangle.$$

We begin with the last term above, because it is the simplest one. Using (13.16) and (13.7), one finds

$$\begin{aligned}
\langle x|\hat{K}_1^P|x\rangle &= \int_0^\tau d\tau' \langle x|\hat{K}_0(\tau-\tau')\hat{P}\hat{K}_0(\tau')|x\rangle \\
&= \int_0^\tau d\tau' d^{2\omega}y d^{2\omega}z \langle x|\hat{K}_0(\tau-\tau')|y\rangle \langle y|\hat{P}|z\rangle \langle z|\hat{K}_0(\tau')|x\rangle
\end{aligned}$$

$$= \int_0^\tau d\tau' \int d^{2\omega}y \, \langle x|\, \hat{K}_0(\tau-\tau')\, |y\rangle \, P(y) \, \langle y|\, \hat{K}_0(\tau')\, |x\rangle$$

$$= \int_0^\tau d\tau' \int d^{2\omega}y \frac{\exp\left[-\frac{(x-y)^2}{4(\tau-\tau')} - \frac{(x-y)^2}{4\tau'}\right]}{[4\pi(\tau-\tau')]^\omega \, [4\pi\tau']^\omega} P(y)$$

$$= \int_0^\tau d\tau' \int d^{2\omega}y \frac{\exp\left[-\frac{\tau}{4(\tau-\tau')\tau'}(x-y)^2\right]}{[4\pi(\tau-\tau')]^\omega \, [4\pi\tau']^\omega} P(y).$$

To convert the last integral to a more convenient form, we note that the exponential factor above is similar to the zeroth-order heat kernel given by (13.16),

$$e^{\tau\Box_x} P(x) = \int d^{2\omega}y \frac{1}{(4\pi\tau)^\omega} \exp\left[-\frac{(x-y)^2}{4\tau}\right] P(y).$$

Therefore, the result can be rewritten in the operator form as

$$\langle x|\, \hat{K}_1^P \, |x\rangle = \frac{1}{(4\pi\tau)^\omega} \int_0^\tau d\tau' \exp\left[\frac{\tau'(\tau-\tau')}{\tau}\Box_x\right] P(x). \tag{13.17}$$

Note that the operator exponential in equation (13.17) has to be understood as a shorthand for the corresponding nonlocal kernel; the function $P(x)$ does not need to be infinitely differentiable.

The remaining diagonal matrix elements for $\hat{K}_1^\Gamma$ and $\hat{K}_1^h$ can be expressed through the nondiagonal matrix element $\langle x|\, \hat{K}_1^P \, |y\rangle$. It is not difficult to compute this matrix element by the same method we used for $\langle x|\, \hat{K}_1^P \, |x\rangle$. The calculation leading to (13.17) needs to be only slightly modified by introducing the Fourier transform,

$$P(z) \equiv \int \frac{d^{2\omega}k}{(2\pi)^\omega} e^{ik\cdot z} p(k).$$

Then we can calculate the Gaussian integral over $d^{2\omega}z$ (see Exercise 13.2):

$$\left\langle x|\hat{K}_1^P|y\right\rangle = \int_0^\tau d\tau' \langle x|\hat{K}_0(\tau-\tau')\hat{P}\hat{K}_0(\tau')|y\rangle \tag{13.18}$$

$$= \int_0^\tau d\tau' \int d^{2\omega}z \int \frac{d^{2\omega}k}{(2\pi)^\omega} \frac{\exp\left[-\frac{(x-z)^2}{4(\tau-\tau')} - \frac{(z-y)^2}{4\tau'} + ik\cdot z\right]}{[4\pi(\tau-\tau')]^\omega [4\pi\tau']^\omega} p(k)$$

$$= \frac{\exp\left[-\frac{(x-y)^2}{4\tau}\right]}{(4\pi\tau)^\omega} \int_0^\tau d\tau' \int \frac{d^{2\omega}k}{(2\pi)^\omega} \exp\left[-\frac{\tau'(\tau-\tau')}{\tau}k^2 \right.$$

$$\left. + \frac{1}{\tau} ik\cdot(x\tau' + y(\tau-\tau'))\right] p(k). \tag{13.19}$$

In the limit $y \to x$ we recover the result (13.17), as expected.

Exercise 13.2
Verify the following Gaussian integral over the 2ω-dimensional Euclidean space,

$$\int d^{2\omega}\mathbf{x} \exp\left[-A|\mathbf{x}-\mathbf{a}|^2 - B|\mathbf{x}-\mathbf{b}|^2 + 2\mathbf{c}\cdot\mathbf{x}\right]$$
$$= \frac{\pi^\omega}{(A+B)^\omega} \exp\left[-\frac{AB|\mathbf{a}-\mathbf{b}|^2}{A+B} + \frac{2\mathbf{c}\cdot(A\mathbf{a}+B\mathbf{b})+|\mathbf{c}|^2}{A+B}\right].$$

Here $A>0$, $B>0$ are constants and **a**, **b**, **c** are fixed 2ω-dimensional vectors. The scalar product of 2ω-dimensional vectors is denoted by $\mathbf{a}\cdot\mathbf{b}$.

We now turn to

$$\hat{K}_1^\Gamma(\tau) \equiv \int_0^\tau d\tau' \hat{K}_0(\tau-\tau')\hat{\Gamma}\hat{K}_0(\tau'),$$

where the operator $\hat{\Gamma}$ is defined by (13.6). The matrix element $\langle x|\,\hat{K}_1^\Gamma(\tau)\,|y\rangle$ can be written as

$$\langle x|\,\hat{K}_1^\Gamma(\tau)\,|y\rangle = \int_0^\tau d\tau' \int d^{2\omega}z\, \langle x|\,\hat{K}_0(\tau-\tau')\,|z\rangle\, h^{\mu\nu}_{,\nu}(z) \frac{\partial}{\partial z^\mu} \langle z|\,\hat{K}_0(\tau')\,|y\rangle$$
$$= -\frac{\partial}{\partial y^\mu} \int_0^\tau d\tau' \int d^{2\omega}z\, \langle x|\,\hat{K}_0(\tau-\tau')\,|z\rangle\, h^{\mu\nu}_{,\nu}(z)\, \langle z|\,\hat{K}_0(\tau')\,|y\rangle\,,$$

where we took into account the fact that $\langle z|\,\hat{K}_0(\tau)\,|y\rangle$ is a function of $(z-y)$ and τ and replaced the derivative ∂_z by $-\partial_y$. The obtained expression looks very similar to (13.18), except that here we have $h^{\mu\nu}_{,\nu}(z)$ instead of $P(z)$. Therefore we can use (13.19) to obtain

$$\langle x|\,\hat{K}_1^\Gamma(\tau)\,|x\rangle = -\lim_{y\to x} \frac{\partial}{\partial y^\mu} \langle x|\,\hat{K}_1^P(\tau)\,|y\rangle \Big|_{P(z)\to h^{\mu\nu}_{,\nu}(z)}$$
$$= -\frac{1}{(4\pi\tau)^\omega} \int_0^\tau d\tau' \exp\left[\frac{\tau'(\tau-\tau')}{\tau}\Box_x\right] \frac{\tau-\tau'}{\tau} h^{\mu\nu}_{,\mu\nu}(x).$$

The diagonal matrix element of the operator $\hat{K}_1^h$ is computed in a similar way.

Exercise 13.3
Verify the formula

$$\langle x|\,\hat{K}_1^h(\tau)\,|x\rangle =$$
$$\int_0^\tau d\tau' \frac{\exp\left[\frac{\tau'(\tau-\tau')}{\tau}\Box_x\right]}{(4\pi\tau)^\omega} \left\{-\frac{\delta_{\mu\nu}h^{\mu\nu}(x)}{2\tau} + \left(\frac{\tau-\tau'}{\tau}\right)^2 h^{\mu\nu}_{,\mu\nu}(x)\right\}.$$

Hint: Follow the strategy used to calculate $\langle x|\,\hat{K}_1^\Gamma\,|y\rangle$.

13.2 Trace of the heat kernel

The trace of the heat kernel is equal to

$$\mathrm{Tr}\,\hat{K}(\tau) = \int d^{2\omega}x \, \langle x| \left(\hat{K}_0 + \hat{K}_1\right) |x\rangle + O\left(h^2\right).$$

Combining the results of the previous calculations, we find

$$\langle x| \hat{K}_1(\tau) |x\rangle = \frac{1}{(4\pi\tau)^\omega} \int_0^\tau d\tau' \exp\left[\frac{\tau'(\tau-\tau')}{\tau}\Box_x\right] \times \left\{P(x) - \frac{1}{2\tau}\delta_{\mu\nu}h^{\mu\nu}(x) - \frac{\tau'(\tau-\tau')}{\tau^2}h^{\mu\nu}_{\ ,\mu\nu}(x)\right\}, \quad (13.20)$$

where we neglected higher-order terms in h and set

$$P(x) = \frac{1}{4}\delta_{\mu\nu}\Box h^{\mu\nu}(x) - V(x) + O\left(h^2\right).$$

The exponential is expanded as

$$\exp\left[\frac{\tau'(\tau-\tau')}{\tau}\Box_x\right] = \hat{1} + \sum_{n=1}^{\infty}\frac{1}{n!}\left(\frac{\tau'(\tau-\tau')}{\tau}\Box_x\right)^n$$

and yields terms such as $\Box^n h^{\mu\nu}$ and $\Box^n V$ with prefactors that can be integrated term by term over $d\tau'$. After some algebra, we can rewrite (13.20) as

$$\begin{aligned}\langle x| \hat{K}_1(\tau) |x\rangle =& \frac{1}{(4\pi\tau)^\omega}\Big\{P(x)\tau - \frac{1}{2}\delta_{\mu\nu}h^{\mu\nu}(x) - \frac{1}{6}\tau h^{\mu\nu}_{\ ,\mu\nu}(x) \\ &+ \frac{\tau}{6}\Box P - \frac{\tau}{12}\delta_{\mu\nu}\Box h^{\mu\nu}(x) - \frac{\tau}{30}\Box h^{\mu\nu}_{\ ,\mu\nu}(x) + \Box^2(\ldots)\Big\} \\ =& \frac{1}{(4\pi\tau)^\omega}\Big\{-\frac{1}{2}\delta_{\mu\nu}h^{\mu\nu}(x) - \tau V(x) \\ &+ \frac{\tau}{6}\left[\delta_{\mu\nu}\Box h^{\mu\nu}(x) - h^{\mu\nu}_{\ ,\mu\nu}(x)\right] + \Box(\ldots)\Big\}.\end{aligned}$$

The covariant volume factor $\sqrt{g}$ and the Ricci scalar R are related to $h^{\mu\nu}$ by

$$\sqrt{g} = 1 - \frac{1}{2}\delta_{\mu\nu}h^{\mu\nu} + O\left(h^2\right), \quad R = \delta_{\mu\nu}\Box h^{\mu\nu} - h^{\mu\nu}_{\ ,\mu\nu} + O\left(h^2\right). \quad (13.21)$$

Using these formulae, we obtain

$$\langle x| \hat{K}_1(\tau) |x\rangle = \frac{\sqrt{g}}{(4\pi\tau)^\omega}\left[-\tau V(x) + \frac{\tau}{6}R(x) + \Box(\ldots) + O(h^2)\right]. \quad (13.22)$$

Exercise 13.4
Derive the relations (13.21).

The final result for the trace of the heat kernel is

$$\operatorname{Tr}\hat{K} = \int d^{2\omega}x \,\langle x| \left(\hat{K}_0 + \hat{K}_1\right) |x\rangle$$
$$= \frac{1}{(4\pi\tau)^{\omega}} \int d^{2\omega}x\sqrt{g}\left[1 + \left(\frac{R}{6} - V\right)\tau + O\left(h^2\right)\right]. \tag{13.23}$$

We have skipped here the terms $\Box(...)$ because they are total derivatives and hence vanish after the integration over $d^{2\omega}x$. The disregarded higher-order terms $O(h^2)$ involve R^2, V^2, VR etc. Expression (13.23) represents the first two terms of the expansion for the trace of the heat kernel in terms of the powers of curvature. It is possible to compute further terms of this expansion, although the formulae rapidly become complicated at higher orders. The second-order terms were found by Barvinsky and Vilkovisky.[1] We state their result without proof:

$$\operatorname{Tr}\hat{K}(\tau) = \int \frac{d^{2\omega}x\sqrt{g}}{(4\pi\tau)^{\omega}} \left\{ 1 + \tau\left[\frac{R}{6} - V\right] \right.$$
$$+ \frac{\tau^2}{2}\left[V - \frac{R}{6}\right] f_1\left(-\tau\Box_g\right) V + \tau^2 V f_2\left(-\tau\Box_g\right) R$$
$$\left. + \tau^2 R f_3\left(-\tau\Box_g\right) R + \tau^2 R_{\mu\nu} f_4\left(-\tau\Box_g\right) R^{\mu\nu} + O(R^3, V^3, ...) \right\}, \tag{13.24}$$

where $\Box_g$ is the covariant Laplacian and the auxiliary functions $f_i(\xi)$ are defined by

$$f_1(\xi) \equiv \int_0^1 e^{-\xi u(1-u)} du, \qquad f_2(\xi) \equiv -\frac{f_1(\xi)}{6} - \frac{f_1(\xi) - 1}{2\xi}, \tag{13.25}$$

$$f_4(\xi) \equiv \frac{f_1(\xi) - 1 + \frac{1}{6}\xi}{\xi^2}, \qquad f_3(\xi) \equiv \frac{f_1(\xi)}{32} + \frac{f_1(\xi) - 1}{8\xi} - \frac{f_4(\xi)}{8}. \tag{13.26}$$

Since the functions $f_i(\xi)$ are analytic and have Taylor expansions that converge uniformly for all $\xi \geq 0$, the operators $f_i(-\tau\Box_g)$ are well-defined. Expressions such as $f_i(-\tau\Box_g)V(x)$ can be also rewritten as integrals of $V(x)$ with nonlocal kernels, but we shall not need their explicit form here. Note that the nonlocal terms such as $f_1(-\tau\Box_g)V$ contain all powers of τ.

[1] A. O. Barvinsky and G. A. Vilkovisky, *Nucl. Phys. B* **333** (1990), 471.

13.3 The Seeley–DeWitt expansion

Instead of developing the perturbation theory in terms of the powers of curvature, one can find a perturbative expansion for the heat kernel in terms of powers of the proper time τ. This expansion is called the *Seeley–DeWitt expansion* and has the form

$$\langle x| \hat{K}(\tau) |x\rangle = \frac{\sqrt{g}}{(4\pi\tau)^{\omega}} \left[1 + a_1(x)\tau + a_2(x)\tau^2 + O\left(\tau^3\right)\right], \quad (13.27)$$

where the *Seeley–DeWitt coefficients* $a_i(x)$ are local, scalar functions of the curvature tensor $R_{\kappa\lambda\mu\nu}$ and $V(x)$. To derive (13.27), one does not need to assume that the curvature is small.

We will omit a direct derivation of the Seeley–DeWitt expansion; instead, we obtain the coefficients a_1 and a_2 by expanding (13.24) in powers of τ and comparing the result with (13.27). Because the terms of order R^3 and higher that were neglected in (13.24) contain at least a third power of the proper time τ, the result in (13.24) accounts for all terms of order τ^2. Expanding the nonlocal operators in (13.24) in powers of τ, one obtains

$$\mathrm{Tr}\, \hat{K}(\tau) = \int \frac{d^{2\omega}x\sqrt{g}}{(4\pi\tau)^{\omega}} \left\{1 + \tau\left[\frac{R}{6} - V\right] + \tau^2\left[\frac{1}{2}V^2 - \frac{1}{6}VR + \frac{1}{120}R^2 + \frac{1}{60}R_{\mu\nu}R^{\mu\nu}\right] + O(\tau^3, R^3, V^3, ...)\right\}. \quad (13.28)$$

Exercise 13.5

To derive the coefficients in the above formula, verify that

$$f_1(0) = 1, \quad f_2(0) = -\frac{1}{12}, \quad f_3(0) = \frac{1}{120}, \quad f_4(0) = \frac{1}{60}$$

for the functions in (13.25)–(13.26).

Comparing (13.28) with (13.27), we conclude that (up to the total derivative terms) the first two Seeley–DeWitt coefficients are

$$a_1 = \frac{1}{6}R - V,$$

$$a_2 = \frac{1}{2}V^2 - \frac{1}{6}VR + \frac{1}{120}R^2 + \frac{1}{60}R_{\mu\nu}R^{\mu\nu}.$$

The heat kernel enters (12.74), where we need to integrate from $\tau = 0$ to $\tau = \infty$. The Seeley–DeWitt expansion (13.27) is valid only for small τ and so cannot be used to compute the zeta function. The behavior of the heat kernel at small τ corresponds to the ultraviolet limit of quantum field theory. This can be informally justified by noting that τ has dimension x^2 and thus small values of τ correspond to small distances. Effects of QFT at small distances, i.e. *local* effects, include vacuum polarization. On the other hand, large values of τ correspond to the infrared limit, which is related to particle production. To obtain the infrared behavior of the heat kernel, one needs a representation valid uniformly for all τ, such as (13.23).

14

Results from effective action

Summary Divergences in the effective action. Renormalization of constants. Nonlocal terms in the renormalized action. Polyakov action in $1+1$ dimensions. Conformal anomaly.

In this chapter we will use the results obtained in the previous chapter to calculate the effective action for a scalar field in a curved background and to determine the expectation value $\langle \hat{T}_{\mu\nu} \rangle$, which accounts for the backreaction of quantum scalar field on the metric.

14.1 Renormalization of the effective action

Let us recall that the effective action can be expressed in terms of the zeta function, defined through the trace of the heat kernel,

$$\zeta(s) \equiv \frac{1}{\Gamma(s)} \int_0^\infty \tau^{s-1} \mathrm{Tr}\, \hat{K}(\tau)\, d\tau, \tag{14.1}$$

as

$$\Gamma_{\mathrm{E}}\left[g_{\mu\nu}\right] = -\frac{1}{2} \left.\frac{d}{ds}\right|_{s=0} \zeta(s). \tag{14.2}$$

For $s=0$ the above integral diverges at $\tau \to 0$. Therefore it is usually assumed that the definition (14.1) is applicable only for those s for which the integral converges. The value of $\zeta(s)$ at $s=0$ is then obtained by analytic continuation. This procedure removes all the divergences in the effective action without any justification and thus does not allow us to reveal the physical meaning of these divergences.

Therefore in this section we will define $\zeta(s)$ for small s by regularizing the integral in (14.1) through an explicit cutoff parameter instead of performing the analytic continuation. This will allow us to reveal the nature of divergences and to justify their removal by applying the renormalization procedure.

To be specific, we consider a minimally coupled massless field ($V = 0$) in the four-dimensional Euclidean space ($\omega = 2$). The zeta function is obtained by substituting (13.24) into (14.1). The integral in (14.1) converges at the upper limit (for large proper time τ), and hence the divergences are due to the behavior of the heat kernel at small τ. Since small proper times correspond to small distances $(x-y) \sim \tau^{1/2}$ or equivalently to large momenta $k \sim \tau^{-1/2}$, these divergences are called *ultraviolet divergences*. For small τ one can use the Seeley–DeWitt expansion (13.28) instead of (13.24). Assuming that τ_1 is sufficiently small, so that the Seeley–DeWitt expansion is applicable for $\tau < \tau_1$, we rewrite the expression for the ζ function as

$$\zeta(s) = \frac{1}{(4\pi)^2\Gamma(s)} \int d^4x\sqrt{g} \left[\int_0^{\tau_1} \tau^{s-3} d\tau + \frac{R}{6} \int_0^{\tau_1} \tau^{s-2} d\tau \right.$$
$$\left. + \left(\frac{1}{120} R^2 + \frac{1}{60} R_{\mu\nu} R^{\mu\nu} \right) \int_0^{\tau_1} \tau^{s-1} d\tau + (\text{finite terms}) \right]. \tag{14.3}$$

The divergences we are interested in arise as $\tau \to 0$ when we set $s = 0$. Further terms of the expansion in τ contain τ^{s+n} with $n \geq 0$ and therefore are finite.

To examine the divergences in (14.3), let us introduce a cutoff at $\tau_0 < \tau_1$ and denote the resulting integrals for brevity by

$$A(\tau_0) \equiv \int_{\tau_0}^{\tau_1} \tau^{s-3} d\tau, \; B(\tau_0) \equiv \int_{\tau_0}^{\tau_1} \tau^{s-2} d\tau, \; C(\tau_0) \equiv \int_{\tau_0}^{\tau_1} \tau^{s-1} d\tau.$$

For $\tau_0 \to 0$ the leading divergences in the ζ function are

$$\zeta(s) = \frac{1}{\Gamma(s)} \int \frac{d^4x\sqrt{g}}{(4\pi)^2} \left[A(\tau_0) + \frac{R}{6} B(\tau_0) \right.$$
$$\left. + \left(\frac{1}{120} R^2 + \frac{1}{60} R_{\mu\nu} R^{\mu\nu} \right) C(\tau_0) + (\text{finite terms}) \right],$$

and for $s = 0$ the functions A, B, C diverge as

$$A(\tau_0) \sim \frac{1}{2} \tau_0^{-2}, \quad B(\tau_0) \sim \tau_0^{-1}, \quad C(\tau_0) \sim |\ln \tau_0|$$

when $\tau_0 \to 0$. For small s the Γ function has the expansion (A1.25),

$$\frac{1}{\Gamma(s)} = s + O(s^2).$$

Therefore the Euclidean effective action can be written as

$$\Gamma_E[g_{\mu\nu}] = -\frac{1}{2}\frac{d\zeta}{ds}\bigg|_{s=0} = -\int \frac{d^4x\sqrt{g}}{32\pi^2}\left[\frac{1}{2\tau_0^2} + \frac{1}{6\tau_0}R\right.$$
$$\left. + \left(\frac{1}{120}R^2 + \frac{1}{60}R_{\mu\nu}R^{\mu\nu}\right)|\ln\tau_0| + (\text{finite terms})\right]. \qquad (14.4)$$

This is the regularized effective action in which the first three terms become infinite when the cutoff parameter τ_0 is set to 0.

The divergent terms in the Lorentzian effective action $\Gamma_L[g_{\mu\nu}]$ can be obtained by a straightforward analytic continuation of (14.4). Since no Green's functions are present in the divergent terms, we only need to replace $\sqrt{g}$ by $\sqrt{-g}$ and R by $-R$.

The backreaction of the quantum field on the gravitational background causes a modification of the Einstein equation. The total action for the gravitational background is a sum of the free gravitational action (5.16) and the (Lorentzian) effective action $\Gamma_L[g_{\mu\nu}]$ induced by the quantum fields.

The classical action of general relativity, $S^{\text{grav}}[g_{\mu\nu}]$, contains the cosmological constant term Λ and the curvature term R that are similar to the divergent terms in (14.4). The renormalization procedure is implemented as follows. We assume that the free gravitational action (without backreaction of quantum fields) contains also the terms quadratic in curvature,

$$S_{\text{bare}}^{\text{grav}}[g_{\mu\nu}] = \int d^4x\sqrt{-g}\left[-\frac{R+2\Lambda_B}{16\pi G_B} + \alpha_B\left(\frac{R^2}{120} + \frac{R_{\mu\nu}R^{\mu\nu}}{60}\right)\right], \qquad (14.5)$$

where Λ_B, G_B, and α_B are called the *bare coupling constants*; these constants are not observable because the quantum fields are always present and cannot be "switched off." The modified action for gravity is thus the sum of the free action and the effective action,

$$S_{\text{bare}}^{\text{grav}}[g_{\mu\nu}] + \Gamma_L[g_{\mu\nu}] = \int d^4x\sqrt{-g}\left\{\left[-\frac{\Lambda_B}{8\pi G_B} - \frac{A(\tau_0)}{32\pi^2}\right]\right.$$
$$-\left[\frac{1}{16\pi G_B} + \frac{B(\tau_0)}{192\pi^2}\right]R$$
$$\left. + \left[\alpha_B - \frac{C(\tau_0)}{32\pi^2}\right]\left[\frac{R^2}{120} + \frac{R_{\mu\nu}R^{\mu\nu}}{60}\right] + (\text{finite terms})\right\}.$$

If the bare constants were finite, the presence of the divergent factors $A(\tau_0)$, $B(\tau_0)$, and $C(\tau_0)$ would make the total action infinite in the limit $\tau_0 \to 0$. The renormalization procedure assumes that the bare constants are *functions of* τ_0

chosen in such a way that they cancel the divergences in the effective action. The renormalized coupling constants are then

$$\frac{\Lambda}{8\pi G} = \frac{\Lambda_B}{8\pi G_B} - \frac{A(\tau_0)}{32\pi^2},$$

$$\frac{1}{16\pi G} = \frac{1}{16\pi G_B} - \frac{B(\tau_0)}{192\pi^2},$$

$$\alpha = \alpha_B - \frac{C(\tau_0)}{32\pi^2}.$$

After removing the cutoff (setting $\tau_0 = 0$), the renormalized constants are equal to the observed values of the constant α, the cosmological constant Λ, and Newton's constant G.

The resulting gravitational action (14.5) coincides with the standard Einstein–Hilbert action (5.16) only if $\alpha = 0$. Generally (without fine-tuning) the action contains two extra terms that are quadratic in the curvature. These terms are necessary to renormalize the backreaction of matter fields on gravity. If the curvature is small ($R \ll 1$ in Planck units), the extra terms are insignificant in comparison with S^{grav}, which is linear in R. In this limit Einstein's general relativity is a good approximation that agrees with the available experiments. When the curvature is large ($R \sim 1$ in Planck units), the extra terms may become significant.

The divergences found in equation (14.4) result from the backreaction of the scalar field. Other fields will give similar contributions, differing only in the numerical coefficients in front of R^2 and $R_{\mu\nu}R^{\mu\nu}$. Therefore in general we need to introduce four independent bare constants into the bare gravitational action, controlling the terms 1, R, R^2, and $R_{\mu\nu}R^{\mu\nu}$.

In dimensions other than four, the divergences contain other powers of τ_0; the leading divergence is

$$\left. \int_{\tau_0}^{\tau_1} \frac{d\tau}{(4\pi\tau)^\omega} \tau^{s-1} \right|_{s=0} \sim \tau_0^{-\omega}.$$

Therefore in 2ω dimensions we expect to find $\omega + 1$ divergent terms: $\tau_0^{-\omega}, \ldots,$ τ_0^{-1}, and $|\ln \tau_0|$.

14.2 Finite terms in the effective action

The Seeley–DeWitt expansion is valid only for small τ and so is not adequate for calculation of the finite nonlocal contributions to the effective action. Therefore we employ (13.24) with $V=0$ and find

$$\zeta(s)=\frac{1}{(4\pi)^{\omega}\Gamma(s)}\int d^{2\omega}x\sqrt{g}\left\{\int_0^{\infty} d\tau\,\tau^{s-1-\omega}\left[1+\frac{\tau}{6}R\right.\right.$$
$$\left.\left.+\tau^2 R f_3\left(-\tau\Box_g\right)R+\tau^2 R_{\mu\nu}f_4\left(-\tau\Box_g\right)R^{\mu\nu}\right]\right\}. \tag{14.6}$$

First we consider the two-dimensional spacetime ($\omega=1$). In this case the Ricci tensor is always proportional to the metric:

$$R_{\mu\nu}=\frac{1}{2}g_{\mu\nu}R. \tag{14.7}$$

Remark The Einstein equation (in vacuum) is identically satisfied in two dimensions due to (14.7). Therefore to obtain a nontrivial two-dimensional theory of gravity, the Einstein–Hilbert action needs to be modified. The renormalized effective action provides one such modification.

The first two terms in (14.6) are local and, as we have explained before, are responsible for the renormalization of the coupling constants. Leaving only finite contributions to (14.6) and using (14.7), we obtain

$$\zeta(s)=\frac{1}{4\pi\Gamma(s)}\int d^2x\sqrt{g}\int_0^{\infty} d\tau\,\tau^s R\left[f_3\left(-\tau\Box_g\right)+\frac{1}{2}f_4\left(-\tau\Box_g\right)\right]R,$$

and the finite contribution to the renormalized effective action is then

$$\Gamma_{\mathrm{E}}\left[g_{\mu\nu}\right]=-\frac{1}{2}\left.\frac{d\zeta}{ds}\right|_{s=0}$$
$$=-\frac{1}{8\pi}\int d^2x\sqrt{g}R\int_0^{\infty} d\tau\left[f_3\left(-\tau\Box_g\right)+\frac{1}{2}f_4\left(-\tau\Box_g\right)\right]R.$$

To compute the integral, we formally change the integration variable from τ to $\xi=-\tau\Box_g$ and obtain the following nonlocal expression,

$$\Gamma_{\mathrm{E}}\left[g_{\mu\nu}\right]=\frac{1}{8\pi}I_0\int d^2x\sqrt{g}R\,\Box_g^{-1}R, \tag{14.8}$$

where I_0 is a constant computed in Exercise 14.1,

$$I_0\equiv\int_0^{\infty} d\xi\left[f_3(\xi)+\frac{1}{2}f_4(\xi)\right]=\frac{1}{12}.$$

The resulting action

$$\Gamma_E[g_{\mu\nu}] = \frac{1}{96\pi}\int d^2x\sqrt{g}R\Box_g^{-1}R \qquad (14.9)$$

$$\equiv \frac{1}{96\pi}\int d^2x\sqrt{g(x)}\,d^2y\sqrt{g(y)}R(x)R(y)G_E(x,y)$$

is called the *Polyakov action.* Here G_E is the (Euclidean) Green's function of the Laplace operator $\Box_g$.

Exercise 14.1*
Verify that

$$I_0 \equiv \int_0^\infty d\xi\left[f_3(\xi)+\frac{1}{2}f_4(\xi)\right] = \frac{1}{12},$$

where the auxiliary functions $f_3(\xi)$ and $f_4(\xi)$ are defined in (13.25)–(13.26).

Hint: Rewrite I_0 as a double integral over ξ and u, regularize the integral over ξ by the factor $\exp(-a\xi)$ with $a>0$, exchange the order of integration and take the limit $a\to 0$ at the end of calculation.

Remark: the four-dimensional result In the four-dimensional case ($\omega = 2$), we omit the calculations and only quote the result,

$$\Gamma_E[g_{\mu\nu}] \sim \int d^4x\sqrt{g}R\ln\left(\frac{-\Box_g}{\mu^2}\right)R + \text{terms with } R_{\mu\nu}\ln\left(\frac{-\Box_g}{\mu^2}\right)R^{\mu\nu},$$

where μ is a mass scale introduced for dimensional reasons (the operator $\Box_g$ has dimension m^2). The logarithm of the Laplace operator is defined by

$$\ln\left(\frac{-\Box_g}{\mu^2}\right) = \int_0^{+\infty} d\left(m^2\right)\left[\frac{1}{\mu^2+m^2} - \frac{1}{-\Box_g+m^2}\right],$$

where the second term in brackets is the Green's function of the operator $-\Box_g + m^2$.

Note that a change of the parameter μ, for example $\mu\to\tilde{\mu}=\mu/b$, would add a term $(\ln b)R^2$ to the action, thus changing the constant α in front of the R^2 term. This constant thus becomes scale dependent and "running." This is a manifestation of the renormalization group properties of the theory. The value of α at a given energy (normalization point) must be determined experimentally and then the dependence of this constant from the energy is obtained by solving the renormalization group equation.

14.2.1 EMT from the Polyakov action

Using the effective action (14.9), we can compute the vacuum expectation value of the energy-momentum tensor of quantum fields.

To obtain the Lorentzian effective action $\Gamma_L[g_{\mu\nu}]$ we have to perform an analytic continuation of $\Gamma_E[g_{\mu\nu}]$ back to the Lorentzian time and substitute the Feynman

Green's functions instead of the Euclidean ones. The vacuum expectation value of the EMT is then

$$\langle 0_{\text{in}} | \hat{T}_{\mu\nu}(x) | 0_{\text{in}} \rangle = \frac{2}{\sqrt{-g(x)}} \frac{\delta \Gamma_L}{\delta g^{\mu\nu}(x)} \bigg|_{G_F \to G_{\text{ret}}}.$$

Before presenting detailed calculations, we quote the final result,

$$\langle 0_{\text{in}} | \hat{T}_{\mu\nu} | 0_{\text{in}} \rangle = \frac{1}{48\pi} \bigg\{ -2\nabla_\mu \nabla_\nu \left(\Box_g^{-1} R\right) + \nabla_\mu \left(\Box_g^{-1} R\right) \nabla_\nu \left(\Box_g^{-1} R\right)$$
$$+ \left[2R - \frac{1}{2} \nabla^\lambda \left(\Box_g^{-1} R\right) \nabla_\lambda \left(\Box_g^{-1} R\right) \right] g_{\mu\nu} \bigg\}. \qquad (14.10)$$

Here the operator $\Box_g^{-1}$ represents the retarded Green's function $G_{\text{ret}}(x, y)$, so that for any $f(x)$

$$\left(\Box_g^{-1} f\right)(x) \equiv \int d^2 y \sqrt{-g}\, f(y) G_{\text{ret}}(x, y).$$

The expression in (14.10) is nonlocal and describes simultaneously the particle production and the vacuum polarization effects.

Derivation of (14.10) First we need to convert the Euclidean effective action $\Gamma_{\text{E}}\left[g_{\mu\nu}\right]$ to the Lorentzian one, $\Gamma_{\text{L}}\left[g_{\mu\nu}\right]$. We recall that the Euclidean metric $g_{\mu\nu}$ entering $\Gamma_{\text{E}}[g_{\mu\nu}]$ is related to the Lorentzian metric by an analytic continuation with an additional sign change:

$$g_{\mu\nu}^{(\text{E})} = - \left. g_{\mu\nu} \right|_{t \to -i\tau}.$$

Therefore we replace R by $-R$ and $\Box$ by $-\Box$ before performing the analytic continuation. It is easy to see that action (14.9) also changes the sign,

$$\Gamma_{\text{E}} = -\frac{1}{96\pi} \int d^2 x^{(\text{E})} \sqrt{g^{(\text{E})}} R[g^{(\text{E})}] \Box_{g^{(\text{E})}}^{-1} R[g^{(\text{E})}].$$

As a result of the analytic continuation, we have $d^2 x^{(\text{E})} = i d^2 x$ and $\sqrt{g^{(\text{E})}} = \sqrt{-g}$, where x and $g_{\mu\nu}$ are now the Lorentzian quantities. Following the strategy outlined above, we have to replace the Euclidean Green's function G_{E} by the Feynman function $\frac{1}{i} G_{\text{F}}$ and obtain the Lorentzian effective action,

$$\Gamma_{\text{L}}\left[g_{\mu\nu}\right] = i\Gamma_{\text{E}}\left[g_{\mu\nu}^{(\text{E})}\right]_{\tau = it}$$
$$= -i \frac{1}{96\pi} \int i d^2 x_1 \int i d^2 x_2 \sqrt{-g(x_1)} R(x_1) \frac{1}{i} G_{\text{F}}(x_1, x_2) \sqrt{-g(x_2)} R(x_2)$$
$$= \frac{1}{96\pi} \int d^2 x \sqrt{-g} R \Box_g^{-1} R, \qquad (14.11)$$

where the symbol $\Box_g^{-1}$ is the Feynman Green's function G_{F} of the D'Alembert operator in the Lorentzian spacetime.

It remains to compute the variation of (14.11) with respect to $g_{\mu\nu}$. The required calculations are summarized in the following exercises.

Exercise 14.2*

(**a**) Verify that

$$\delta\Gamma^{\alpha}_{\mu\nu} = \frac{1}{2} g^{\alpha\beta} \left(\nabla_\mu \delta g_{\beta\nu} + \nabla_\nu \delta g_{\beta\mu} - \nabla_\beta \delta g_{\mu\nu} \right),$$

where ∇_μ is the covariant derivative defined according to the metric $g_{\mu\nu}$.

(**b**) Show that the variation of $\Box_g \phi$, where ϕ is a scalar function, is

$$\delta\Box_g \phi = \left(\delta g^{\mu\nu}\right) \nabla_\mu \nabla_\nu \phi - g^{\mu\nu} \left(\delta\Gamma^{\alpha}_{\mu\nu}\right) \nabla_\alpha \phi,$$

and the variation of the inverse D'Alembert operator is

$$\delta\Box_g^{-1}\phi = -\Box_g^{-1} \left(\delta\Box_g\right) \Box_g^{-1}\phi.$$

(**c**) Derive the variation of the Riemann tensor in the form

$$\delta R^{\alpha}{}_{\beta\mu\nu} = \nabla_\mu \delta\Gamma^{\alpha}_{\beta\nu} - \nabla_\nu \delta\Gamma^{\alpha}_{\mu\beta}.$$

Hint: Perform all calculations in a locally inertial frame where $\Gamma^{\alpha}_{\mu\nu} = 0$, and then generalize to arbitrary coordinates.

Exercise 14.3*

Compute the variation of the Polyakov action (14.11) with respect to $g^{\mu\nu}$ and derive (14.10).

14.3 Conformal anomaly

We have shown in Section 5.5 that the trace of the energy-momentum tensor vanishes for a classical conformally invariant field, that is, $T^{\mu}_{\mu} \equiv T_{\mu\nu} g^{\mu\nu} = 0$. On the other hand, the vacuum expectation value $\langle \hat{T}^{\mu}_{\mu} \rangle$ is in general nonzero even for a conformally invariant quantum field. This nonvanishing trace is called the *conformal anomaly* or *trace anomaly*. For a "free" conformal field in an arbitrary gravitational background the trace can be calculated *exactly*. In this section we will show that the trace anomaly is local and it is simply proportional to the corresponding coefficient in the Seeley–DeWitt expansion.

First we consider a minimally coupled scalar field in two dimensions. In this case, the expectation value of the energy-momentum tensor was calculated above in (14.10). Using the identities

$$g^{\mu\nu} \nabla_\mu \nabla_\nu \left(\Box_g^{-1} R\right) = \Box_g \left(\Box_g^{-1} R\right) = R,$$

$$g^{\mu\nu} g_{\mu\nu} = 2,$$

we immediately find

$$\langle 0|\, \hat{T}^{\mu}_{\mu}(x)\, |0\rangle = \frac{R(x)}{24\pi} = \frac{a_1(x)}{4\pi}, \tag{14.12}$$

where a_1 is the first Seeley–DeWitt coefficient. It is clear that the trace of the EMT does not vanish if $R \neq 0$.

The reason for conformal symmetry to be broken can be understood from the path integral formulation of QFT. The quantum theory would be conformally invariant if a path integral such as

$$\int \mathcal{D}\phi\, e^{-S_{\mathrm{E}}[\phi, g_{\mu\nu}]}$$

were invariant under conformal transformations. However, this is impossible because one cannot choose the integration measure $\mathcal{D}\phi$ to be simultaneously conformally invariant and generally covariant. For instance, the generally covariant integration measure (12.62) is not conformally invariant because the eigenvalues and the eigenfunctions are not preserved under conformal transformations.

The trace anomaly can be calculated exactly in an arbitrary number of dimensions even in those cases when the energy-momentum tensor itself can be determined only perturbatively. To demonstrate how this can be done, we present for simplicity a direct calculation of the trace anomaly in two dimensions. The generalization of the presented calculations to an arbitrary number of dimensions is largely straightforward. The idea is to calculate the variation of the effective action under an infinitesimal conformal transformation and thus infer the expectation value of $\hat{T}^{\mu}_{\mu}$.

First we recall that the effective action is expressed through the zeta function as

$$\Gamma_{\mathrm{E}}\left[g_{\mu\nu}\right] = -\frac{1}{2}\left.\frac{d\zeta_g}{ds}\right|_{s=0},$$

where

$$\zeta_g(s) = \mathrm{Tr}\left[\hat{O}_g^{-s}\right].$$

Under an infinitesimal conformal transformation

$$g_{\alpha\beta} \to \tilde{g}_{\alpha\beta} = \Omega^2 g_{\alpha\beta} = (1+\delta\Omega)^2 g_{\alpha\beta},$$

the operator $\hat{O}_g$ from (12.70) with $V = 0$ is transformed as

$$\begin{aligned}\hat{O}_{\Omega^2 g} &= \Omega_g^{-1}\hat{O}\Omega^{-1} = (1-\delta\Omega)\hat{O}_g(1-\delta\Omega) + O\left(\delta\Omega^2\right) \\ &= \hat{O}_g - \delta\Omega\hat{O}_g - \hat{O}_g\delta\Omega + O\left(\delta\Omega^2\right).\end{aligned}$$

The transformed zeta function is then

$$\begin{aligned}\zeta_{g+\delta g}(s) &= \mathrm{Tr}\left[\left(\hat{O}_g - \delta\Omega\hat{O}_g - \hat{O}_g\delta\Omega\right)^{-s}\right] \\ &= \mathrm{Tr}\left[\hat{O}_g^{-s} + s\hat{O}_g^{-s-1}\left(\delta\Omega\hat{O}_g + \hat{O}_g\delta\Omega\right)\right] \\ &= \mathrm{Tr}\hat{O}_g^{-s} + 2s\mathrm{Tr}\left[\delta\Omega\hat{O}_g^{-s}\right],\end{aligned}$$

up to terms of order $O\left(\delta\Omega^2\right)$. (The order of the operators under the trace can be exchanged.) Therefore we obtain

$$\Gamma_\mathrm{E}\left[g_{\mu\nu} + \delta g_{\mu\nu}\right] = -\frac{1}{2}\left.\frac{d\zeta_{g+\delta g}}{ds}\right|_{s=0} = \Gamma_\mathrm{E}\left[g_{\mu\nu}\right] - \lim_{s\to 0}\mathrm{Tr}\left(\delta\Omega\hat{O}^{-s}\right). \quad (14.13)$$

Note that the limit $s \to 0$ must be taken *after* computing the trace. Using the coordinate basis, one gets

$$\begin{aligned}\mathrm{Tr}\left(\delta\Omega\hat{O}^{-s}\right) &= \int d^2x\, d^2y\, \langle x|\, \delta\Omega\, |y\rangle\, \langle y|\, \hat{O}^{-s}\, |x\rangle \\ &= \int d^2x\, \delta\Omega(x)\, \langle x|\, \hat{O}^{-s}\, |x\rangle\, .\end{aligned} \quad (14.14)$$

To calculate the matrix element $\langle x|\, \hat{O}^{-s}\, |x\rangle$, we note that

$$\hat{O}^{-s} = \frac{1}{\Gamma(s)}\int_0^{+\infty} d\tau\, \tau^{s-1} e^{-\tau\hat{O}} = \frac{1}{\Gamma(s)}\int_0^{+\infty} d\tau\, \tau^{s-1}\hat{K}(\tau),$$

and hence

$$\begin{aligned}\langle x|\, \hat{O}^{-s}\, |x\rangle &= \frac{1}{\Gamma(s)}\int_0^{+\infty} d\tau\, \tau^{s-1}\, \langle x|\, \hat{K}(\tau)\, |x\rangle \\ &= \frac{\sqrt{g}}{4\pi\Gamma(s)}\int_0^{+\infty} d\tau\, \tau^{s-2}\left[1 + a_1(x)\tau + a_2(x)\tau^2 + O\left(\tau^3\right)\right],\end{aligned} \quad (14.15)$$

where we have substituted the Seeley–DeWitt expansion (13.27) for $\langle x|\, \hat{K}\, |x\rangle$. The integral in (14.15) diverges at both the upper and the lower limits. However, the upper limit divergence is spurious because the Seeley–DeWitt expansion is valid only for small τ. As we have seen, the heat kernel actually decays for large τ and hence the integral converges as $\tau \to \infty$. The contribution to the trace anomaly comes from small τ and therefore can be calculated using the Seeley–DeWitt expansion. However, to avoid the spurious ultraviolet divergence one must regularize the integral at large τ. To this end, we simply multiply the integrand by $\exp(-\alpha\tau)$ with $\alpha > 0$, compute the limit $s \to 0$ at fixed α, and then set $\alpha = 0$. The integrals in the expansion (14.15) are

$$\frac{1}{\Gamma(s)} \int_0^{+\infty} d\tau\, \tau^{s-2} e^{-\alpha\tau} = \frac{\alpha^{1-s}\Gamma(s-1)}{\Gamma(s)} = \frac{\alpha^{1-s}}{s-1};$$

$$\frac{a_1}{\Gamma(s)} \int_0^{+\infty} d\tau\, \tau^{s-1} e^{-\alpha\tau} = a_1 \alpha^{-s}; \quad \frac{a_2}{\Gamma(s)} \int_0^{+\infty} d\tau\, \tau^{s} e^{-\alpha\tau} = a_2 s \alpha^{-s-1}.$$

For $s \to 0$ at fixed $\alpha > 0$, only the first two terms in expansion (14.15) give nonvanishing contributions to $\langle x | \hat{O}^{-s} | x \rangle$. Notice that all higher order in τ terms also vanish in this limit. Therefore the result,

$$\lim_{\alpha \to +0} \left(\lim_{s \to +0} \langle x | \hat{O}^{-s} | x \rangle \right) = \frac{1}{4\pi} \sqrt{g}\, a_1(x), \tag{14.16}$$

is *exact.* It follows from (14.13) and (14.14) that the variation of the effective action under infinitesimal conformal transformation is

$$\delta\Gamma_{\mathrm{E}} = \Gamma_{\mathrm{E}} \left[g_{\mu\nu} + \delta g_{\mu\nu} \right] - \Gamma_{\mathrm{E}} \left[g_{\mu\nu} \right] = -\frac{1}{4\pi} \int d^2x \sqrt{g}\, a_1 \delta\Omega(x). \tag{14.17}$$

When we convert the Euclidean action Γ_{E} into the Lorentzian Γ_{L}, the variation (14.17) is transformed via the replacements $R \to -R$, $d^2x^{(\mathrm{E})} = i d^2x$, and $\delta\Gamma_{\mathrm{L}} = i\delta\Gamma_{\mathrm{E}}$. Since (14.17) does not contain any Green's functions, it is clear that the variation of the Lorentzian effective action after a conformal transformation is

$$\delta\Gamma_{\mathrm{L}} = -\frac{1}{4\pi} \int d^2x \sqrt{-g}\, a_1 \delta\Omega(x). \tag{14.18}$$

On the other hand, the variation $\delta\Gamma_{\mathrm{L}}$ is related to the expectation value of the energy-momentum tensor as

$$\delta\Gamma_{\mathrm{L}} = \int d^2x \frac{\delta\Gamma_L}{\delta g^{\mu\nu}} \delta g^{\mu\nu} = -\int d^2x \sqrt{-g} \langle \hat{T}_{\mu\nu} \rangle g^{\mu\nu} \delta\Omega(x). \tag{14.19}$$

Comparing (14.18) and (14.19), we obtain

$$\langle \hat{T}^{\mu}_{\mu} \rangle = \frac{a_1}{4\pi}.$$

In a 2ω-dimensional space, the heat kernel is proportional to $\tau^{-\omega}$. Thus, it is clear from (14.15) that the nonvanishing contribution to the trace anomaly comes from the a_ω-term in the Seeley–DeWitt expansion. Therefore, in 2ω dimensions the trace anomaly is proportional to a_ω.

Finally, let us show that the Polyakov action for the minimally coupled scalar field in a two-dimensional (Euclidean) space can be derived by integrating the trace anomaly. Since every metric in a two-dimensional space is conformally flat, there exists a coordinate system (*conformal gauge*) in which the metric can be written in a conformally flat form,

$$ds^2 = g_{\alpha\beta}dx^\alpha dx^\beta = e^{2\sigma}\delta_{\alpha\beta}dx^\alpha dx^\beta, \tag{14.20}$$

where $\delta_{\alpha\beta}$ is the Euclidean metric of the two-dimensional flat space and σ is a function of the coordinates. In this coordinate system, we have

$$\sqrt{g} = e^{2\sigma}, \quad R = -2\Box_g\sigma = -2e^{-2\sigma}\Box_f\sigma, \tag{14.21}$$

where R is the scalar curvature, $\Box_g$ is the two-dimensional Laplacian in the space with metric $g_{\alpha\beta}$, and $\Box_f$ is the Laplacian in flat space with metric $\delta_{\alpha\beta}$. The trace anomaly for a massless scalar field,

$$\left\langle \hat{T}^\mu_\mu \right\rangle = \frac{2}{\sqrt{g}}g^{\mu\nu}\frac{\delta\Gamma}{\delta g^{\mu\nu}} = \frac{a_1}{4\pi} = \frac{R}{24\pi},$$

can be rewritten in the conformal gauge in the form

$$\frac{\delta\Gamma}{\delta\sigma} = \frac{1}{12\pi}\Box_f\sigma.$$

This equation can be easily integrated:

$$\Gamma(\sigma) - \Gamma(\sigma = 0) = \frac{1}{24\pi}\int d^2x\sigma\Box_f\sigma. \tag{14.22}$$

Since $\sigma = 0$ corresponds to the flat space, while the effective action for a scalar field in a two-dimensional Euclidean flat space is equal to zero, we have $\Gamma(\sigma = 0) = 0$. Using (14.21), we find

$$\sigma = -\frac{1}{2\Box_g}R, \quad \Box_f\sigma = -\frac{1}{2}e^{2\sigma}R = -\frac{1}{2}\sqrt{g}R.$$

Substituting these expressions in (14.22), we finally obtain

$$\Gamma[g] = \frac{1}{96\pi}\int d^2x\sqrt{g}R\frac{1}{\Box_g}R,$$

in agreement with (14.9).

Note that the effective action can be unambiguously recovered from the trace anomaly only if the conformally invariant part of the effective action vanishes, i.e. if $\Gamma(\sigma = 0) = 0$. Otherwise, there are infinitely many conformally equivalent effective actions that reproduce the same trace anomaly but can lead to different energy-momentum tensors.[1]

[1] For more details, see the paper by V. Mukhanov, A. Wipf, and A. Zelnikov, *Phys. Lett. B* **332** (1994), 283 (preprint `arxiv:hep-th/9403018`).

Appendix 1

Mathematical supplement

A1.1 Functionals and distributions (generalized functions)

This appendix is an informal introduction to functionals and distributions.

Functionals

A *functional* is a map from a space of functions into numbers. If a functional S maps a function $q(t)$ into a number a, we write $S[q] = a$ or $S[q(t)] = a$. This notation is intended to show that the value $S[q]$ depends on the behavior of $q(t)$ at all t, not only at one particular t.

Some functionals can be written as integrals,

$$A[q(t)] = \int_{t_1}^{t_2} F(q(t))\, dt,$$

where $F(q)$ is an ordinary function applied to the value of q. For example, the functional

$$A[q(t)] = \int_0^1 [q(t)]^2\, dt$$

yields $A[t^n] = (2n+1)^{-1}$ and $A[\sin t] = \frac{1}{2} - \frac{1}{4}\sin 2$.

A functional may not be well-defined on all functions. For example, the above functional $A[q]$ can be applied only to functions $q(t)$ that are square-integrable on the interval $[0, 1]$. Together with a functional one always implies a suitable space of functions on which the functional is well-defined. Functions from this space are called *base functions* of a given functional.

Distributions

Not all functionals are expressible in the form of an integral. For example, the *delta function* denoted by $\delta(t - t_0)$ is by definition a functional that returns the value of a function at the point t_0, i.e.

$$\delta(t-t_0)\,[f(t)] \equiv f(t_0).$$

This functional cannot be written as an integral because there exists no function $F(t,f)$ such that for any continuous function $f(t)$,

$$f(t_0) = \int F(t, f(t))\, dt.$$

However, it is very convenient to be able to represent such functionals as integrals. So one writes

$$\delta(t-t_0)\,[f(t)] = f(t_0) \equiv \int_{-\infty}^{+\infty} f(t)\delta(t-t_0)\, dt \tag{A1.1}$$

even though $\delta(t-t_0)$ is not a function with numeric values (it is a "generalized function") and the integration is purely symbolic. This notation is a convenient shorthand because one can manipulate expressions linear in the δ function as if they were normal functions; for instance,

$$\int [a_1\delta(x-x_1) + a_2\delta(x-x_2)]\, f(x)\, dx = a_1 f(x_1) + a_2 f(x_2).$$

However, expressions such as $\sqrt{\delta(t)}$ or $\exp[\delta(t)]$ are undefined.

Note that the functional $\delta(t-t_0)$ is well-defined only on functions that are continuous at $t=t_0$. If we need to work with these functionals, we usually restrict the base functions to be everywhere continuous.

As an example, consider the functional

$$B[q(t)] \equiv 3\sqrt{q(1)} + \sin[q(2)],$$

where $q(1)$ and $q(2)$ are the values of the function $q(t)$. This functional depends only on the values of $q(t)$ at $t=1$ and $t=2$ and can be written in an integral form as

$$\begin{aligned} B[q(t)] &= 3\sqrt{q(1)} + \sin[q(2)] \\ &= \int_{-\infty}^{+\infty} dt \left\{ 3\delta(t-1)\sqrt{q(t)} + \delta(t-2)\sin[q(t)] \right\}. \end{aligned} \tag{A1.2}$$

Generalized function and *distribution* are other names for "a linear functional on a suitable space of functions." A functional is *linear* if

$$S[f(t) + cg(t)] = S[f] + cS[g]$$

for arbitrary base functions f, g and an arbitrary constant c. It is straightforward to verify that $\delta(t-t_0)$ is a linear functional.

The application of a linear functional A to a function $f(x)$ is written symbolically as an integral

$$A[f] \equiv \int f(x)A(x)dx, \qquad \text{(A1.3)}$$

where $A(x)$ is the *integration kernel* which represents the functional. Note that there may be no actual integration in equation (A1.3) because $A(x)$ is not necessarily an ordinary function. For instance, there is no real integration performed in equations (A1.1) and (A1.2).

Remark The δ function is sometimes "defined" by the conditions $\delta(x) = 0$ for $x \neq 0$ and $\delta(0) = +\infty$, while $\int \delta(x)dx = 1$. However, these contradictory requirements cannot be satisfied by any function with numeric values. It is more consistent to say that $\delta(x - x_0)$ is not really a function of x and to treat equation (A1.1) as a purely symbolic relation.

Distributions defined on a certain space of base functions build a linear space. An ordinary function $a(x)$ naturally defines a functional

$$a[f(x)] \equiv \int a(x)f(x)dx$$

and thus also belongs to the space of distributions if the integral converges for all base functions $f(x)$. For example, the function $a(x) \equiv 1$ defines a distribution on the base space of integrable functions on $[-\infty, +\infty]$, although $a(x)$ itself does not belong to the base space.

Distributions can be multiplied by ordinary functions, and the result is a distribution. For example, suppose $A(x)$ is a distribution and $a(x)$ is an ordinary function, then the action of Aa on a base function $f(x)$ is

$$A(x)a(x)[f] \equiv \int A(x)a(x)f(x)dx \equiv A(x)[af].$$

Sometimes two distributions can be multiplied, e.g. $\delta(x - x_0)\delta(y - y_0)$ is defined on continuous functions $f(x, y)$ and yields the value $f(x_0, y_0)$.

Two distributions are equal when they give equal results for all base functions. For instance, one can easily show that in the space of distributions $(x - x_0)\delta(x - x_0) = 0$ when applied to continuous base functions.

Derivatives of the δ function are defined as functionals that yield the value of the derivative of a function at a fixed point. If $\delta(x - x_0)$ were a normal function, one would expect the following identity to hold,

$$\int f(x)\delta'(x - x_0)\,dx = -\int f'(x)\delta(x - x_0)\,dx = -f'(x_0).$$

Therefore one *defines* the distribution $\delta'(x - x_0)$ as the functional

$$\delta'(x - x_0)[f(x)] \equiv -f'(x_0).$$

More generally,

$$\frac{d^n \delta(x-x_0)}{dx^n}[f] = (-1)^n \left. \frac{d^n f}{dx^n} \right|_{x=x_0} .$$

Derivatives of the δ function are functionals defined on sufficiently smooth base functions.

Principal value integrals

Not all distributions arise from combinations of δ functions. Another important example is the principal value integral.

If the space of base functions includes all continuous functions, then the distribution

$$a(x) = \frac{1}{x-x_0}$$

is undefined on some base functions because the integral with a function $f(x)$ diverges at the pole $x = x_0$ if $f(x_0) \neq 0$. The Cauchy principal value prescription helps to define $a[f]$ in such cases.

Definition

For integrals $\int_A^B F(x)dx$ where $F(x)$ has a pole at $x = x_0$ within the interval (A, B), one defines the *principal value* denoted by $\mathcal{P}\int$ as

$$\mathcal{P}\int_A^B F(x)dx \equiv \lim_{\varepsilon\to+0}\left[\int_A^{x_0-\varepsilon} F(x)dx + \int_{x_0+\varepsilon}^B F(x)dx\right],$$

when the limit exists. The idea is to cut out a neighborhood of the pole symmetrically at both sides. If the integrand contains several poles, the same limit procedure is applied to each pole separately; if there are no poles, the usual integration is performed. For example,

$$\mathcal{P}\int_{-\infty}^{+\infty} \frac{dx}{x^3} = 0; \quad \mathcal{P}\int_0^M \frac{dx}{x^2-1} = \frac{1}{2}\ln\frac{M-1}{M+1}, \; M > 1.$$

We write

$$\mathcal{P}\frac{1}{x-x_0}$$

to denote the distribution that acts by applying the principal value prescription to the integral, i.e.

$$\left(\mathcal{P}\frac{1}{x-x_0}\right)[f(x)] \equiv \mathcal{P}\int_A^B \frac{f(x)dx}{x-x_0}.$$

This integral converges in a neighborhood of $x = x_0$ if $f(x)$ is continuous there.

It is almost always the case that one cannot use the ordinary function $1/x$ as a distribution and *must* use $\mathcal{P}\frac{1}{x}$ instead, because the base functions are typically such that the ordinary integral $\int \frac{dx}{x} f(x)$ would diverge.

Example calculation with residues

A typical example is the principal value integral

$$\mathcal{P}\int_{-\infty}^{+\infty} \frac{e^{-ikx}}{x} dx \equiv \lim_{\varepsilon\to+0} \left[\int_{-\infty}^{-\varepsilon} \frac{e^{-ikx}}{x} dx + \int_{\varepsilon}^{+\infty} \frac{e^{-ikx}}{x} dx \right]. \qquad \text{(A1.4)}$$

Since the indefinite integral

$$\int \frac{e^{-ikx}}{x} dx \qquad \text{(A1.5)}$$

cannot be computed, we need to use the method of residues. First we assume that $\operatorname{Re} k > 0$ and consider the contour C in the complex x plane that goes around the pole at $x = 0$ along a semicircle of radius ε (see Fig. A1.1). The contour may be closed in the lower half-plane since $\operatorname{Re} k > 0$. The integral around the contour C is found from the residue at $x = 0$ which is equal to 1, so

$$\oint_C \frac{e^{-ikx}}{x} dx = -2\pi i.$$

This integral differs from that of equation (A1.4) only by the contribution of the semicircle. The function near the pole is nearly equal to $1/x$ and one can easily show by an explicit calculation that in the limit $\varepsilon \to +0$ the integral around the

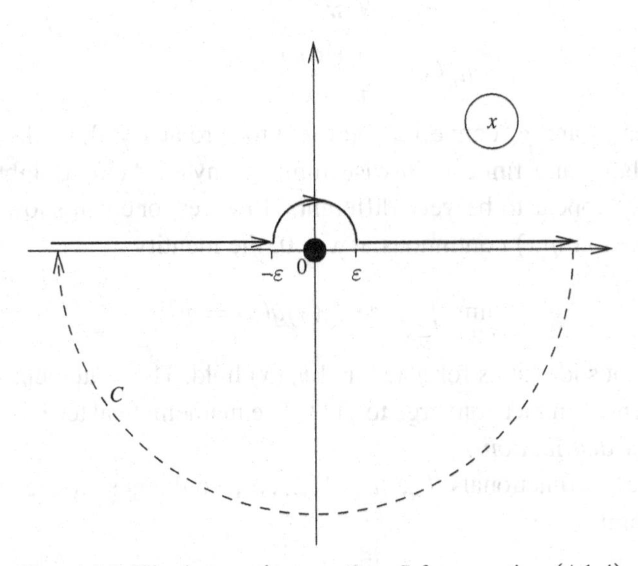

Fig. A1.1 The integration contour C for equation (A1.4).

semicircle is equal to $-\pi i$ times the residue (a half of the integral over the full circle). Therefore,

$$\mathcal{P}\int_{-\infty}^{+\infty} \frac{e^{-ikx}}{x} dx = -2\pi i - (-\pi i) = -\pi i, \quad \operatorname{Re} k > 0.$$

Analogous calculations give the opposite sign for $k < 0$ and the final result is

$$\mathcal{P}\int_{-\infty}^{+\infty} \frac{e^{-ikx}}{x} dx = -i\pi \operatorname{sign} k.$$

We could have chosen another contour instead of C; a very similar calculation yields the same answer for the contour with the semicircle in the opposite direction. We would like to emphasize that the choice of a contour is a purely technical issue inherent in the method of residues. The principal value integral is well-defined regardless of any integration in the complex plane; one would not need to choose any contours if we could compute the indefinite integral (A1.5) or if there existed another method for evaluating the two integrals in Eq. (A1.4) separately.

Convergence in the distributional sense

The δ function may be approximated by certain sequences of functions, for example (here $n = 1, 2, \ldots$)

$$f_n(x) = \begin{cases} 0, & |x| > \frac{1}{2n}, \\ n, & |x| < \frac{1}{2n}; \end{cases}$$

$$g_n(x) = \sqrt{\frac{n}{\pi}} \exp\left[-nx^2\right];$$

$$h_n(x) = \frac{1}{\pi} \frac{\sin nx}{x}.$$

The sequences f_n and g_n converge pointwise to zero at $x \neq 0$, while the sequence h_n does not have any finite pointwise limit at any x. At first sight these three sequences may appear to be very different. However, one can show that for any integrable function $q(x)$ continuous at $x = 0$, the identity

$$\lim_{n\to\infty} \int_{-\infty}^{+\infty} dx\, f_n(x) q(x) = q(0)$$

and the analogous identities for $g_n(x)$ and $h_n(x)$ hold. This statement suggests that all three sequences in fact converge to $\delta(x)$. The mathematical term is *convergence in the sense of distributions.*

A sequence of functionals F_n, $n = 1, 2, \ldots$, converges in the distributional sense if the limit

$$\lim_{n\to\infty} F_n[q]$$

exists for all base functions $q(x)$. It is clear from our example that a sequence of functions may converge in the distributional sense even if it has no pointwise limits.

Various statements concerning the δ function can (and should) be verified by calculations with explicit sequences of ordinary functions that converge to the δ function in the distributional sense.

The Sokhotsky formula

Another example of convergence to a distribution is the family of functions

$$a_\varepsilon(x) \equiv \frac{1}{x+i\varepsilon}, \quad \varepsilon > 0.$$

As $\varepsilon \to 0$, the functions $a_\varepsilon(x)$ converge pointwise to $1/x$ everywhere except at $x=0$.

The Sokhotsky formula is the limit (understood in the distributional sense)

$$\lim_{\varepsilon\to+0} \frac{1}{x+i\varepsilon} = -i\pi\delta(x) + \mathcal{P}\frac{1}{x}. \tag{A1.6}$$

This formula is derived by integrating $a_\varepsilon(x)f(x)$, where $f(x)$ is an arbitrary continuous base function,

$$\int_{-\infty}^{+\infty} \frac{1}{x+i\varepsilon} f(x)dx = \int_{-\infty}^{+\infty} \frac{xf(x)dx}{x^2+\varepsilon^2} - i\int_{-\infty}^{+\infty} \frac{\varepsilon f(x)dx}{x^2+\varepsilon^2}, \tag{A1.7}$$

and showing that in the limit $\varepsilon \to 0$ the two terms in the right-hand side converge to

$$\mathcal{P}\int_{-\infty}^{+\infty} \frac{f(x)}{x} dx - i\pi f(0).$$

We omit the detailed proof.

Distributional convergence of integrals

The concept of convergence in the distributional sense applies also to integrals. For example, consider the ordinarily divergent integral

$$a(x) \equiv \int_0^{+\infty} dk \sin kx. \tag{A1.8}$$

If we take equation (A1.8) at face value as an equality of functions, then $a(x)$ would be undefined for any x except $x=0$ where $a(0)=0$. However, if we interpret equation (A1.8) in the distributional sense, it yields a certain well-defined distribution $a(x)$.

To demonstrate this, we attempt to define the functional

$$a[f] \equiv \int dx\, f(x) a(x) = \int dx\, f(x) \int_0^{+\infty} dk \sin kx.$$

This expression is meaningless because of the divergent integral over k. If we now formally reverse the order of integrations, we get a meaningful formula

$$a[f] \equiv \int_0^{+\infty} dk \int dx\, f(x) \sin kx. \tag{A1.9}$$

The integrations performed in this order do converge as long as $f(x)$ is sufficiently well-behaved (continuous and decaying at infinity). Therefore it is reasonable to *define* the functional $a[f]$ by equation (A1.9).

We can now reduce the well-defined functional $a[f]$ to a simpler form. To transform the expression (A1.9), it is useful to be able to interchange the order of integrations. However, this can be done for uniformly convergent integrals, while the double integral (A1.9) converges non-uniformly. Therefore we temporarily introduce a cutoff into the integral over dk at the upper limit (large k). At the end of the calculation we shall remove the cutoff and obtain the final result. Now we show the details of this procedure for the functional (A1.9).

A simple way to introduce a cutoff is to multiply the integrand by $\exp(-\alpha k)$ where $\alpha > 0$ is a real parameter; the original integral is restored when $\alpha = 0$. It is clear that for any sufficiently well-behaved function $f(x)$,

$$\lim_{\alpha \to +0} \int_0^{+\infty} dk \int dx\, f(x) e^{-\alpha k} \sin kx = \int_0^{+\infty} dk \int dx\, f(x) \sin kx \equiv a[f].$$

The double integral

$$\int_0^{+\infty} dk \int dx\, f(x) e^{-\alpha k} \sin kx$$

converges uniformly in k and x, so we can reverse the order of integrations *before* evaluating the limit $\alpha \to 0$. In the inner integral we obtain the family of functions

$$a_\alpha(x) \equiv \int_0^{+\infty} dk \sin kx \exp(-\alpha k) = \frac{x}{\alpha^2 + x^2}.$$

At this point we can impose the limit $\alpha \to 0$, use Eqs. (A1.6)–(A1.7) and find

$$a[f] = \lim_{\alpha \to +0} \int dx\, a_\alpha(x) f(x) = \lim_{\alpha \to +0} \int \frac{x f(x) dx}{x^2 + \alpha^2} = \mathcal{P} \int dx \frac{f(x)}{x}.$$

This holds for any base function $f(x)$, therefore we obtain the following equality of distributions,

$$a(x) \equiv \int_0^{+\infty} dk \sin kx = \lim_{\alpha \to +0} a_\alpha(x) = \mathcal{P}\frac{1}{x}. \tag{A1.10}$$

We may say that the integral (A1.8) diverges in the usual sense but converges in the distributional sense.

The distributional limit of a divergent integral is usually found by regularizing the integral with a convenient factor such as $\exp(-\alpha k)$ and by removing the cutoff after the integration. The way to introduce the cutoff in k is of course not unique. For instance, we could multiply the integrand by $\exp\left(-\alpha k^2\right)$ or simply replace the infinite upper limit in equation (A1.8) by a parameter $k_{\max}$ and then evaluate the limit $k_{\max} \to +\infty$. The calculations are somewhat less transparent in that case but the result is the same. We are free to choose a cutoff in any form, as long as the cutoff allows us to reverse the order of integration.

Remark We should keep in mind that there must be some base functions $f(x)$ to which both sides of equation (A1.10) are applied as linear functionals. Only then the manipulations with the artificial cutoff become well-defined operations in the space of distributions. Although it is tempting to treat $a(x)$ as an ordinary function equal to $1/x$, it would be an abuse of notation since e.g.

$$a(2) = \int_0^{+\infty} dk \sin 2k = \frac{1}{2} \quad ???$$

is a meaningless statement. Expressions such as equation (A1.8) usually appear as inner integrals in calculations, for example,

$$\int_{-\infty}^{+\infty} dx \int_0^{+\infty} dk\, x e^{-x^2} \sin kx,$$

which looks like an application of the distribution $a(x)$ to the base function xe^{-x^2}. In such cases we are justified to treat the inner integral as the distribution (A1.10).

Fourier representations of distributions

A well-known integral representation of the δ function is

$$\delta\left(x - x_0\right) = \int_{-\infty}^{+\infty} \frac{dk}{2\pi} e^{ik(x-x_0)}. \tag{A1.11}$$

The integral in equation (A1.11) diverges for all x and must be understood in the distributional sense, similarly to the integral (A1.8).

Distributions often turn up in calculations when we use Fourier transforms. If $\tilde{f}(k)$ is a Fourier transform of $f(x)$, so that

$$f(x) = \int \tilde{f}(k) e^{ikx} dk,$$

then $\tilde{f}(k)$ may well be a distribution since the only way it is connected with real functions is through integration. We shall see examples of this in Appendix A1.2.

Solving equations for distributions

Distributions may be added together, multiplied by ordinary functions, or differentiated to yield other distributions. For example, the distribution $\mathcal{P}\frac{1}{x^2}$ multiplied by the function $2x$ yields the distribution $\mathcal{P}\frac{2}{x}$. Although such calculations are in most cases intuitively obvious, they need to be verified more formally by analyzing explicit distributional limits.

A curious phenomenon occurs when solving algebraic equations that involve distributions, e.g.

$$(x - x_0)\, a(x) = 1. \tag{A1.12}$$

Note that $(x - x_0)\, \delta(x - x_0) = 0$. So the solution of equation (A1.12) in terms of distributions is

$$a(x) = \mathcal{P}\frac{1}{x - x_0} + A\delta(x - x_0),$$

where the constant A is an arbitrary number. This shows that one should be careful when doing arithmetic with distributions. For instance, dividing a distribution by x is possible but the result contains the term $A\delta(x)$ with an arbitrary constant A.

A1.2 Green's functions, boundary conditions, and contours

Green's functions are used to solve linear differential equations. The typical problem involves a linear differential operator $\hat{L}_x$ such as

$$\hat{L}_x = \frac{d^2}{dx^2} + a^2. \tag{A1.13}$$

A *Green's function* of the operator $\hat{L}$ is a distribution $G(x, x')$ that solves the equation

$$\hat{L}_x G(x, x') = \delta(x - x'). \tag{A1.14}$$

Because of this relation which can be symbolically represented by $\hat{L}_x \hat{G} = \hat{1}$, the Green's function is frequently written as the "inverse" of the operator $\hat{L}_x$, i.e. $\hat{G} = \hat{L}_x^{-1}$. However, one should keep in mind that this notation is symbolic and the operator such as $\hat{L}_x$ does not actually have an inverse operator.

The Green's function must also satisfy a set of boundary conditions imposed usually at $|x| = \infty$ or perhaps at some finite boundary points, according to the particular problem. For example, the causal boundary condition in one dimension (real x) is

$$G_{\text{ret}}(x, x') = 0 \text{ for } x < x'. \tag{A1.15}$$

This condition specifies the retarded Green's function.

Green's functions can be used to solve equations of the form

$$\hat{L}_x f(x) = s(x),$$

where $s(x)$ is a known "source" function. The general solution of the above equation can be written as

$$f(x) = f_0(x) + \int G(x, x')s(x')dx',$$

where $f_0(x)$ is a general solution of the homogeneous equation, $\hat{L}f_0 = 0$.

Equation (A1.14) defines a Green's function only up to a solution of the homogeneous equation. Boundary conditions are needed to fix the Green's function uniquely. Green's functions obtained with different boundary conditions differ by a solution of the homogeneous equation.

Using Fourier transforms

To find a Green's function, it is often convenient to use Fourier transforms, especially when $G(x, x') = G(x - x')$. In that case we can use the Fourier representation in n dimensions,

$$G(x - x') \equiv G(\Delta x) = \int \frac{d^n k}{(2\pi)^n} g(k) e^{ik\cdot\Delta x}. \tag{A1.16}$$

However, it often turns out that the Fourier transform of a Green's function is a *distribution* and not an ordinary function.

As an example, we consider the Green's function $G(x - x')$ of the one-dimensional operator (A1.13). The Fourier image $g(k)$ of $G(\Delta x)$ defined by equation (A1.16) satisfies

$$\left(a^2 - k^2\right) g(k) = 1. \tag{A1.17}$$

Here we are forced to treat $g(k)$ as a distribution because the ordinary solution

$$G(\Delta x) = \int_{-\infty}^{+\infty} \frac{dk}{2\pi} \frac{1}{a^2 - k^2} e^{ik\Delta x} \quad ???$$

involves a meaningless divergent integral. In the space of distributions, the general solution of equation (A1.17) is

$$g(k) = \mathcal{P}\frac{1}{a^2 - k^2} + g_1\delta(k - a) + g_2\delta(k + a), \tag{A1.18}$$

with arbitrary complex constants $g_{1,2}$. Then the general form of the Green's function is found from equation (A1.16) with $n = 1$,

$$G(\Delta x) = \mathcal{P}\int_{-\infty}^{+\infty} \frac{dk}{2\pi} \frac{e^{ik\Delta x}}{a^2 - k^2} + \frac{g_1}{2\pi} e^{ia\Delta x} + \frac{g_2}{2\pi} e^{-ia\Delta x}.$$

The last two terms correspond to the general solution of the homogeneous equation,

$$\hat{L}_x \left(g_1 e^{ia\Delta x} + g_2 e^{-ia\Delta x} \right) = 0.$$

The constants $g_{1,2}$ can be found from the particular boundary conditions after computing the principal value integral. We find, for instance, that the boundary conditions (A1.15) require

$$g_1 = \frac{\pi}{2ia}, \quad g_2 = -\frac{\pi}{2ia} \tag{A1.19}$$

(see equation (A4.35) in the solution to Exercise 12.2) and the retarded Green's function is expressed by equation (A4.36).

Contour integration and boundary conditions

We have shown that the boundary condition for $G(x, x')$ determines the choice of the constants $g_{1,2}$ which parametrize the general solution of the homogeneous oscillator equation, while the nontrivial part of Green's function (a special solution of the inhomogeneous equation) is equal to a certain principal value integral. Instead of using the principal value prescription, we could select a contour C in the complex k plane and express the Green's function as

$$G(\Delta x) = \int_C \frac{dk}{2\pi} \frac{e^{ik\Delta x}}{a^2 - k^2} + \frac{\tilde{g}_1}{2\pi} e^{ia\Delta x} + \frac{\tilde{g}_2}{2\pi} e^{-ia\Delta x}. \tag{A1.20}$$

In effect we replaced the principal value prescription $\mathcal{P}\int$ by a certain choice of the contour. This alternative prescription adds some residue terms at the poles $k = \pm a$, so the constants $\tilde{g}_{1,2}$ differ from those in equation (A1.19). The resulting Green's function is of course the same because the change in the constants $g_{1,2} \to \tilde{g}_{1,2}$ cancels the extra residue terms.

Example calculation with a contour

Let us select the contour C shown in Fig. A1.2, where both semicircles are *arbitrarily* chosen to lie in the upper half-plane. The contour C must be closed in the lower half-plane if $\Delta x < 0$ and in the upper half-plane if $\Delta x > 0$. The integral along each semicircle is equal to $-\pi i$ times the residue at the corresponding pole. Therefore the integral along the contour C is

$$\int_C \frac{e^{ik\Delta x}}{a^2 - k^2} dk = \begin{cases} 0, & \Delta x > 0; \\ \frac{2\pi}{a} \sin(a\Delta x), & \Delta x < 0. \end{cases}$$

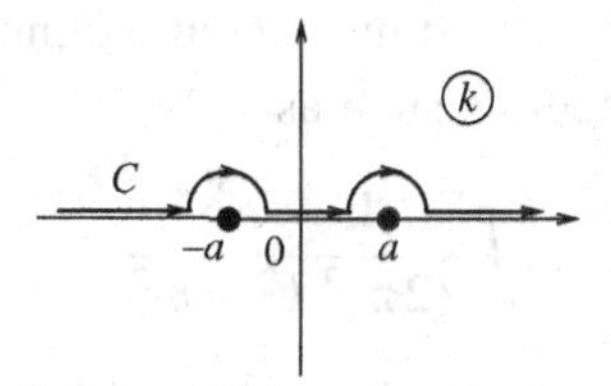

Fig. A1.2 Alternative integration contour for Green's function, equation (A1.20).

To satisfy the boundary conditions (A1.15), we must choose the constants as

$$g_{1,2} = \pm\frac{\pi}{ia}.$$

Note that this differs from equation (A1.19). The resulting Green's function is

$$G_{\text{ret}}(\Delta x) = \theta(x - x')\frac{\sin a\Delta x}{a},$$

which coincides with equation (A4.36). The same result is obtained from any other choice of the contour in equation (A1.20) when the constants $g_{1,2}$ are chosen correctly.

Choosing the contour as in Fig. A1.2 is equivalent to considering the limit

$$\lim_{\varepsilon\to+0}\int_{-\infty}^{+\infty}\frac{dk\, e^{ik\Delta x}}{a^2 - k^2 - ik\varepsilon},$$

since a replacement $k \to k + \frac{1}{2}i\varepsilon$ under the integral corresponds to shifting the integration line upwards.

We could choose the contour of integration in a clever way to make $g_{1,2} = 0$. This is achieved if both semicircles in Fig. A1.2 are turned upside-down. This is the calculation often presented in textbooks, where one is instructed to rewrite the integral as

$$\lim_{\varepsilon\to+0}\int_{-\infty}^{+\infty}\frac{dk\, e^{ik\Delta x}}{a^2 - k^2 \pm i\varepsilon} \quad \text{or} \quad \lim_{\varepsilon\to+0}\int_{-\infty}^{+\infty}\frac{dk\, e^{ik\Delta x}}{a^2 - k^2 \pm ik\varepsilon} \tag{A1.21}$$

with small real $\varepsilon > 0$ and to take the limit $\varepsilon \to +0$. As we have seen, such limits with a prescription for inserting ε into the denominator are equivalent to particular choices of contours in the complex k plane. It is difficult to remember the correct prescription of the contour or the specific ansatz with ε that one needs for each operator $\hat{L}_x$ and for each set of boundary conditions. These tricks are unnecessary if one treats the Fourier image $g(k)$ as a distribution with unknown constants, as in equation (A1.18). One is then free to choose either a principal value prescription or an arbitrary contour in the complex k plane, as long as one determines the relevant constants from boundary conditions.

So far we considered only one-dimensional examples. In higher-dimensional spaces, one often obtains integrals such as

$$\int \frac{d^3\mathbf{k}}{(2\pi)^3} \frac{e^{i\mathbf{k}\cdot\mathbf{x}}}{k^2 - m^2}$$

in which the kernel $1/(k^2 - m^2)$ must be understood as a distribution and rewritten as

$$\text{“}\frac{1}{k^2 - m^2}\text{”} = \mathcal{P}\frac{1}{k^2 - m^2} + h(\mathbf{k})\,\delta\left(k^2 - m^2\right),$$

where $h(\mathbf{k})$ is an arbitrary function of the vector $\mathbf{k}$. To obtain an explicit principal value formulation of such integrals, one first separates the divergent integration over a scalar variable (in this case over dk),

$$\mathcal{P}\int \frac{d^3\mathbf{k}}{(2\pi)^3} \frac{e^{i\mathbf{k}\cdot\mathbf{x}}}{k^2 - m^2} = \int_0^{\pi} d\theta \sin\theta\, \mathcal{P}\int_0^{+\infty} \frac{k^2 dk}{(2\pi)^2} \frac{e^{ikx\cos\theta}}{k^2 - m^2},$$

and then uses the principal value prescription. (In this particular case the integration over $d\theta$ can be performed first.) The relevant arbitrary functions such as $h(\mathbf{k})$ must be determined from the appropriate boundary conditions.

A1.3 Euler's gamma function and analytic continuations

Euler's gamma function $\Gamma(x)$ is a transcendental function that generalizes the factorial $n!$ from natural n to complex numbers. We shall now summarize some of its standard properties.

The usual definition is

$$\Gamma(x) = \int_0^{+\infty} t^{x-1} e^{-t} dt. \qquad \text{(A1.22)}$$

The integral (A1.22) converges for real $x > 0$ (and also for complex x such that $\operatorname{Re} x > 0$) and defines an analytic function. It is easy to check that $\Gamma(n) = (n-1)!$ for integer $n \geq 1$; in particular, $\Gamma(1) = 1$. The gamma function can be analytically continued to all complex x.

Analytic continuations

If an analytic function $f(x)$ is defined only for some x, an analytic continuation can be used to obtain values for other x.

A familiar case of analytic continuation is the geometric series,

$$f(x) = \sum_{n=0}^{\infty} x^n, \quad |x| < 1.$$

The series converges only for $|x| < 1$, yielding

$$f(x) = \frac{1}{1-x}, \quad |x| < 1, \tag{A1.23}$$

which defines the function $f(x)$ for all $x \neq 1$ and coincides with the old definition for $|x| < 1$. Therefore, equation (A1.23) provides the analytic continuation of $f(x)$ to the entire complex plane (except for the pole at $x = 1$).

If a function $f(x)$ is defined by an integral relation such as

$$f(x) = \int F(x, y) dy,$$

where the integral converges only for some x, one might be able to transform the specific integral until one obtains some other formula for $f(x)$ that is valid for a wider range of x. According to a standard theorem of complex calculus, two analytic functions that coincide in some region of the complex plane must coincide in the entire plane (perhaps after branch cuts). Therefore any formula for $f(x)$ defines the same analytic function. The hard part is to obtain a better formula out of the original definition. Unfortunately, there is no general method to perform the analytic continuation. One has to apply tricks that are suitable to the problem at hand.

The gamma function for all x

The analytic continuation of $\Gamma(x)$ can be performed as follows. Integrating equation (A1.22) by parts, one obtains the identity

$$x\Gamma(x) = \Gamma(x+1), \quad x > 0. \tag{A1.24}$$

This formula determines $\Gamma(x)$ for $\mathrm{Re}\, x > -1$, because $\Gamma(x+1)$ is well-defined and one can write

$$\Gamma(x) \equiv \frac{\Gamma(x+1)}{x}, \quad \mathrm{Re}\, x > -1.$$

The point $x = 0$ is clearly a pole of $\Gamma(x)$, but at $x \neq 0$ the function is finite. Subsequently we define $\Gamma(x)$ for $\mathrm{Re}\, x > -2$ by

$$\Gamma(x) \equiv \frac{\Gamma(x+2)}{x(x+1)}, \quad \mathrm{Re}\, x > -2,$$

for $\mathrm{Re}\, x > -3$ and so on. (Thus $\Gamma(x)$ has poles at $x = 0, -1, -2, \ldots$) The resulting analytic function coincides with the original integral for $\mathrm{Re}\, x > 0$.

Series expansions

One can expand the gamma function in power series as

$$\Gamma(1+\varepsilon) = 1 - \gamma\varepsilon + O\left(\varepsilon^2\right),$$

where

$$\gamma \equiv -\int_0^{+\infty} dt\, e^{-t} \ln t \approx 0.5772$$

is Euler's constant. From the above series it is easy to deduce the asymptotic behavior at the poles, for instance

$$\Gamma(x \to 0) = \frac{\Gamma(x+1)}{x} = \frac{1}{x} - \gamma + O(x). \tag{A1.25}$$

Product identity

A convenient identity connects $\Gamma(x)$ and $\Gamma(1-x)$:

$$\Gamma(x)\Gamma(1-x) = \frac{\pi}{\sin \pi x}. \tag{A1.26}$$

This identity holds for all (complex) x; for instance, it follows that $\Gamma\left(\frac{1}{2}\right) = \sqrt{\pi}$. One can also obtain the formula

$$\Gamma(ix)\Gamma(-ix) = |\Gamma(ix)|^2 = \frac{\pi}{x \sinh \pi x}. \tag{A1.27}$$

Finally, equation (A1.26) allows one to express $\Gamma(x)$ for $\operatorname{Re} x \le 0$ through $\Gamma(1-x)$, which is another way to define the analytic continuation of $\Gamma(x)$ to all complex x.

Derivation of equation (A1.26) We first derive the identity for $0 < \operatorname{Re} x < 1$. Using equation (A1.22), we have

$$\Gamma(x)\Gamma(1-x) = \int_0^{+\infty} ds \int_0^{+\infty} dt\, s^{x-1} t^{-x} e^{-(s+t)},$$

where the integrals are convergent if $0 < \operatorname{Re} x < 1$. After a change of the variables $(s, t) \to (u, v)$,

$$u \equiv s+t, \; v \equiv \ln \frac{s}{t}, \quad ds\, dt = \frac{u e^v}{(e^v+1)^2} du\, dv,$$

where $0 < u < +\infty$ and $-\infty < v < +\infty$, the integral over u is elementary and we get

$$\Gamma(x)\Gamma(1-x) = \int_{-\infty}^{+\infty} \frac{e^{vx} dv}{e^v+1}.$$

The integral converges for $0 < \operatorname{Re} x < 1$ and is evaluated using residues by shifting the contour to $v = 2\pi i + \tilde{v}$ which multiplies the integral by $\exp(2\pi i x)$. The residue at $v = i\pi$ is equal to $-\exp(i\pi x)$. We find

$$\left(1 - e^{2\pi i x}\right)\Gamma(x)\Gamma(1-x) = -2\pi i e^{\pi i x},$$

from which equation (A1.26) follows for $0 < \operatorname{Re} x < 1$.

To show that the identity holds for all x, we use equation (A1.24) to find for integer $n \geq 1$

$$\Gamma(x-n)\Gamma(1-(x-n)) = \frac{\Gamma(x)\Gamma(1-x)}{(x-1)\ldots(x-n)}(1-x)\ldots(n-x)$$
$$= (-1)^n \frac{\pi}{\sin \pi x} = \frac{\pi}{\sin \pi(x-n)}, \quad 0 < \operatorname{Re} x < 1.$$

Expressing integrals through the gamma function

Some transcendental integrals such as

$$\int_0^{+\infty} x^{s-1} e^{-bx} dx \tag{A1.28}$$

are expressed through the gamma function after a change of variable $y = bx$,

$$\int_0^{+\infty} x^{s-1} e^{-bx} dx = b^{-s}\Gamma(s).$$

However, complications arise when s and b are complex numbers, because of the ambiguity of the phase of b. For example, i^i is an inherently ambiguous expression since one may write

$$i^i = \left[\exp\left(\frac{i\pi}{2} + 2\pi i n\right)\right]^i = \exp\left(-\frac{\pi}{2} - 2\pi n\right), \quad n \in \mathbb{Z}.$$

We consider equation (A1.28) with a complex b such that $\operatorname{Re} b > 0$. (The integral diverges if $\operatorname{Re} b < 0$, converges conditionally when $\operatorname{Re} b = 0$, $b \neq 0$, and $0 < \operatorname{Re} s < 1$, while for other s the limit $\operatorname{Re} b \to +0$ may be taken only in the distributional sense.) The integrand is rewritten as

$$x^{s-1} e^{-bx} = \exp\left[-bx + (s-1)\ln x\right].$$

The contour of integration may be rotated to the half-line $x = e^{i\phi} y$, with a fixed angle $|\phi| < \frac{\pi}{2}$, and y varying in the interval $0 < y < +\infty$. Therefore, if $\operatorname{Re} s > 0$ we can change the variable $bx \equiv y$ as long as $\operatorname{Re} b > 0$. Then we should select the branch of the complex logarithm function covering the region $-\frac{\pi}{2} < \phi < \frac{\pi}{2}$,

$$\ln(A + iB) \equiv \ln|A + iB| + i\,(\operatorname{sign} B) \arctan \frac{|B|}{A}, \quad A > 0. \tag{A1.29}$$

With this definition of the logarithm, the integral (A1.28) is transformed to

$$\int_0^{+\infty} x^{s-1} e^{-bx} dx = b^{-s} \int_0^{+\infty} y^{s-1} e^{-y} dy = \exp(-s \ln b) \Gamma(s). \tag{A1.30}$$

In the calculations for the Unruh effect (Section 8.5) we encountered the following integral,

$$F(\omega, \Omega) \equiv \int_{-\infty}^{+\infty} \frac{du}{2\pi} \exp\left(i\Omega u + i\frac{\omega}{a} e^{-au}\right).$$

This integral can be expressed through the gamma function. Changing the variable to $x \equiv e^{-au}$, we obtain

$$F(\omega, \Omega) = \frac{1}{2\pi a} \int_0^{+\infty} dx\, x^{-\frac{i\Omega}{a} - 1} e^{\frac{i\omega}{a} x} \tag{A1.31}$$

which is of the form (A1.28) with

$$b = -\frac{i\omega}{a}, \quad s = -\frac{i\Omega}{a}.$$

Since $\mathrm{Re}\, s = 0$, the integral in Eq. (A1.31) diverges at $x = 0$. To obtain the distributional limit of this integral, we need to take the limit of s having a vanishing positive real part. Since b also must satisfy $\mathrm{Re}\, b > 0$, we choose

$$b = -\frac{i\omega}{a} + \varepsilon, \; s = -\frac{i\Omega}{a} + \varepsilon, \quad \varepsilon > 0,$$

and take the limit of $\varepsilon \to +0$. Then we can use equation (A1.30) in which we must evaluate

$$\ln b = \lim_{\varepsilon \to +0} \ln\left(-\frac{i\omega}{a} + \varepsilon\right) = \ln\left|\frac{\omega}{a}\right| - i\frac{\pi}{2} \mathrm{sign}\left(\frac{\omega}{a}\right).$$

Substituting into equation (A1.30), we find

$$F(\omega, \Omega) = \frac{1}{2\pi a} \exp\left[\frac{i\Omega}{a} \ln\left|\frac{\omega}{a}\right| + \frac{\pi\Omega}{2a} \mathrm{sign}\left(\frac{\omega}{a}\right)\right] \Gamma\left(-\frac{i\Omega}{a}\right).$$

Now it is straightforward to obtain the relation

$$F(\omega, \Omega) = F(-\omega, \Omega) \exp\left(\frac{\pi\Omega}{a}\right), \quad \omega > 0,\, \Omega > 0.$$

Finally, we derive an explicit formula for the quantity

$$|\beta_{\omega\Omega}|^2 = \frac{\Omega}{\omega} |F(-\omega, \Omega)|^2$$

which is related to the mean particle number by equation (8.43). Using equation (A1.27), we get

$$\begin{aligned}|\beta_{\omega\Omega}|^2 &= \frac{\Omega}{4\pi^2 a^2 \omega} \exp\left(-\frac{\pi\Omega}{a}\right) \left|\Gamma\left(-\frac{i\Omega}{a}\right)\right|^2 \\ &= \frac{1}{2\pi\omega a}\left[\exp\left(\frac{2\pi\Omega}{a}\right) - 1\right]^{-1}.\end{aligned}$$

Appendix 2

Backreaction derived from effective action

In this appendix, we derive the backreaction of a quantum system on a classical background, starting from a fully quantized theory rather than from heuristic considerations as in Section 12.1.7. We follow the paper by A. Barvinsky and D. Nesterov, *Nucl. Phys. B* **608** (2001), 333, preprint `arxiv:gr-qc/0008062`.

We are interested in describing the backreaction of a quantum system $\hat{q}$ on a classical background B. Denote by $S[q, B]$ the classical action describing the complete system. For simplicity, we treat q and B as systems with one degree of freedom each; the considerations are straightforwardly generalized to more realistic cases. Working in the Heisenberg picture, we consider a fully quantized system $(\hat{q}(t), \hat{B}(t))$ in a (time-independent) quantum state $|\psi\rangle$. We are interested in a state $|\psi\rangle$ such that the variable $\hat{B}$ is approximately classical, i.e. it has a large expectation value and small quantum fluctuations around it, while $\hat{q}$ is essentially in the vacuum (ground) state. This assumption can be formulated mathematically as

$$\langle\psi|\,\hat{B}(t)\,|\psi\rangle \equiv B_{\mathrm{c}}(t), \quad \langle\psi|\,\hat{q}(t)\,|\psi\rangle \approx 0,$$
$$\hat{B}(t) \equiv B_{\mathrm{c}}(t) + \hat{b}(t),$$

where $B_{\mathrm{c}}(t)$ is the "classical" expectation value of $\hat{B}$ in the state $|\psi\rangle$. It is implied that the "quantum fluctuation" $\hat{b}(t)$ is small in comparison with $B_{\mathrm{c}}(t)$. By definition, $\langle\psi|\,\hat{b}\,|\psi\rangle = 0$. Below, we shall simply neglect any terms involving $\hat{b}(t)$. (In a more complete treatment, these terms can be retained.)

The quantum Heisenberg equation for $\hat{B}(t)$, which is $\delta S[\hat{q}, \hat{B}]/\delta B(t) = 0$, can be expanded around $\hat{B} = B_{\mathrm{c}}$, $\hat{q} = 0$ in powers of $\hat{b}(t)$ and $\hat{q}(t)$ as follows,

$$\frac{\delta S[\hat{q}, \hat{B}]}{\delta B(t)} = \frac{\delta S[0, B_{\mathrm{c}}]}{\delta B(t)} + \int dt_1 \frac{\delta^2 S[0, B_{\mathrm{c}}]}{\delta B(t)\delta q(t_1)} \hat{q}(t_1) + \int dt_1 \frac{\delta^2 S[0, B_{\mathrm{c}}]}{\delta B(t)\delta B(t_1)} \hat{b}(t_1)$$
$$+ \frac{1}{2} \int dt_1 dt_2 \frac{\delta^3 S[0, B_{\mathrm{c}}]}{\delta B(t)\delta q(t_1)\delta q(t_2)} \hat{q}(t_1)\hat{q}(t_2)$$

$$+\frac{1}{2}\int dt_1 dt_2 \frac{\delta^3 S[0, B_c]}{\delta B(t)\delta q(t_1)\delta B(t_2)}\hat{q}(t_1)\hat{b}(t_2)$$

$$+\frac{1}{2}\int dt_1 dt_2 \frac{\delta^3 S[0, B_c]}{\delta B(t)\delta B(t_1)\delta B(t_2)}\hat{b}(t_1)\hat{b}(t_2)+\ldots=0.$$

Here, the arguments $[0, B_c]$ indicate that the derivatives of the action are computed at $\hat{q}=0$ and $\hat{B}=B_c$. Computing the expectation value of the above equation in the state $|\psi\rangle$ and disregarding terms of higher order in the fluctuations, as well as terms containing $\hat{b}$, we obtain an effective equation for $B_c(t)$,

$$\frac{\delta S[0, B_c]}{\delta B(t)}+\frac{1}{2}\int dt_1 dt_2 \frac{\delta^3 S[0, B_c]}{\delta B(t)\delta q(t_1)\delta q(t_2)}\langle\psi|\,\hat{q}(t_1)\hat{q}(t_2)\,|\psi\rangle=0.$$

The resulting equation can be rewritten as

$$\frac{\delta S[0, B_c]}{\delta B(t)}+\frac{1}{2}\int dt_1 dt_2 \frac{\delta^3 S[0, B_c]}{\delta B(t)\delta q(t_1)\delta q(t_2)}\langle 0|\,\hat{q}(t_1)\hat{q}(t_2)\,|0\rangle=0, \tag{A2.1}$$

where we have replaced $|\psi\rangle$ by the vacuum state $|0\rangle$ of the variable $\hat{q}$, according to the assumption that $|\psi\rangle$ is the ground state of $\hat{q}$.

Equation (A2.1) would coincide with equation (12.47) if the integral term involving $\delta^3 S/(\delta B\delta q\delta q)$ were equal to $\delta\Gamma[B_c]/\delta B(t)$. At this point, equation (A2.1) involves the expectation value $\langle 0|\,\hat{q}(t_1)\hat{q}(t_2)\,|0\rangle$ that is not even expressed as a functional of $B_c(t)$ alone. Let us now show how this can be done.

To achieve a heuristic insight, consider the case where $\hat{q}(t)$ is a harmonic oscillator. The quantity $\langle 0|\,\hat{q}(t_1)\hat{q}(t_2)\,|0\rangle$ was computed in Exercise 3.4(a) on page 41 for the harmonic oscillator driven by an external force $J(t)$. In that case, a nonzero external force J creates a nonzero expectation value of $\hat{q}$ in the vacuum state $|0\rangle$. To describe the present situation, where $\langle 0|\,\hat{q}\,|0\rangle=0$, we can use the result of that exercise with $J=0$,

$$\langle 0|\,\hat{q}(t_1)\hat{q}(t_2)\,|0\rangle=\frac{1}{2\omega}e^{i\omega(t_2-t_1)}.$$

Comparing this with equations (3.24) and (3.27), we obtain the relation

$$\langle 0|\,\hat{q}(t_1)\hat{q}(t_2)\,|0\rangle=\frac{1}{i}G_F(t_2, t_1)-\frac{1}{i}G_{ret}(t_2, t_1).$$

In fact, this relation applies much more generally (we omit the derivation). In the general case, the above Green's functions are inverses of the operator $-\hat{A}$, where we defined an auxiliary operator

$$\hat{A}\equiv\frac{\delta^2 S[q=0, B_c]}{\delta q(t_1)\delta q(t_2)},$$

acting on functions $f(t)$ as

$$\left(\hat{A}f\right)(t) \equiv \int dt_1 \frac{\delta^2 S[0, B_c]}{\delta q(t)\delta q(t_1)} f(t_1).$$

Since linear operators are equivalent to integration kernels (in the sense of distributions), we shall write $\hat{A}$ and $A(t_1, t_2)$ interchangeably. Note that $\hat{A} = -(\Box + V)\,\delta(x_1 - x_2)$ in the case of a scalar field considered in Section 12.3, while

$$\hat{A} = -\left[m\frac{d^2}{dt_1^2} + m\omega^2\right]\delta(t_1 - t_2)$$

in the case of a harmonic oscillator.

Next, we note that equation (A2.1) contains $\langle 0|\,\hat{q}(t_1)\hat{q}(t_2)\,|0\rangle$ only in the combination

$$\int dt_1 dt_2 \frac{\delta^3 S[0, B_c]}{\delta B(t)\delta q(t_1)\delta q(t_2)} \langle 0|\,\hat{q}(t_1)\hat{q}(t_2)\,|0\rangle\,.$$

In most cases, the third-order functional derivative contains only local terms of the form $\delta(t_1 - t_2)$. Recalling that $G_{\mathrm{ret}}(t_1, t_1) = 0$, we find that the term containing G_{ret} can be omitted. Thus, Eq. (A2.1) becomes

$$\frac{\delta S[0, B_c]}{\delta B(t)} + \frac{1}{2i}\int dt_1 dt_2 \frac{\delta^3 S[0, B_c]}{\delta B(t)\delta q(t_1)\delta q(t_2)} G_F(t_2, t_1) = 0.$$

The last step is to demonstrate that the above equation can be obtained from the one-loop effective action Γ defined by equation (12.63) in Sec. 12.3. This is shown by a formal calculation with functional determinants. We use the general formula for the variation of the determinant of the operator $\hat{A}$ (see equation (A4.13) on page 229),

$$\delta(\det \hat{A}) = (\det \hat{A})\ \mathrm{Tr}\ (\hat{A}^{-1}\delta\hat{A}).$$

We assume that this formula holds even for infinite-dimensional operators, after suitable renormalizations of the determinant and the trace. Note that the inverse operator $-\hat{A}^{-1}$ must be understood as the *Feynman* Green's function $\hat{G}_F$ because the functional determinant is defined using the "in-out" boundary conditions (see Section 12.3). In other words,

$$\delta(\det \hat{A}) = (\det \hat{A})\ \mathrm{Tr}\ (-\hat{G}_F\delta\hat{A})$$
$$\equiv -(\det \hat{A})\int dt_1 dt_2 G_F(t_2, t_1)\delta A(t_1, t_2).$$

We now transform equation (12.63) into the (Lorentzian) one-loop effective action and rewrite it as

$$\Gamma[B_c] = i\Gamma_E[B_c] = \frac{i}{2}\ln\det\frac{\delta^2 S[q=0, B_c]}{\delta q(t_1)\delta q(t_2)} \equiv \frac{i}{2}\ln\det\hat{A}.$$

Finally, we compute

$$\frac{\delta\Gamma[B_c]}{\delta B_c(t)} = \frac{i}{2}\frac{1}{\det\hat{A}}\det\hat{A}\ \mathrm{Tr}\left(\hat{A}^{-1}\frac{\delta\hat{A}}{\delta B_c(t)}\right) = -\frac{i}{2}\mathrm{Tr}\left(\hat{G}_F\frac{\delta^3 S[0, B_c]}{\delta B_c\delta q\delta q}\right)$$

$$= \frac{1}{2i}\int dt_1 dt_2 G_F(t_2, t_1)\frac{\delta^3 S}{\delta B_c(t)\delta q(t_1)\delta q(t_2)}.$$

Therefore, the effective equation of motion for $J_c(t)$,

$$\frac{\delta S[0, B_c]}{\delta B_c(t)} + \frac{\delta\Gamma[B_c]}{\delta B_c(t)} = 0,$$

is indeed equivalent to equation (A2.1).

Appendix 3

Mode expansions cheat sheet

We present a list of formulae relevant to mode expansions of free, real scalar fields. This should help resolve any confusion about the signs **k** and −**k** or similar technicalities.

All equations (except commutation relations) hold for operators as well as for classical quantities. The formulae for a field quantized in a box are obtained by replacing the factors $(2\pi)^3$ in the denominators with the volume V of the box. (Note that this replacement changes the physical dimension of the modes $\phi_{\mathbf{k}}$.)

$$\phi(\mathbf{x},t) = \int \frac{d^3\mathbf{k}\, e^{i\mathbf{k}\cdot\mathbf{x}}}{(2\pi)^{3/2}} \phi_{\mathbf{k}}(t); \qquad \phi_{\mathbf{k}}(t) = \int \frac{d^3\mathbf{x}\, e^{-i\mathbf{k}\cdot\mathbf{x}}}{(2\pi)^{3/2}} \phi(\mathbf{x},t)$$

$$a_{\mathbf{k}}^-(t) = \sqrt{\frac{\omega_k}{2}}[\phi_{\mathbf{k}} + \frac{i}{\omega_k}\pi_{\mathbf{k}}]; \quad a_{\mathbf{k}}^+(t) = \sqrt{\frac{\omega_k}{2}}[\phi_{-\mathbf{k}} - \frac{i}{\omega_k}\pi_{-\mathbf{k}}]$$

$$\phi_{\mathbf{k}}(t) = \frac{a_{\mathbf{k}}^-(t) + a_{-\mathbf{k}}^+(t)}{\sqrt{2\omega_k}}; \qquad \pi_{\mathbf{k}}(t) = \sqrt{\frac{\omega_k}{2}} \frac{a_{\mathbf{k}}^-(t) - a_{-\mathbf{k}}^+(t)}{i}$$

Time-independent creation and annihilation operators $\hat{a}_{\mathbf{k}}^{\pm}$ are defined by

$$\hat{a}_{\mathbf{k}}^{\pm}(t) \equiv \hat{a}_{\mathbf{k}}^{\pm} \exp(\pm i\omega_k t)$$

Note that all $a_{\mathbf{k}}^{\pm}$ below are time-independent.

$$\phi^\dagger(x) = \phi(x); \quad (\phi_{\mathbf{k}})^\dagger = \phi_{-\mathbf{k}}; \quad (a_{\mathbf{k}}^-)^\dagger = a_{\mathbf{k}}^+$$

$$\pi(\mathbf{x},t) = \frac{d}{dt}\phi(\mathbf{x},t); \quad \pi_{\mathbf{k}}(t) = \frac{d}{dt}\phi_{\mathbf{k}}(t)$$

$$\left[\hat{\phi}(\mathbf{x},t), \hat{\pi}(\mathbf{x}',t)\right] = i\delta(\mathbf{x}-\mathbf{x}')$$

$$\left[\hat{\phi}_{\mathbf{k}}(t), \hat{\pi}_{\mathbf{k}'}(t)\right] = i\delta(\mathbf{k}+\mathbf{k}')$$

$$\left[\hat{a}_{\mathbf{k}}^{-}, \hat{a}_{\mathbf{k}'}^{+}\right] = \delta\left(\mathbf{k}-\mathbf{k}'\right)$$

$$\hat{\phi}\left(\mathbf{x}, t\right) = \int \frac{d^3\mathbf{k}}{(2\pi)^{3/2}} \frac{1}{\sqrt{2\omega_k}} \left[\hat{a}_{\mathbf{k}}^{-} e^{-i\omega_k t + i\mathbf{k}\cdot\mathbf{x}} + \hat{a}_{\mathbf{k}}^{+} e^{i\omega_k t - i\mathbf{k}\cdot\mathbf{x}}\right]$$

Mode expansions may use anisotropic mode functions $v_{\mathbf{k}}(t)$. Isotropic mode expansions use scalar k instead of vector $\mathbf{k}$ because $v_{\mathbf{k}} \equiv v_k$ for all $|\mathbf{k}| = k$.

$$\hat{\phi}\left(\mathbf{x}, t\right) = \int \frac{d^3\mathbf{k}}{(2\pi)^{3/2}} \frac{1}{\sqrt{2}} \left[\hat{a}_{\mathbf{k}}^{-} v_{\mathbf{k}}^{*}(t) e^{i\mathbf{k}\cdot\mathbf{x}} + \hat{a}_{\mathbf{k}}^{+} v_{\mathbf{k}}(t) e^{-i\mathbf{k}\cdot\mathbf{x}}\right]$$

(Note: the factor $\sqrt{2}$ and the choice of $v_{\mathbf{k}}^{*}$ instead of $v_{\mathbf{k}}$ are for consistency with literature. This could have been chosen differently.)

$$v_{-\mathbf{k}} = v_{\mathbf{k}} \neq v_{\mathbf{k}}^{*}; \quad \ddot{v}_{\mathbf{k}} + \omega_k^2(t) v_{\mathbf{k}} = 0; \quad \dot{v}_{\mathbf{k}} v_{\mathbf{k}}^{*} - v_{\mathbf{k}} \dot{v}_{\mathbf{k}}^{*} = 2i$$

$$\phi_{\mathbf{k}}(t) = \frac{1}{\sqrt{2}} \left[a_{\mathbf{k}}^{-} v_{\mathbf{k}}^{*}(t) + a_{-\mathbf{k}}^{+} v_{\mathbf{k}}(t)\right]; \quad \pi_{\mathbf{k}}(t) = \frac{1}{\sqrt{2}} \left[a_{\mathbf{k}}^{-} \dot{v}_{\mathbf{k}}^{*}(t) + a_{-\mathbf{k}}^{+} \dot{v}_{\mathbf{k}}(t)\right]$$

Here the $a_{\mathbf{k}}^{\pm}$ are time-independent although $v_{\mathbf{k}}$ and $\phi_{\mathbf{k}}$, $\pi_{\mathbf{k}}$ depend on time:

$$a_{\mathbf{k}}^{-} = \frac{1}{i\sqrt{2}} \left[\dot{v}_{\mathbf{k}}(t)\phi_{\mathbf{k}}(t) - v_{\mathbf{k}}(t)\pi_{\mathbf{k}}(t)\right]$$

$$a_{\mathbf{k}}^{+} = \frac{i}{\sqrt{2}} \left[\dot{v}_{\mathbf{k}}^{*}(t)\phi_{-\mathbf{k}}(t) - v_{\mathbf{k}}^{*}(t)\pi_{-\mathbf{k}}(t)\right]$$

Free scalar field mode functions in the flat space:

$$v_k(t) = \frac{1}{\sqrt{\omega_k}} e^{i\omega_k t}$$

Bogolyubov transformations

Note: $\hat{a}_{\mathbf{k}}^{\pm}$ are defined by $v_{\mathbf{k}}(\eta)$ and $\hat{b}_{\mathbf{k}}^{\pm}$ are defined by $u_{\mathbf{k}}(\eta)$.

$$u_{\mathbf{k}}(\eta) = \alpha_{\mathbf{k}} v_{\mathbf{k}}(\eta) + \beta_{\mathbf{k}} v_{\mathbf{k}}^{*}(\eta); \quad |\alpha_{\mathbf{k}}|^2 - |\beta_{\mathbf{k}}|^2 = 1$$

$$\hat{b}_{\mathbf{k}}^{-} = \alpha_{\mathbf{k}} \hat{a}_{\mathbf{k}}^{-} - \beta_{\mathbf{k}} \hat{a}_{-\mathbf{k}}^{+}, \quad \hat{b}_{\mathbf{k}}^{+} = \alpha_{\mathbf{k}}^{*} \hat{a}_{\mathbf{k}}^{+} - \beta_{\mathbf{k}}^{*} \hat{a}_{-\mathbf{k}}^{-}$$

$$\alpha_{\mathbf{k}} = \alpha_{-\mathbf{k}}, \quad \beta_{\mathbf{k}} = \beta_{-\mathbf{k}}$$

$$\hat{a}_{\mathbf{k}}^{-} = \alpha_{\mathbf{k}}^{*} \hat{b}_{\mathbf{k}}^{-} + \beta_{\mathbf{k}} \hat{b}_{-\mathbf{k}}^{+}, \quad \hat{a}_{\mathbf{k}}^{+} = \alpha_{\mathbf{k}} \hat{b}_{\mathbf{k}}^{+} + \beta_{\mathbf{k}}^{*} \hat{b}_{-\mathbf{k}}^{-}$$

Appendix 4

Solutions to exercises

Chapter 1

Exercise 1.1 (p. 6)

For a field $\phi(\mathbf{x})$ which is a function only of space, the mode $\phi_{\mathbf{k}}$ is

$$\phi_{\mathbf{k}} = \int \frac{d^3\mathbf{x}}{(2\pi)^{3/2}} e^{-i\mathbf{k}\cdot\mathbf{x}} \phi(\mathbf{x}).$$

Substituting into equation (1.10), we get

$$I = \int \frac{d^3\mathbf{x} d^3\mathbf{y} d^3\mathbf{k}}{(2\pi)^3} e^{i\mathbf{k}\cdot(\mathbf{y}-\mathbf{x})} \phi(\mathbf{x}) \phi(\mathbf{y}) \sqrt{k^2+m^2}.$$

Therefore

$$K(\mathbf{x},\mathbf{y}) = \int \frac{d^3\mathbf{k}}{(2\pi)^3} e^{i\mathbf{k}\cdot(\mathbf{y}-\mathbf{x})} \sqrt{k^2+m^2}.$$

This integral does not converge and should be understood in the distributional sense (see Appendix A1.1). Compare

$$\int \frac{d^3\mathbf{k}}{(2\pi)^3} e^{i\mathbf{k}\cdot\mathbf{x}} = \delta(\mathbf{x}); \int \frac{d^3\mathbf{k}}{(2\pi)^3} \mathbf{k} e^{i\mathbf{k}\cdot\mathbf{x}} = -i\nabla\delta(\mathbf{x}).$$

Exercise 1.2 (p. 8)

We substitute the Fourier transform of $\phi(\mathbf{x})$ into the integral over the cube-shaped region,

$$\phi_L = \frac{1}{L^3}\int_{L^3} \phi(\mathbf{x})\, d^3\mathbf{x} = \frac{1}{L^3}\int_{L^3} d^3\mathbf{x} \int \frac{d^3\mathbf{k}}{(2\pi)^{3/2}} e^{i\mathbf{k}\cdot\mathbf{x}} \phi_{\mathbf{k}}.$$

The integral over $d^3\mathbf{x}$ can be easily computed using the formula

$$\int_{-L/2}^{L/2} dx\, e^{ik_x x} = \frac{2}{k_x L} \sin \frac{k_x L}{2} \equiv f(k_x).$$

Then the expectation value of ϕ_L^2 is

$$\langle \phi_L^2 \rangle = \int \frac{d^3\mathbf{k} d^3\mathbf{k}'}{(2\pi)^3} \langle \phi_\mathbf{k} \phi_{\mathbf{k}'} \rangle f(k_x) f(k_y) f(k_z) f(k'_x) f(k'_y) f(k'_z). \quad \text{(A4.1)}$$

If $\delta\phi_\mathbf{k}$ is the given "typical amplitude of fluctuations" in the mode $\phi_\mathbf{k}$, then the expectation value of $\langle \phi_\mathbf{k} \phi_{\mathbf{k}'} \rangle$ in the vacuum state is

$$\langle \phi_\mathbf{k} \phi_{\mathbf{k}'} \rangle = (\delta\phi_\mathbf{k})^2 \delta(\mathbf{k} + \mathbf{k}').$$

So the integral over $\mathbf{k}, \mathbf{k}'$ in equation (A4.1) reduces to a single integral over $\mathbf{k}$,

$$\langle \phi_L^2 \rangle = \int \frac{d^3\mathbf{k}}{(2\pi)^3} (\delta\phi_\mathbf{k})^2 \left[f(k_x) f(k_y) f(k_z) \right]^2. \quad \text{(A4.2)}$$

The function $f(k)$ is of order 1 for $|kL| \lesssim 1$ but very small for $|kL| \gg 1$. Therefore the integration in equation (A4.2) selects the vector values $\mathbf{k}$ of magnitude $|\mathbf{k}| \lesssim L^{-1}$. As a qualitative estimate, we may take $\delta\phi_\mathbf{k}$ to be constant throughout the effective region of integration in $\mathbf{k}$ and obtain

$$\langle \phi_L^2 \rangle \sim \int_{|\mathbf{k}|<L^{-1}} d^3\mathbf{k} (\delta\phi_\mathbf{k})^2 \sim k^3 (\delta\phi_\mathbf{k})^2 \Big|_{k=L^{-1}}.$$

Exercise 1.3 (p. 11)

The problem is similar to the Schrödinger equation with a step-like potential barrier between two free regions. The general solution in the tunneling region $0 < t < T$ is

$$q(t) = A \cosh \Omega_0 t + B \sinh \Omega_0 t. \quad \text{(A4.3)}$$

The matching condition at $t = 0$ selects $A = 0$ and $B = q_1 \omega_0 / \Omega_0$. The general solution in the region $t > T$ is

$$q(t) = q_2 \sin [\omega_0 (t - T) + \alpha].$$

The constant q_2 is determined by the matching conditions at $t = T$: the values $q(T)$, $\dot{q}(T)$ must match $q_2 \sin\alpha$ and $q_2 \omega_0 \cos\alpha$. Therefore q_2 is found as

$$q_2^2 = [q(T)]^2 + \left[\frac{\dot{q}(T)}{\omega_0} \right]^2.$$

Substituting the values from equation (A4.3), we have

$$q_2^2 = q_1^2 \left[1 + \left(1 + \frac{\omega_0^2}{\Omega_0^2} \right) \sinh^2 \Omega_0 T \right]. \tag{A4.4}$$

For $\Omega_0 T \gg 1$ we can approximate this exact answer by

$$q_2 \approx q_1 \frac{\exp(\Omega_0 T)}{2} \sqrt{1 + \frac{\omega_0^2}{\Omega_0^2}}.$$

Exercise 1.4 (p. 11)

The "number of particles" is formally estimated using the energy of the oscillator. A state with an amplitude q_0 has energy

$$E = \frac{1}{2}\left(\dot{q}^2 + \omega_0^2 q^2\right) = \frac{1}{2} q_0^2 \omega_0^2.$$

Therefore the number of particles is related to the amplitude by

$$n = \frac{q_0^2 \omega_0 - 1}{2}. \tag{A4.5}$$

If the oscillator was initially in the ground state, then $q_1 = \omega_0^{-1/2}$ and equation (A4.4) gives

$$n = \frac{1}{2}\left(1 + \frac{\omega_0^2}{\Omega_0^2} \right) \sinh^2 \Omega_0 T.$$

There are no produced particles if $T = 0$; the number of particles is exponentially large in $\Omega_0 T$.

Exercise 1.5 (p. 12)

To find the strongest currently available electric field, one can perform an Internet search for descriptions of Schwinger effect experiments. The electric field of strongest lasers available in 2003 was $\sim 10^{11}$ V/m. There was a proposed X-ray laser experiment where the radiation is focused, yielding peak fields of order 10^{17}–10^{18} V/m. (See A. Ringwald, *Phys. Lett. B* **510** (2001), 107; preprint `arxiv:hep-ph/0103185`.)

Rewriting equation (1.16) in SI units, we get

$$P = \exp\left(-\frac{m_e c^3}{\hbar e E} \right).$$

For the electric field of a laser, $E = 10^{11}$V/m, the result is

$$P \approx \exp\left(-\frac{(9.11\cdot 10^{-31})^2 (3.00\cdot 10^8)^3}{(1.05\cdot 10^{-34})(1.60\cdot 10^{-19})(10^{11})} \right) \sim e^{-10^7}.$$

Thus, even the strongest laser field gives no measurable particle production. For the proposed focusing experiment, P is between 10^{-11} and 10^{-2}, and some particle production could be observed.

Exercise 1.6 (p. 12)

We need to express all quantities in SI units. The equation $T = a/(2\pi)$ becomes

$$kT = \frac{\hbar}{c}\frac{a}{2\pi},$$

where $k \approx 1.38\cdot 10^{-23}$J/K is Boltzmann's constant. The boiling point of water is $T = 373$K, so the required acceleration is $a \sim 10^{22}$m/s^2, which is clearly beyond any practical possibility.

Chapter 2

Exercise 2.1 (p. 15)

We choose the general solution of equation (2.7) as

$$q(t) = A\cos\omega (t - t_1) + B\sin\omega (t - t_1).$$

The initial condition at $t = t_1$ gives $A = q_1$. The final condition at $t = t_2$ gives

$$B = \frac{q_2 - q_1\cos\omega (t_2 - t_1)}{\sin\omega (t_2 - t_1)}.$$

The classical trajectory exists and is unique if $\sin\omega (t_2 - t_1) \neq 0$. Otherwise we need to consider two possibilities: either $q_1 = q_2$ or not. If $q_1 = q_2$, the value of B remains undetermined (there are infinitely many classical trajectories). If $q_1 \neq q_2$, the value of B is formally infinite; this indicates that the action does not have a minimum (there is no classical trajectory).

Exercise 2.2 (p. 17)

The first functional derivative is

$$\frac{\delta S}{\delta q(t_1)} = \frac{\partial L}{\partial q} - \frac{d}{dt}\frac{\partial L}{\partial \dot{q}}\bigg|_{q(t_1)} = -\left[\ddot{q}(t_1) + \omega^2 q(t_1)\right]. \qquad \text{(A4.6)}$$

As expected, it vanishes "on-shell" (i.e. on solutions). To evaluate the second functional derivative, we need to rewrite equation (A4.6) as an integral of some function over time, e.g.

$$\ddot{q}(t_1) + \omega^2 q(t_1) = \int \left[\ddot{q}(t) + \omega^2 q(t)\right] \delta(t - t_1)\, dt. \tag{A4.7}$$

For an expression of the form $\int q(t) f(t) dt$, the functional derivative with respect to $q(t_2)$ is $f(t_2)$. We can rewrite equation (A4.7) in this form:

$$\ddot{q}(t_1) + \omega^2 q(t_1) = \int \left[\delta''(t - t_1) + \omega^2 \delta(t - t_1)\right] q(t) dt.$$

Therefore the answer is

$$\frac{\delta^2 S}{\delta q(t_1)\, \delta q(t_2)} = -\delta''(t_2 - t_1) - \omega^2 \delta(t_2 - t_1).$$

Exercise 2.3 (p. 19)

(a) The Hamilton action functional

$$S[q(t), p(t)] = \int [p\dot{q} - H(p, q)]\, dt$$

is extremized when

$$\frac{\delta S}{\delta q(t)} = 0, \quad \frac{\delta S}{\delta p(t)} = 0.$$

Computing the functional derivatives, we obtain the Hamilton equations (2.17).

When computing $\delta S / \delta p(t)$, we did not have to integrate by parts because S does not depend on $\dot{p}$. Therefore the variation $\delta p(t)$ is not constrained at the boundary points. However, to compute $\delta S / \delta q(t)$ we need to integrate by parts, which yields a boundary term

$$p(t) \delta q(t)|_{t_1}^{t_2}.$$

This boundary term must vanish. Therefore an appropriate extremization problem is to specify $q(t_1)$ and $q(t_2)$ without restricting $p(t)$. Alternatively, one might specify $p(t_1) = 0$ and fix $q(t_2)$, or vice versa.

(b) A simple calculation using equation (2.17) gives

$$\frac{dH}{dt} = \frac{\partial H}{\partial q}\dot{q} + \frac{\partial H}{\partial p}\dot{p} = 0.$$

(c) The Hamiltonian H is defined as $p\dot{q} - L$, where $\dot{q}$ is replaced by a function of p determined by equation (2.13). This equation is equivalent to the first of the Hamilton equations (2.17). Therefore the function $p\dot{q} - H$ is equal to L on the classical paths.

Exercise 2.4 (p. 19)

We use the elementary identity

$$[A, BC] = B[A, C] + [A, B]C. \tag{A4.8}$$

Computing the commutator

$$[\hat{q}, \hat{p}^2] = [\hat{q}, \hat{p}]\hat{p} + \hat{p}[\hat{q}, \hat{p}] = 2i\hbar\hat{p},$$

we then obtain the result,

$$\hat{q}\hat{p}^2\hat{q} - \hat{p}^2\hat{q}^2 = [\hat{q}, \hat{p}^2]\hat{q} = 2i\hbar\hat{p}\hat{q}.$$

Exercise 2.5 (p. 22)

The "bra"-vector $\langle\psi|$ corresponding to the "ket"-vector $\alpha|v\rangle + \beta|w\rangle$ is defined as a linear map acting on vectors $|u\rangle$ as

$$\langle\psi|u\rangle \equiv (\alpha|v\rangle + \beta|w\rangle, |u\rangle).$$

Since

$$(\alpha|v\rangle, |u\rangle) = (|u\rangle, \alpha|v\rangle)^* = \alpha^*(|u\rangle, |v\rangle)^* = \alpha^*(|v\rangle, |u\rangle),$$

we have

$$(\alpha|v\rangle + \beta|w\rangle, |u\rangle) = \alpha^*(|v\rangle, |u\rangle) + \beta^*(|w\rangle, |u\rangle).$$

This is obviously equal to the action of the "bra"-vector $\alpha^*\langle v| + \beta^*\langle w|$ on the "ket"-vector $|u\rangle$.

Exercise 2.6 (p. 29)

We insert the decomposition of unity, $\int |q\rangle\langle q|\, dq$, into the normalization condition $\langle p_1|p_2\rangle = \delta(p_1 - p_2)$ and obtain

$$\delta(p_1 - p_2) = \int \langle p_1|q\rangle\langle q|p_2\rangle\, dq. \tag{A4.9}$$

Since from our earlier calculations we know that

$$\langle p|q\rangle = C\exp\left(-\frac{ipq}{\hbar}\right),$$

we now substitute this into equation (A4.9) and find the condition for C,

$$\delta(p_1 - p_2) = |C|^2 \int_{-\infty}^{+\infty} dq \exp\left[-\frac{i(p_1 - p_2)q}{\hbar}\right]$$
$$= 2\pi\hbar|C|^2\delta(p_1 - p_2).$$

From this we obtain $|C| = (2\pi\hbar)^{-1/2}$. Note that C is determined up to an irrelevant phase factor.

Exercise 2.7 (p. 30)

(a) See the solution for Exercise 2.4. First we find

$$[\hat{q}, \hat{p}^n] = i\hbar\hat{p}^{n-1} + \hat{p}\left[\hat{q}, \hat{p}^{n-1}\right].$$

Then we use induction to prove that

$$[\hat{q}, \hat{p}^n] = i\hbar n\hat{p}^{n-1}$$

for $n = 1, 2, \ldots$ The statement of the problem follows since $[\hat{q}, \hat{q}^m] = 0$.

The analogous relation with $\hat{p}$ is obtained automatically if we interchange $\hat{q} \leftrightarrow \hat{p}$ and change the sign of the commutator ($i\hbar$ to $-i\hbar$).

(b) We can generalize the result of part (a) to terms of the form $\hat{q}^a\hat{p}^b\hat{q}^c$ by using equation (A4.8),

$$\left[\hat{q}, \hat{q}^a\hat{p}^b\hat{q}^c\right] = i\hbar b\hat{q}^a\hat{p}^{b-1}\hat{q}^c \equiv i\hbar\frac{\partial}{\partial p}\hat{q}^a\hat{p}^b\hat{q}^c. \tag{A4.10}$$

Here it is implied that the derivative $\partial/\partial p$ acts only on $\hat{p}$ where it appears in the expression; the operator ordering should remain unchanged. To prove equation (A4.10), it suffices to demonstrate that for any two terms $f(\hat{p}, \hat{q})$ and $g(\hat{p}, \hat{q})$ of the form $\hat{q}^a\hat{p}^b\hat{q}^c$ that satisfy equation (2.39), the product fg will also satisfy that equation.

An analytic function $f(p, q)$ is expanded into a sum of terms of the form $\ldots\hat{q}^a\hat{p}^b\hat{q}^c\hat{p}^d\ldots$ and the relation (A4.10) can be generalized to terms of this form. Each term of the expansion of $f(\hat{p}, \hat{q})$ satisfies the relation; therefore the sum will also satisfy the relation.

Exercise 2.8 (p. 30)

Note that $\hat{q}$ does not commute with $d\hat{q}/dt$ (coordinates cannot be measured together with velocities). So the time derivative of e.g. $\hat{q}^3$ must be written as

$$\frac{d}{dt}\hat{q}^3 = \hat{q}^2\dot{\hat{q}} + \hat{q}\dot{\hat{q}}\hat{q} + \dot{\hat{q}}\hat{q}^2.$$

It is easy to show that for any operators $\hat{A}$, $\hat{B}$, $\hat{H}$ (not necessarily Hermitian) that satisfy

$$\frac{\partial}{\partial t}\hat{A} = [\hat{A}, \hat{H}], \quad \frac{\partial}{\partial t}\hat{B} = [\hat{B}, \hat{H}],$$

the following properties hold:

$$\frac{\partial}{\partial t}(\hat{A}+\hat{B}) = [\hat{A}+\hat{B}, \hat{H}]; \quad \frac{\partial}{\partial t}(\hat{A}\hat{B}) = \frac{\partial\hat{A}}{\partial t}\hat{B} + \hat{A}\frac{\partial\hat{B}}{\partial t} = [\hat{A}\hat{B}, \hat{H}].$$

By induction, starting from $\hat{p}$ and $\hat{q}$, we prove the same property for arbitrary terms of the form $\ldots\hat{q}^a\hat{p}^b\hat{q}^c\hat{p}^d\ldots$ and their linear combinations. Any observable

$A(p, q)$ that can be approximated by such polynomial terms will satisfy the same equation (2.40).

Chapter 3

Exercise 3.1 (p. 34)

The result follows by simple algebra.

More generally, if $\hat{A}(t)$ and $\hat{B}(t)$ are operators satisfying the equation

$$\frac{d}{dt}\hat{A} = [\hat{A}, \hat{B}]$$

and $\hat{A}(t_0)$ is a c-number, i.e. $\hat{A}(t_0) = A_0\hat{1}$, then $\hat{A}(t) = A_0\hat{1}$ for all other t. This follows because all derivatives $d^n\hat{A}/dt^n$, $n \geq 1$, vanish at $t = t_0$. Therefore it suffices to verify the commutator $\left[\hat{a}^-(t), \hat{a}^+(t)\right] = 1$ at *one* value of t.

Exercise 3.2 (p. 34)

The differential equation

$$\frac{dy}{dx} = f(x)y + g(x)$$

with the initial condition $y(x_0) = y_0$ has the following solution,

$$y(x) = y_0 \exp\left(\int_{x_0}^{x} f(x')dx'\right) + \int_{x_0}^{x} dx' g(x') \exp\left(\int_{x'}^{x} f(x'')dx''\right).$$

(This can be easily derived using the method of variation of constants.) The solution for the driven harmonic oscillator is a special case of this formula with $f(x) = -i\omega$ and $g(x) = J$.

Exercise 3.3 (p. 37)

First we compute the matrix element

$$0 = \langle n_{\text{out}}|\, \hat{a}_{\text{in}}^- \,|0_{\text{in}}\rangle = \langle n_{\text{out}}|\, \hat{a}_{\text{in}}^- \left(\sum_{k=0}^{\infty} \Lambda_k \,|k_{\text{out}}\rangle\right).$$

Since $\hat{a}_{\text{in}}^- = \hat{a}_{\text{out}}^- - C$ and $\hat{a}_{\text{out}}^- \,|k_{\text{out}}\rangle = \sqrt{k}\,|k-1_{\text{out}}\rangle$, we obtain

$$0 = -C\Lambda_n + \sqrt{n+1}\Lambda_{n+1}.$$

Exercise 3.4 (p. 41)

(a) Expanding $\hat{q}(t_1)$ for $t_1 \geq T$ in the "in" creation and annihilation operators $\hat{a}^{\pm}_{\text{in}}$, we find

$$\hat{q}(t_1) = \frac{1}{\sqrt{2\omega}} \left(\hat{a}^{-}_{\text{in}} e^{-i\omega t_1} + \hat{a}^{-}_{\text{in}} e^{-i\omega t_1} + 2\text{Re}\left(J_0 e^{-i\omega t_1} \right) \right)$$

and then we have

$$\langle 0_{\text{in}} | \hat{q}(t_2) \hat{q}(t_1) | 0_{\text{in}} \rangle = \frac{1}{2\omega} e^{-i\omega(t_2 - t_1)} + \frac{2}{\omega} \text{Re}\left(J_0 e^{-i\omega t_1} \right) \text{Re}\left(J_0 e^{-i\omega t_2} \right).$$

The first term is the expectation value without the external force. The second term can be written as a double integral of $J(t)$ as required, since

$$J_0 = \frac{i}{\sqrt{2\omega}} \int_0^T e^{i\omega t} J(t) dt.$$

(b) To compute this matrix element, we expand $\hat{q}(t_1)$ in the operators $\hat{a}^{\pm}_{\text{out}}$ and $\hat{q}(t_2)$ in the operators $\hat{a}^{\pm}_{\text{in}}$. Then we need to compute the following matrix elements:

$$\langle 0_{\text{out}} | \hat{a}^{-}_{\text{out}} | 0_{\text{in}} \rangle = J_0 \langle 0_{\text{out}} | 0_{\text{in}} \rangle,$$

$$\langle 0_{\text{out}} | \hat{a}^{-}_{\text{out}} \hat{a}^{+}_{\text{in}} | 0_{\text{in}} \rangle = \left(1 - |J_0|^2 \right) \langle 0_{\text{out}} | 0_{\text{in}} \rangle.$$

The final result is

$$\frac{\langle 0_{\text{out}} | \hat{q}(t_2) \hat{q}(t_1) | 0_{\text{in}} \rangle}{\langle 0_{\text{out}} | 0_{\text{in}} \rangle} = \frac{1}{2\omega} e^{-i\omega(t_2 - t_1)} + \frac{1}{2\omega} J_0^2 e^{-i\omega(t_1 + t_2)}.$$

Again, the last term

$$\frac{1}{2\omega} J_0^2 e^{-i\omega(t_1 + t_2)}$$

is rewritten as a double integral as required.

Chapter 4

Exercise 4.1 (p. 42)

From linear algebra it is known that a positive-definite symmetric matrix M_{ij} can be diagonalized using an orthogonal basis v^i_α with positive eigenvalues ω^2_α, $\alpha = 1, \ldots, N$. In other words, there exists a nondegenerate matrix $v_{i\alpha}$ such that

$$\sum_i M_{ij} v_{i\alpha} = \omega_\alpha v_{j\alpha}, \quad \sum_i v_{i\alpha} v_{i\beta} = \delta_{\alpha\beta}.$$

Here we do *not* use the Einstein summation convention but write all sums explicitly.

Transforming q_i into a new set of variables $\tilde{q}_\alpha$ by

$$q_i \equiv \sum_\alpha v_{i\alpha} \tilde{q}_\alpha,$$

we rewrite the quadratic term in the action as

$$\sum_{ij} q_i M_{ij} q_j = \sum_{\alpha\beta ij} \tilde{q}_\alpha v_{i\alpha} M_{ij} v_{j\beta} \tilde{q}_\beta = \sum_{\alpha\beta} \omega_\beta \delta_{\alpha\beta} \tilde{q}_\alpha \tilde{q}_\beta = \sum_\alpha \omega_\alpha \tilde{q}_\alpha^2.$$

This provides the required diagonalized form of the action.

Exercise 4.2 (p. 44)

We compute the action of the transformed field $\tilde{\phi}(x)$ after a Lorentz transformation with a matrix Λ^μ_ν. Since the determinant of Λ is equal to 1, we may change the variables of integration d^4x to the transformed variables $d^4\tilde{x}$ with the corresponding Jacobian equal to 1. The action (4.5) has two terms, one with ϕ^2 and the other with derivatives of ϕ. The field values ϕ do not change under the Lorentz transformation, therefore the integral over $d^4\tilde{x}$ of $\tilde{\phi}^2$ is the same as the integral of ϕ^2 over d^4x. However, the values of the field derivatives $\partial_\mu \phi$ do change,

$$\partial_\mu \phi \to \Lambda^\nu_\mu \partial_\nu \phi.$$

The action contains the scalar term $m^2\phi^2$ that does not change, and also the term

$$\eta^{\mu\nu} \left(\partial_\mu \phi\right) \left(\partial_\nu \phi\right)$$

that transforms according to the Lorentz transformation of the field derivatives,

$$\eta^{\mu\nu} \left(\partial_\mu \phi\right) \left(\partial_\nu \phi\right) \to \eta^{\mu\nu} \Lambda^{\mu'}_\mu \left(\partial_{\mu'} \phi\right) \Lambda^{\nu'}_\nu \left(\partial_{\nu'} \phi\right).$$

But the Lorentz transformation leaves the metric unchanged [see equation (4.6)]. Therefore this term in the action is unchanged as well. We obtain the invariance of the action under Lorentz transformations.

Exercise 4.3 (p. 44)

Solution with explicit variation From the action (4.5) we obtain the variation δS with respect to a small change $\delta\phi(x)$ of the field, assuming that $\delta\phi$ vanishes at spatial and temporal infinities:

$$\begin{aligned} \delta S &= \int d^4x \left[\eta^{\mu\nu} \left(\partial_\mu \phi\right) \left(\partial_\nu \delta\phi\right) - m^2 \phi \delta\phi\right] \\ &= \int d^4x \left[-\eta^{\mu\nu} \left(\partial_\nu \partial_\mu \phi\right) - m^2 \phi\right] \delta\phi \end{aligned}$$

(the second line follows by Gauss's theorem). The expression in square brackets must vanish for the action to be extremized, so the equation of motion is

$$-\eta^{\mu\nu}\left(\partial_\nu\partial_\mu\phi\right)-m^2\phi=-\ddot{\phi}+\Delta\phi-m^2\phi=0.$$

Solution with functional derivatives The equation of motion is $\delta S/\delta\phi=0$. To compute the functional derivative, we rewrite the action in an explicit integral form with some function $M(x, y)$,

$$S[\phi]=\frac{1}{2}\int d^4x d^4y\phi(x)\phi(y)M(x,y). \tag{A4.11}$$

(The factor 1/2 is for convenience.) Integrating by parts, we find

$$M(x,y)=-m^2\delta(x-y)+\eta^{\mu\nu}\frac{\partial}{\partial x^\mu}\frac{\partial}{\partial y^\nu}\delta(x-y). \tag{A4.12}$$

Thus the functional derivative of the action (A4.11) is

$$\frac{\delta S}{\delta\phi(x)}=\int d^4y\phi(y)M(x,y).$$

Substituting M from equation (A4.12), we find

$$\frac{\delta S}{\delta\phi(x)}=-m^2\phi(x)-\eta^{\mu\nu}\frac{\partial^2}{\partial x^\mu\partial x^\nu}\phi(x)$$

as required.

Exercise 4.4 (p. 45)

If $\phi(\mathbf{x})$ is a real-valued function, then

$$(\phi_{\mathbf{k}})^*=\int\frac{d^3\mathbf{x}}{(2\pi)^{3/2}}e^{i\mathbf{k}\cdot\mathbf{x}}\phi(\mathbf{x})=\phi_{-\mathbf{k}}.$$

Exercise 4.5 (p. 47)

We use the relations

$$\hat{\phi}_{\mathbf{k}}=\frac{1}{\sqrt{2\omega_k}}\left(\hat{a}_{\mathbf{k}}^-e^{-i\omega_k t}+\hat{a}_{-\mathbf{k}}^+e^{i\omega_k t}\right),\ \hat{\pi}_{\mathbf{k}}=i\sqrt{\frac{\omega_k}{2}}\left(\hat{a}_{-\mathbf{k}}^+e^{i\omega_k t}-\hat{a}_{\mathbf{k}}^-e^{-i\omega_k t}\right).$$

(Here $\hat{a}_{\mathbf{k}}^{\pm}$ are time-independent operators.) Then we find

$$\frac{1}{2}\left(\hat{\pi}_{\mathbf{k}}\hat{\pi}_{-\mathbf{k}}+\omega_k^2\hat{\phi}_{\mathbf{k}}\hat{\phi}_{-\mathbf{k}}\right)=\frac{\omega_k}{2}\left(\hat{a}_{\mathbf{k}}^-\hat{a}_{\mathbf{k}}^++\hat{a}_{-\mathbf{k}}^+\hat{a}_{-\mathbf{k}}^-\right).$$

Since we integrate over all $\mathbf{k}$, we may change the integration variable $\mathbf{k} \to -\mathbf{k}$ when needed. Therefore we may write the Hamiltonian as

$$\hat{H} = \int d^3\mathbf{k} \frac{\omega_k}{2} \left(\hat{a}_{\mathbf{k}}^+ \hat{a}_{\mathbf{k}}^- + \hat{a}_{\mathbf{k}}^- \hat{a}_{\mathbf{k}}^+ \right).$$

Chapter 5

Exercise 5.1 (p. 60)

The computation is split into three parts: (1) the variation of the determinant $\sqrt{-g}$ with respect to $g^{\alpha\beta}$; (2) the variation of the action with respect to $\Gamma^{\lambda}_{\rho\sigma}$; (3) the variation of the action with respect to $g^{\alpha\beta}$.

(1) To find the variation of the determinant $\sqrt{-g}$, we need to compute the derivative of $g \equiv \det g_{\alpha\beta}$ with respect to a parameter. We can use the matrix identity (for finite-dimensional matrices A)

$$\det A = \exp\left(\operatorname{Tr}\ln A\right).$$

Choosing $A \equiv A(s)$ as a matrix that depends on some parameter s, we get

$$\frac{d}{ds}\det A = \frac{d}{ds}\exp\left(\operatorname{Tr}\ln A\right) = (\det A)\operatorname{Tr}\left(A^{-1}\frac{dA}{ds}\right). \tag{A4.13}$$

Here A^{-1} is the inverse matrix. We now set $A \equiv g_{\mu\nu}$ (the covariant metric tensor in some basis) and $s \equiv g_{\alpha\beta}$ with *fixed* α and β. Then $g \equiv \det A$ and

$$\frac{\partial g}{\partial s} \equiv \frac{\partial g}{\partial g_{\alpha\beta}} = g\, g^{\mu\nu}\frac{\partial g_{\mu\nu}}{\partial g_{\alpha\beta}} = g\, g^{\mu\nu}\delta^{\alpha}_{\mu}\delta^{\beta}_{\nu} = g\, g^{\alpha\beta}.$$

The derivative with respect to components of the inverse matrix $g^{\mu\nu}$ is computed quickly if we recall that the determinant of $g^{\mu\nu}$ is g^{-1}:

$$\frac{\partial g}{\partial g^{\alpha\beta}} = -g\, g_{\alpha\beta}; \qquad \frac{\partial\sqrt{-g}}{\partial g^{\alpha\beta}} = -\frac{1}{2}\sqrt{-g}g_{\alpha\beta}. \tag{A4.14}$$

(Note that $\sqrt{-g}\sqrt{-g} = -g > 0$.)

We may also consider $\sqrt{-g(x')}$ with a *fixed* x' to be a functional of $g^{\alpha\beta}(x)$. The functional derivative of this functional is

$$\frac{\delta\sqrt{-g(x')}}{\delta g^{\alpha\beta}(x)} = \frac{\partial\sqrt{-g}}{\partial g^{\alpha\beta}}\delta(x-x') = -\frac{\sqrt{-g}}{2}g_{\alpha\beta}\delta(x-x').$$

As a by-product, we also find the spatial derivatives of the determinant:

$$\partial_\mu g = g\, g_{\alpha\beta,\mu}g^{\alpha\beta}; \quad \partial_\mu\sqrt{-g} = \frac{\sqrt{-g}}{2}g^{\alpha\beta}g_{\alpha\beta,\mu}. \tag{A4.15}$$

(2) To compute $\delta S/\delta\Gamma^{\lambda}_{\rho\sigma}$, we rewrite the action as an integral of $\Gamma^{\lambda}_{\rho\sigma}$ times some function. This requires some reshuffling of indices and integrations by parts. For example,

$$\int \sqrt{-g}d^4x\, g^{\alpha\beta}\Gamma^{\mu}_{\alpha\beta,\mu} = -\int d^4x\Gamma^{\lambda}_{\rho\sigma}\left(\sqrt{-g}g^{\rho\sigma}\right)_{,\lambda},$$

$$-\int \sqrt{-g}d^4x\, g^{\alpha\beta}\Gamma^{\mu}_{\alpha\mu,\beta} = \int d^4x\Gamma^{\lambda}_{\rho\sigma}\left(\sqrt{-g}g^{\rho\beta}\delta^{\sigma}_{\lambda}\right)_{,\beta}.$$

The functional derivatives of these terms with respect to $\Gamma^{\lambda}_{\rho\sigma}$ can be read off from these integrals. The terms bilinear in Γ need to be rewritten twice, with $\Gamma^{\lambda}_{\rho\sigma}$ at the first place or at the second place:

$$\Gamma^{\mu}_{\alpha\beta}\Gamma^{\nu}_{\mu\nu} = \Gamma^{\lambda}_{\rho\sigma}\Gamma^{\nu}_{\lambda\nu}\delta^{\rho}_{\alpha}\delta^{\sigma}_{\beta} = \Gamma^{\rho}_{\alpha\beta}\Gamma^{\lambda}_{\rho\sigma}\delta^{\sigma}_{\lambda},$$

$$\Gamma^{\nu}_{\alpha\mu}\Gamma^{\mu}_{\beta\nu} = \Gamma^{\lambda}_{\rho\sigma}\Gamma^{\sigma}_{\beta\lambda}\delta^{\rho}_{\alpha} = \Gamma^{\sigma}_{\alpha\lambda}\Gamma^{\lambda}_{\rho\sigma}\delta^{\rho}_{\beta}.$$

The functional derivatives of these terms are then computed by omitting $\Gamma^{\lambda}_{\rho\sigma}$ from the above expressions:

$$\frac{\delta}{\delta\Gamma^{\lambda}_{\rho\sigma}}\left(\int \Gamma^{\mu}_{\alpha\beta}\Gamma^{\nu}_{\mu\nu}\sqrt{-g}g^{\alpha\beta}d^4x\right) = \sqrt{-g}g^{\alpha\beta}\left(\Gamma^{\nu}_{\lambda\nu}\delta^{\rho}_{\alpha}\delta^{\sigma}_{\beta} + \Gamma^{\rho}_{\alpha\beta}\delta^{\sigma}_{\lambda}\right),$$

$$\frac{\delta}{\delta\Gamma^{\lambda}_{\rho\sigma}}\left(-\int \Gamma^{\nu}_{\alpha\mu}\Gamma^{\mu}_{\beta\nu}\sqrt{-g}g^{\alpha\beta}d^4x\right) = -\sqrt{-g}g^{\alpha\beta}\left(\Gamma^{\sigma}_{\beta\lambda}\delta^{\rho}_{\alpha} + \Gamma^{\sigma}_{\alpha\lambda}\delta^{\rho}_{\beta}\right).$$

Therefore the equation of motion for $\Gamma^{\lambda}_{\rho\sigma}$ is

$$0 = \frac{\delta S}{\delta\Gamma^{\lambda}_{\rho\sigma}} = -\left(\sqrt{-g}g^{\rho\sigma}\right)_{,\lambda} + \left(\sqrt{-g}g^{\rho\beta}\delta^{\sigma}_{\lambda}\right)_{,\beta}$$
$$+\left(\Gamma^{\nu}_{\lambda\nu}g^{\rho\sigma} + \Gamma^{\rho}_{\alpha\beta}g^{\alpha\beta}\delta^{\sigma}_{\lambda} - \Gamma^{\rho}_{\alpha\lambda}g^{\sigma\alpha} - \Gamma^{\sigma}_{\alpha\lambda}g^{\rho\alpha}\right)\sqrt{-g}.$$

It is now convenient to convert the upper indices ρ, σ into lower indices μ, ν by multiplying both parts by $g_{\mu\rho}g_{\nu\sigma}$ (before doing this, we rename the mute index ν above into α). The derivatives of $\sqrt{-g}$ are shown in equation (A4.15). The common factor $\sqrt{-g}$ is canceled. We obtain the following equation for $\Gamma^{\mu}_{\alpha\beta}$:

$$\Gamma^{\alpha}_{\lambda\alpha}g_{\mu\nu} + \Gamma^{\rho}_{\alpha\beta}g_{\mu\rho}g_{\lambda\nu}g^{\alpha\beta} - \Gamma^{\rho}_{\nu\lambda}g_{\mu\rho} - \Gamma^{\sigma}_{\mu\lambda}g_{\nu\sigma}$$
$$= \frac{1}{2}g_{\lambda\nu}g^{\alpha\beta}\left(2g_{\alpha\mu,\beta} - g_{\alpha\beta,\mu}\right) + \frac{1}{2}g_{\mu\nu}g^{\alpha\beta}g_{\alpha\beta,\lambda} - g_{\mu\nu,\lambda}.$$

This is a complicated (although linear) equation that needs to be solved for Γ. One way is to separate the terms on both sides by their index symmetry and by their dependence on $g_{\alpha\beta}$. To make the symmetry in the indices easier to use, we lower the index μ in $\Gamma^{\mu}_{\alpha\beta}$ to obtain the auxiliary quantity $\Gamma_{\mu\alpha\beta}$ defined by

$$\Gamma^{\mu}_{\alpha\beta} \equiv g^{\mu\nu}\Gamma_{\nu\alpha\beta}.$$

Then we find

$$\begin{aligned} &g_{\lambda\nu}g^{\alpha\beta}\Gamma_{\mu\alpha\beta} + g_{\mu\nu}g^{\alpha\beta}\Gamma_{\beta\lambda\alpha} - \left(\Gamma_{\mu\nu\lambda} + \Gamma_{\nu\mu\lambda}\right) \\ &= \frac{1}{2}g_{\lambda\nu}g^{\alpha\beta}\left(2g_{\alpha\mu,\beta} - g_{\alpha\beta,\mu}\right) + \frac{1}{2}g_{\mu\nu}g^{\alpha\beta}g_{\alpha\beta,\lambda} - g_{\mu\nu,\lambda}. \end{aligned} \tag{A4.16}$$

Now we note that there are three pairs of terms at each side: terms with free $g_{\lambda\nu}$, terms with free $g_{\mu\nu}$, and terms without a free (undifferentiated) $g_{\mu\nu}$. Moreover, the second and the third pair of terms are symmetric in μ, ν. Therefore, the first pair of terms, which is not symmetric under $\mu \leftrightarrow \nu$, must match separately:

$$g_{\lambda\nu}g^{\alpha\beta}\Gamma_{\mu\alpha\beta} = \frac{1}{2}g_{\lambda\nu}g^{\alpha\beta}\left(2g_{\alpha\mu,\beta} - g_{\alpha\beta,\mu}\right).$$

This equation is obviously solved by

$$\Gamma_{\mu\alpha\beta} = \frac{1}{2}\left(g_{\alpha\mu,\beta} + g_{\beta\mu,\alpha} - g_{\alpha\beta,\mu}\right), \tag{A4.17}$$

which is equivalent to equation (5.18). [Here we identically rewrote

$$2g^{\alpha\beta}g_{\alpha\mu,\beta} = g^{\alpha\beta}\left(g_{\alpha\mu,\beta} + g_{\beta\mu,\alpha}\right),$$

to make $\Gamma_{\mu\alpha\beta}$ symmetric in α, β.] Then we need to check that the other two pairs of terms also cancel. With the above choice of $\Gamma_{\mu\alpha\beta}$ we find

$$\Gamma_{\mu\nu\lambda} + \Gamma_{\nu\mu\lambda} = g_{\mu\nu,\lambda},$$

$$g^{\alpha\beta}\Gamma_{\beta\lambda\alpha} = \frac{1}{2}g^{\alpha\beta}g_{\alpha\beta,\lambda}.$$

Therefore equation (5.18) is a solution.

Finally, we must show that this solution is unique. If there are two solutions $\Gamma_{\mu\alpha\beta}$ and $\Gamma'_{\mu\alpha\beta}$, their difference $D_{\mu\alpha\beta}$ satisfies the homogeneous equation

$$g_{\lambda\nu}g^{\alpha\beta}D_{\mu\alpha\beta} + g_{\mu\nu}g^{\alpha\beta}D_{\beta\lambda\alpha} - \left(D_{\mu\nu\lambda} + D_{\nu\mu\lambda}\right) = 0. \tag{A4.18}$$

We need to show that this equation has no solutions except $D_{\mu\alpha\beta} = 0$ when $g_{\alpha\beta}$ is a nondegenerate matrix. First we antisymmetrize in μ, ν and find $g_{\lambda[\nu}D_{\mu]\alpha\beta}g^{\alpha\beta} = 0$. If we define $u_\mu \equiv D_{\mu\alpha\beta}g^{\alpha\beta}$ and raise the index λ, we find that u_μ satisfies $\delta^\lambda_\nu u_\mu = \delta^\lambda_\mu u_\nu$. The only solution of this is $u_\mu = 0$ (set $\nu = \lambda \neq \mu$ to prove this). So the first term of equation (A4.18) vanishes. Then we contract equation (A4.18) with $g^{\mu\nu}$ and find $g^{\alpha\beta}D_{\beta\lambda\alpha} = 0$. Therefore equation (A4.18) is reduced to $D_{\mu\nu\lambda} + D_{\nu\mu\lambda} = 0$. But a tensor $D_{\mu\nu\lambda}$ which is antisymmetric in the first two indices but symmetric in the last two indices must necessarily vanish. Therefore the solution $\Gamma_{\mu\alpha\beta}$ of equation (A4.16) is unique.

(3) The variation of $R\sqrt{-g}$ with respect to $g^{\alpha\beta}$ is now easy to find. We write

$$R\sqrt{-g} = g^{\mu\nu} R_{\mu\nu}\sqrt{-g},$$

where $R_{\mu\nu}$ is treated as independent of $g^{\alpha\beta}$ since it is a combination of the Γ symbols. Then

$$\frac{\delta}{\delta g^{\alpha\beta}}\left(\int g^{\mu\nu} R_{\mu\nu}\sqrt{-g}d^4x\right)$$
$$= R_{\alpha\beta}\sqrt{-g} + \int g^{\mu\nu} R_{\mu\nu}\frac{\delta\sqrt{-g(x')}}{\delta g^{\alpha\beta}(x)}d^4x' = \left(R_{\alpha\beta} - \frac{1}{2}g_{\alpha\beta}R\right)\sqrt{-g}.$$

The last line gives the required expression.

Remark: other solutions Here we solved for $\Gamma^{\mu}_{\alpha\beta}$ straightforwardly by extremizing the action, without choosing a special coordinate system. Another way to obtain the Einstein equation is to vary the action directly with respect to $g^{\mu\nu}$; direct calculations are cumbersome unless one uses a locally inertial coordinate system.

Chapter 6

Exercise 6.1 (p. 65)

We use the spacetime coordinates $(\mathbf{x}, \eta)$ and note that $\sqrt{-g} = a^4$ and $g^{\alpha\beta} = a^{-2}\eta^{\alpha\beta}$. Then

$$\sqrt{-g}\, m^2\phi^2 = m^2a^2\chi^2,$$
$$\sqrt{-g}\, g^{\alpha\beta}\phi_{,\alpha}\phi_{,\beta} = a^2\left(\phi'^2 - (\nabla\phi)^2\right).$$

Substituting $\phi = \chi/a$, we get

$$a^2\phi'^2 = \chi'^2 - 2\chi\chi'\frac{a'}{a} + \chi^2\left(\frac{a'}{a}\right)^2 = \chi'^2 + \chi^2\frac{a''}{a} - \left[\chi^2\frac{a'}{a}\right]'.$$

The total time derivative term can be omitted from the action, and we obtain the required expression.

Exercise 6.2 (p. 66)

The standard result $dW/dt = 0$ follows if we use the oscillator equation to express $\ddot{x}_{1,2}$ through $x_{1,2}$,

$$\frac{d}{dt}\left(\dot{x}_1x_2 - x_1\dot{x}_2\right) = \ddot{x}_1x_2 - x_1\ddot{x}_2 = \omega^2x_1x_2 - x_1\omega^2x_2 = 0.$$

The solutions $x_1(t)$ and $x_2(t)$ are linearly *dependent* if there exists a constant λ such that $x_2(t) = \lambda x_1(t)$ for all t. It immediately follows that $W[x_1, x_2] = \dot{x}_1 \lambda x_1 - x_1 \lambda \dot{x}_1 = 0$. Conversely, $W[x_1, x_2] = 0$ means that the matrix

$$\begin{pmatrix} \dot{x}_1(t) & x_1(t) \\ \dot{x}_2(t) & x_2(t) \end{pmatrix}$$

is degenerate for each t. Thus, at a fixed time $t = t_0$ there exists λ_0 such that $x_2(t_0) = \lambda_0 x_1(t_0)$ and $\dot{x}_2(t_0) = \lambda_0 \dot{x}_1(t_0)$. The solution of the Cauchy problem with initial conditions $x(t_0) = \lambda_0 x_1(t_0)$, $\dot{x}(t_0) = \lambda_0 \dot{x}_1(t_0)$ is unique; one such solution is $x_2(t)$ and another is $\lambda_0 x_1(t)$; therefore $x_2(t) = \lambda_0 x_1(t)$ for all t.

To derive an expression for $a_{\mathbf{k}}^-$ in terms of $\chi_{\mathbf{k}}(\eta)$ and $v_{\mathbf{k}}(\eta)$, we use equation (6.13) and its time derivative:

$$\chi_{\mathbf{k}}(\eta) = \frac{1}{\sqrt{2}} \left[a_{\mathbf{k}}^- v_k^*(\eta) + a_{-\mathbf{k}}^+ v_k(\eta) \right],$$

$$\chi_{\mathbf{k}}'(\eta) = \frac{1}{\sqrt{2}} \left[a_{\mathbf{k}}^- v_k^{*\prime}(\eta) + a_{-\mathbf{k}}^+ v_k'(\eta) \right].$$

Multiplying the first equation above by v_k', the second equation by v_k, and subtracting, we find

$$\chi_{\mathbf{k}} v_k' - v_k \chi_{\mathbf{k}}' = \frac{1}{\sqrt{2}} a_{\mathbf{k}}^- \left(v_k^* v_k' - v_k^{*\prime} v_k \right).$$

The desired formula for $a_{\mathbf{k}}^-$ follows.

Exercise 6.3 (p. 68)

We compute the commutation relations between $\hat{\chi}(\mathbf{x}, \eta)$ and $\hat{\pi}(\mathbf{x}, \eta)$ using the mode expansion (6.20) and the commutation relations for $\hat{a}_{\mathbf{k}}^{\pm}$:

$$[\hat{\chi}(\mathbf{x}, \eta), \hat{\pi}(\mathbf{y}, \eta)] = \int \frac{d^3\mathbf{k}}{(2\pi)^3} \frac{v_{\mathbf{k}}' v_{\mathbf{k}}^* - v_{\mathbf{k}} v_{\mathbf{k}}'^*}{2} e^{i\mathbf{k}\cdot(\mathbf{x}-\mathbf{y})}.$$

From the known identity

$$\delta(\mathbf{x} - \mathbf{y}) = \int \frac{d^3\mathbf{k}}{(2\pi)^3} e^{i\mathbf{k}\cdot(\mathbf{x}-\mathbf{y})}$$

it follows that equation (6.22) must hold for all $\mathbf{k}$.

Exercise 6.4 (p. 68)

We suppress the index k for brevity and write the normalization condition for $u(\eta)$, expressing u through v using equation (6.24),

$$u^*u' - uu'^* = \left(|\alpha|^2 - |\beta|^2\right)\left(v^*v' - vv'^*\right).$$

It follows that the normalization of $v(\eta)$ and $u(\eta)$ is equivalent to the condition (6.25).

Exercise 6.5 (p. 70)

First we consider the quantum state of one mode $\hat{\phi}_{\mathbf{k}}$. The b-vacuum $\left|{}_{(b)}0_{\mathbf{k},-\mathbf{k}}\right\rangle$ is expanded as the linear combination

$$\left|{}_{(b)}0_{\mathbf{k},-\mathbf{k}}\right\rangle = \sum_{m,n=0}^{\infty} c_{mn}\left|{}_{(a)}m_{\mathbf{k}}, n_{-\mathbf{k}}\right\rangle, \tag{A4.19}$$

where the state $\left|{}_{(a)}m_{\mathbf{k}}, n_{-\mathbf{k}}\right\rangle$ is the result of acting on the a-vacuum state with m creation operators $\hat{a}_{\mathbf{k}}^+$ and n creation operators $\hat{a}_{-\mathbf{k}}^+$,

$$\left|{}_{(a)}m_{\mathbf{k}}, n_{-\mathbf{k}}\right\rangle = \frac{\left(\hat{a}_{\mathbf{k}}^+\right)^m \left(\hat{a}_{-\mathbf{k}}^+\right)^n}{\sqrt{m!n!}}\left|{}_{(a)}0_{\mathbf{k},-\mathbf{k}}\right\rangle. \tag{A4.20}$$

The unknown coefficients c_{mn} may be found after a somewhat long calculation by substituting equation (A4.19) into

$$\left.\begin{aligned}\left(\alpha_{\mathbf{k}}\hat{a}_{\mathbf{k}}^- - \beta_{\mathbf{k}}\hat{a}_{-\mathbf{k}}^+\right)\left|{}_{(b)}0_{\mathbf{k},-\mathbf{k}}\right\rangle = 0,\\ \left(\alpha_{\mathbf{k}}\hat{a}_{-\mathbf{k}}^- - \beta_{\mathbf{k}}\hat{a}_{\mathbf{k}}^+\right)\left|{}_{(b)}0_{\mathbf{k},-\mathbf{k}}\right\rangle = 0.\end{aligned}\right\} \tag{A4.21}$$

We use a faster and more elegant method. Equation (A4.20) implies that the b-vacuum state is a result of acting on the a-vacuum by a combination of the creation operators. We denote this combination by $f(\hat{a}_{\mathbf{k}}^+, \hat{a}_{-\mathbf{k}}^+)$ where $f(x, y)$ is an unknown function. Then from equation (A4.21) we get two equations for $\hat{f}$,

$$\left(\alpha_{\mathbf{k}}\hat{a}_{\mathbf{k}}^- - \beta_{\mathbf{k}}\hat{a}_{-\mathbf{k}}^+\right)\hat{f}\left|{}_{(a)}0_{\mathbf{k},-\mathbf{k}}\right\rangle = 0, \tag{A4.22}$$

$$\left(\alpha_{\mathbf{k}}\hat{a}_{-\mathbf{k}}^- - \beta_{\mathbf{k}}\hat{a}_{\mathbf{k}}^+\right)\hat{f}\left|{}_{(a)}0_{\mathbf{k},-\mathbf{k}}\right\rangle = 0. \tag{A4.23}$$

We know from Exercise 2.7b (p. 30) that the commutator $\left[\hat{a}_{\mathbf{k}}^-, \hat{f}\right]$ is equal to the derivative of $\hat{f}$ with respect to $\hat{a}_{\mathbf{k}}^+$. Therefore equation (A4.22) gives

$$\left(\alpha_{\mathbf{k}}\frac{\partial \hat{f}}{\partial a_{\mathbf{k}}^+} - \beta_{\mathbf{k}}\hat{a}_{-\mathbf{k}}^+\hat{f}\right)\left|{}_{(a)}0_{\mathbf{k},-\mathbf{k}}\right\rangle = 0.$$

Since the function $\hat{f}$ contains only creation operators, it must satisfy

$$\alpha_{\mathbf{k}} \frac{\partial \hat{f}}{\partial a_{\mathbf{k}}^{+}} - \beta_{\mathbf{k}} \hat{a}_{-\mathbf{k}}^{+} \hat{f} = 0.$$

This differential equation has the general solution

$$f\left(\hat{a}_{\mathbf{k}}^{+}, \hat{a}_{-\mathbf{k}}^{+}\right) = C\left(\hat{a}_{-\mathbf{k}}^{+}\right) \exp\left(\frac{\beta_{\mathbf{k}}}{\alpha_{\mathbf{k}}} \hat{a}_{\mathbf{k}}^{+} \hat{a}_{-\mathbf{k}}^{+}\right),$$

where C is an arbitrary function of $\hat{a}_{-\mathbf{k}}^{+}$. To determine this function, we use equation (A4.23) to derive the analogous relation for $\partial f / \partial a_{-\mathbf{k}}^{+}$ and find that C must be a constant. Therefore the b-vacuum is expressed as

$$\left|{}_{(b)}0_{\mathbf{k},-\mathbf{k}}\right\rangle = C \sum_{n=0}^{\infty} \left(\frac{\beta_{\mathbf{k}}}{\alpha_{\mathbf{k}}}\right)^{n} \left|{}_{(a)}n_{\mathbf{k}}, n_{-\mathbf{k}}\right\rangle.$$

The value of C is fixed by normalization,

$$\left\langle {}_{(b)}0_{\mathbf{k},-\mathbf{k}} \middle| {}_{(b)}0_{\mathbf{k},-\mathbf{k}} \right\rangle = 1 \Rightarrow C = \sqrt{1 - \frac{|\beta_{\mathbf{k}}|^{2}}{|\alpha_{\mathbf{k}}|^{2}}} = \frac{1}{|\alpha_{\mathbf{k}}|}.$$

Since $|\beta_{\mathbf{k}}| < |\alpha_{\mathbf{k}}|$, the value of C as given above is always real and nonzero. The final expression for the b-vacuum state is

$$\left|{}_{(b)}0_{\mathbf{k},-\mathbf{k}}\right\rangle = \frac{1}{|\alpha_{\mathbf{k}}|} \sum_{n=0}^{\infty} \left(\frac{\beta_{\mathbf{k}}}{\alpha_{\mathbf{k}}}\right)^{n} \left|{}_{(a)}n_{\mathbf{k}}, n_{-\mathbf{k}}\right\rangle.$$

The vacuum state $\left|{}_{(b)}0\right\rangle$ is the tensor product of the vacuum states $\left|{}_{(b)}0_{\mathbf{k},-\mathbf{k}}\right\rangle$ of all modes. Since each pair $\hat{\phi}_{\mathbf{k}}, \hat{\phi}_{-\mathbf{k}}$ is counted twice in the product over all $\mathbf{k}$, we need to take the square root of the whole expression:

$$\begin{aligned}
|0\rangle &= \left[\prod_{\mathbf{k}} \frac{1}{|\alpha_{\mathbf{k}}|} \sum_{n=0}^{\infty} \left(\frac{\beta_{\mathbf{k}}}{\alpha_{\mathbf{k}}}\right)^{n} \frac{\left(\hat{a}_{\mathbf{k}}^{+} \hat{a}_{-\mathbf{k}}^{+}\right)^{n}}{n!}\right]^{1/2} |0\rangle \\
&= \prod_{\mathbf{k}} \frac{1}{\sqrt{|\alpha_{\mathbf{k}}|}} \exp\left(\frac{\beta_{\mathbf{k}}}{2\alpha_{\mathbf{k}}} \hat{a}_{\mathbf{k}}^{+} \hat{a}_{-\mathbf{k}}^{+}\right) |0\rangle .
\end{aligned}$$

Exercise 6.6 (p. 72)

Similarly to the calculation in Section 4.5, we perform a Fourier transform to find

$$\hat{H} = \frac{1}{2} \int d^{3}\mathbf{k} \left(\hat{\chi}_{\mathbf{k}}' \hat{\chi}_{-\mathbf{k}}' + \omega_{k}^{2}(\eta) \hat{\chi}_{\mathbf{k}} \hat{\chi}_{-\mathbf{k}}\right).$$

Now we expand the operators $\hat{\chi}_{\mathbf{k}}$ through the mode functions and use the identity $v_{\mathbf{k}}(\eta) = v_{-\mathbf{k}}(\eta)$ and equation (6.20). For example, the term $\hat{\chi}'_{\mathbf{k}}\hat{\chi}'_{-\mathbf{k}}$ gives

$$\frac{1}{2}\int d^3\mathbf{k}\hat{\chi}'_{\mathbf{k}}\hat{\chi}'_{-\mathbf{k}} = \int \frac{d^3\mathbf{k}}{4}\left(v'^*_{\mathbf{k}}\hat{a}^-_{\mathbf{k}} + v'_{\mathbf{k}}\hat{a}^+_{-\mathbf{k}}\right)\left(v'^*_{\mathbf{k}}\hat{a}^-_{-\mathbf{k}} + v'_{\mathbf{k}}\hat{a}^+_{\mathbf{k}}\right)$$

$$= \int \frac{d^3\mathbf{k}}{4}\left[v'^2_{\mathbf{k}}\hat{a}^+_{\mathbf{k}}\hat{a}^+_{-\mathbf{k}} + \left(v'^*_{\mathbf{k}}\right)^2\hat{a}^-_{\mathbf{k}}\hat{a}^-_{-\mathbf{k}} + \left|v'_{\mathbf{k}}\right|^2\left(\hat{a}^-_{\mathbf{k}}\hat{a}^+_{\mathbf{k}} + \hat{a}^+_{-\mathbf{k}}\hat{a}^-_{-\mathbf{k}}\right)\right].$$

Since we are integrating over all $\mathbf{k}$, we may exchange $\mathbf{k}$ and $-\mathbf{k}$ in the integrand. After some straightforward algebra we obtain the required result.

Exercise 6.7 (p. 78)

Using the mode expansion and the commutation relations for $\hat{a}^{\pm}_{\mathbf{k}}$, we find

$$\langle 0|\,\hat{\chi}(\mathbf{x},\eta)\,\hat{\chi}(\mathbf{y},\eta)\,|0\rangle = \langle 0|\int \frac{d^3\mathbf{k}}{(2\pi)^3}\frac{1}{2}e^{i\mathbf{k}\cdot(\mathbf{x}-\mathbf{y})}v^*_k v_k\,|0\rangle$$

$$= \int_0^\infty \frac{k^2 dk}{4\pi^2}\int_{-1}^{1} d(\cos\theta)e^{ikL\cos\theta}\frac{|v_k(\eta)|^2}{2}$$

$$= \frac{1}{4\pi^2}\int_0^\infty k^2 dk\,|v_k(\eta)|^2\,\frac{\sin kL}{kL}.$$

Exercise 6.8 (p. 82)

We need to compute the mode function $v_k^{(\mathrm{in})}(\eta)$ at $\eta > \eta_1$ and represent it as a sum of $v_k^{(\mathrm{out})}$ and $v_k^{(\mathrm{out})*}$. To simplify the notation, we rename $v_k^{(\mathrm{in})} \equiv v_k$ and $v_k^{(\mathrm{out})} \equiv u_k$. The mode function v_k and its derivative v_k need to be matched at points $\eta = 0$ and $\eta = \eta_1$. To simplify the matching, we use the ansatz

$$f(t) = A\cos\omega\,(t - t_0) + \frac{B}{\omega}\sin\omega\,(t - t_0)$$

to match $f(t_0) = A$, $f'(t_0) = B$. We find for $0 < \eta < \eta_1$,

$$v_k(\eta) = \frac{1}{\sqrt{\omega_k}}\cos\Omega_k\eta + \frac{i\sqrt{\omega_k}}{\Omega_k}\sin\Omega_k\eta.$$

Then the conditions at $\eta = \eta_1$ are

$$v_k(\eta_1) = \frac{1}{\sqrt{\omega_k}}\cos\Omega_k\eta_1 + \frac{i\sqrt{\omega_k}}{\Omega_k}\sin\Omega_k\eta_1,$$

$$v'_k(\eta_1) = -\frac{\Omega_k}{\sqrt{\omega_k}}\sin\Omega_k\eta_1 + i\sqrt{\omega_k}\cos\Omega_k\eta_1.$$

Hence, for $\eta > \eta_1$ the mode function is

$$v_k(\eta) = \frac{e^{i\omega_k(\eta-\eta_1)}}{\sqrt{\omega_k}}\left[\cos\Omega_k\eta_1 + \frac{i}{2}\left(\frac{\omega_k}{\Omega_k}+\frac{\Omega_k}{\omega_k}\right)\sin\Omega_k\eta_1\right]$$

$$+\frac{e^{-i\omega_k(\eta-\eta_1)}}{\sqrt{\omega_k}}\left(\frac{\omega_k}{\Omega_k}-\frac{\Omega_k}{\omega_k}\right)\frac{i\sin\Omega_k\eta_1}{2}$$

$$=\alpha_k^*\frac{e^{i\omega_k(\eta-\eta_1)}}{\sqrt{\omega_k}}+\beta_k^*\frac{e^{-i\omega_k(\eta-\eta_1)}}{\sqrt{\omega_k}}.$$

The required expressions for α_k and β_k follow after a regrouping of the complex exponentials.

Exercise 6.9 (p. 83)*

The energy density is given by the integral

$$\varepsilon_0 = m_0^4\int_0^{k_{\max}} 4\pi k^2 dk\sqrt{m_0^2+k^2}\frac{\left|\sin\eta_1\sqrt{k^2-m_0^2}\right|^2}{\left|k^4-m_0^4\right|}.$$

We introduce the dimensionless variable $s \equiv k/m_0$ and obtain

$$\frac{\varepsilon_0}{m_0^4} = 4\pi\int_0^{s_{\max}} s^2 ds\frac{\left|\sin A\sqrt{s^2-1}\right|^2}{\left|s^2-1\right|\sqrt{1+s^2}}, \tag{A4.24}$$

where $A \equiv m_0\eta_1 \gg 1$ is a (dimensionless) parameter. The integral over s in equation (A4.24) contains contributions from the intervals $0 < s < 1$ and from $1 < s < s_{\max}$,

$$\frac{\varepsilon_0}{m_0^4} = 4\pi\int_0^1 s^2 ds\frac{\sinh^2 A\sqrt{1-s^2}}{\left(1-s^2\right)\sqrt{1+s^2}} + 4\pi\int_1^{s_{\max}} s^2 ds\frac{\sin^2 A\sqrt{s^2-1}}{\left(s^2-1\right)\sqrt{1+s^2}}. \tag{A4.25}$$

Since this integral cannot be computed exactly, we shall perform an asymptotic estimate for large values of A. The integrand in the first term in equation (A4.25) is exponentially large for most s,

$$\sinh^2 A \approx \frac{1}{4}\exp(2A),$$

while the second term gives only a power-law growth in A,

$$\frac{\sin^2 A\sqrt{s^2-1}}{s^2-1} \le A^2, \quad s \ge 1.$$

(Note that $\left|\frac{\sin x}{x}\right| \leq 1$ for all $x \geq 0$.) This suggests that the first term is the asymptotically dominant one for $A \gg 1$. Now we consider the two integrals in equation (A4.25) in more detail and obtain their asymptotics for $A \to \infty$.

(1) The first integral in equation (A4.25) can be asymptotically estimated in the following way. We rewrite the integrand as a product of quickly varying and slowly varying functions,

$$4\pi \int_0^1 s^2 ds \frac{\sinh^2 A\sqrt{1-s^2}}{(1-s^2)\sqrt{1+s^2}}$$
$$= \pi \int_0^1 ds \left[s^2 e^{2A\sqrt{1-s^2}} \right] \frac{1-2e^{-2A\sqrt{1-s^2}} + e^{-4A\sqrt{1-s^2}}}{(1-s^2)\sqrt{1+s^2}}. \tag{A4.26}$$

The quickly varying expression in the square brackets has the maximum at $s = s_0$ where

$$s_0^2 A = \sqrt{1-s_0^2} \Rightarrow s_0 \approx \frac{1}{\sqrt{A}} \ll 1.$$

This maximum gives the dominant contribution to the integral. Near $s = s_0$ the slowly varying factor is of order $1 + O\left(A^{-1}\right)$ and can be neglected in the calculation of the leading asymptotics. By changing the variable $s\sqrt{A} = u$, we find

$$\int_0^1 s^2 ds \exp\left(2A\sqrt{1-s^2}\right) = e^{2A} \int_0^1 s^2 ds\, e^{-As^2 + O(s^4)}$$
$$= A^{-3/2} e^{2A} \int_0^{\sqrt{A}} u^2 e^{-u^2} du \left(1 + O\left(A^{-1}\right)\right)$$
$$= \frac{\sqrt{\pi}}{4} A^{-3/2} e^{2A} \left(1 + O\left(A^{-1}\right)\right).$$

In the last integral we have approximated

$$\int_0^{\sqrt{A}} u^2 e^{-u^2} du \approx \int_0^{\infty} u^2 e^{-u^2} du = \frac{\sqrt{\pi}}{4},$$

since the difference is exponentially small, of order $\exp\left(-\frac{\text{const}}{A}\right)$, whereas we have already neglected terms of order A^{-1}. Therefore the first contribution to ε_0 is

$$\frac{m_0^4}{4} \left(\frac{\pi}{m_0 \eta_1}\right)^{3/2} e^{2m_0\eta_1} \left(1 + O\left(\frac{1}{m_0 \eta_1}\right)\right).$$

(2) It remains to prove that the first integral in equation (A4.25) gives the dominant contribution for $A \gg 1$. This can be shown by finding an upper bound

for the second integral. We split the range $1 < s < s_{max}$ into two ranges $1 < s < s_1$ and $s_1 < s < s_{max}$, where s_1 is the first point after $s = 1$ where

$$\sin A\sqrt{s_1^2 - 1} = 0.$$

Then

$$s_1 = \sqrt{1 + \frac{\pi^2}{4A^2}} \approx 1 + \frac{\pi^2}{8A^2},$$

and the integrand can be bounded from above on each of the ranges using

$$\frac{\sin^2 A\sqrt{s^2-1}}{s^2-1} \leq A^2, \quad \frac{s^2}{\sqrt{1+s^2}} < 1 \quad \text{for } 0 \leq s \leq s_1,$$

$$\sin^2 A\sqrt{s^2-1} \leq 1, \quad \frac{s^2}{s^2-1} < 1 + \frac{4A^2}{\pi^2} \quad \text{for } s \geq s_1.$$

So the integral satisfies the inequalities

$$\int_1^{s_{max}} s^2 ds \frac{\sin^2 A\sqrt{s^2-1}}{(s^2-1)\sqrt{1+s^2}} < A^2 \int_1^{s_1} ds + \left(1 + \frac{4A^2}{\pi^2}\right) \int_{s_1}^{s_{max}} \frac{ds}{\sqrt{1+s^2}}$$

$$= A^2 (s_1 - 1) + \left(1 + \frac{4A^2}{\pi^2}\right) \left(\sinh^{-1} s_{max} - \sinh^{-1} s_1\right)$$

$$< \frac{\pi^2}{8} + \left(1 + \frac{4A^2}{\pi^2}\right) \ln\left(s_{max} + \sqrt{s_{max}^2 + 1}\right) \sim A^2 \ln s_{max}, \quad A \gg 1.$$

Therefore at large A and fixed s_{max} the contribution of the second integral is subdominant to that of the first integral.

Chapter 7

Exercise 7.1 (p. 89)

We introduce the variable $s \equiv k|\eta|$ and express the mode function through the new function $f(s)$ by

$$v_k(\eta) \equiv \sqrt{s} f(s) = \sqrt{k|\eta|} f(k|\eta|).$$

Then the equation for $f(s)$ is the Bessel equation,

$$s^2 f'' + s f' + (s^2 - n^2) f = 0, \quad n \equiv \sqrt{\frac{9}{4} - \frac{m^2}{H^2}},$$

with the general solution $f(s) = AJ_n(s) + BY_n(s)$, where A and B are arbitrary constants and J_n, Y_n are the Bessel functions. Therefore the mode function $v_k(\eta)$ is

$$v_k(\eta) = \sqrt{k|\eta|} \left[AJ_n(k|\eta|) + BY_n(k|\eta|)\right]. \qquad \text{(A4.27)}$$

The asymptotics of the Bessel functions are known; see e.g. *The Handbook of Mathematical Functions*, eds. M. Abramowitz and I. Stegun (National Bureau of Standards, Washington D.C., 1974):

$$J_n(s) \sim \begin{cases} \frac{1}{\Gamma(n+1)} \left(\frac{s}{2}\right)^n, & s \to 0, \\ \sqrt{\frac{2}{\pi s}} \cos\left(s - \frac{n\pi}{2} - \frac{\pi}{4}\right), & s \to \infty; \end{cases}$$

$$Y_n(s) \sim \begin{cases} -\frac{1}{\pi}\Gamma(n) \left(\frac{2}{s}\right)^n, & s \to 0, \\ \sqrt{\frac{2}{\pi s}} \sin\left(s - \frac{n\pi}{2} - \frac{\pi}{4}\right), & s \to \infty. \end{cases}$$

Since by assumption $m \ll H$, the parameter n is real and $n > 0$. So the mode function $v_k(\eta)$ defined by equation (A4.27) has the following asymptotics:

$$v_k(\eta) \sim \begin{cases} B\frac{1}{\pi} 2^n \Gamma(n) \left(k\,|\eta|\right)^{\frac{1}{2}-n}, & k\,|\eta| \to 0, \\ \sqrt{\frac{2}{\pi}} \left[A \cos\lambda + B \sin\lambda\right], & k\,|\eta| \to +\infty. \end{cases}$$

Here we denoted

$$\lambda \equiv k\,|\eta| - \frac{n\pi}{2} - \frac{\pi}{4}.$$

It is clear that the choice

$$A = \sqrt{\frac{\pi}{2k}}, \qquad B = -iA$$

will result in the asymptotics at early times $k\,|\eta| \to \infty$ of the form

$$v_k(\eta) = \frac{1}{\sqrt{k}} \exp\left(ik\eta + \frac{in\pi}{2} + \frac{i\pi}{4}\right).$$

This coincides with the Minkowski mode function (up to a phase).

Chapter 8

Exercise 8.1 (p. 99)

Consider a boost that transforms one inertial frame to the other and suppose v_r is the relative velocity. The coordinates in the two frames are related by the standard formulae

$$\tilde{t} = \frac{t - v_r x}{\sqrt{1 - v_r^2}},$$

$$\tilde{x} = \frac{x - v_r t}{\sqrt{1 - v_r^2}}.$$

Hence, the lightcone coordinates are related by

$$\tilde{u} = u\frac{1-v_r}{\sqrt{1-v_r^2}} \equiv \alpha u, \quad \tilde{v} = v\frac{1+v_r}{\sqrt{1-v_r^2}} = \alpha^{-1} v.$$

This gives

$$\alpha = \left[\frac{1-v_r}{1+v_r}\right]^{1/2}; \quad v_r = \frac{1-\alpha^2}{1+\alpha^2}.$$

Exercise 8.2 (p. 106)

We substitute the expression for $\hat{b}^{\pm}_{\Omega}$ into the commutation relation and find

$$\begin{aligned}
\delta(\Omega-\Omega') &= \left[\hat{b}^-_{\Omega}, \hat{b}^+_{\Omega'}\right] \\
&= \left[\int d\omega \left(\alpha_{\omega\Omega}\hat{a}^-_{\omega} - \beta_{\omega\Omega}\hat{a}^+_{\omega}\right), \int d\omega' \left(\alpha^*_{\omega'\Omega'}\hat{a}^+_{\omega'} - \beta^*_{\omega'\Omega'}\hat{a}^-_{\omega'}\right)\right] \\
&= \int d\omega d\omega' \left(\alpha_{\omega\Omega}\alpha^*_{\omega'\Omega'}\delta(\omega-\omega') - \beta_{\omega\Omega}\beta^*_{\omega'\Omega'}\delta(\omega-\omega')\right) \\
&= \int d\omega \left(\alpha_{\omega\Omega}\alpha^*_{\omega\Omega'} - \beta_{\omega\Omega}\beta^*_{\omega\Omega'}\right).
\end{aligned}$$

Exercise 8.3 (p. 107)

Let us define the auxiliary function $F(\omega, \Omega)$ by[1]

$$F(\omega, \Omega) \equiv \int_{-\infty}^{+\infty} \frac{du}{2\pi} e^{i\Omega u - i\omega\bar{u}} = \int_{-\infty}^{+\infty} \frac{du}{2\pi} \exp\left[i\Omega u + i\frac{\omega}{a}e^{-au}\right]. \quad \text{(A4.28)}$$

This function can be reduced to Euler's Γ function by changing the variable $u \to t$,

$$t \equiv -\frac{i\omega}{a}e^{-au}.$$

The result is

$$F(\omega, \Omega) = \frac{1}{2\pi a} \exp\left(i\frac{\Omega}{a}\ln\frac{\omega}{a} + \frac{\pi\Omega}{2a}\right)\Gamma\left(-\frac{i\Omega}{a}\right), \quad \omega > 0,\ a > 0.$$

We will need to transform this expression under the replacement $\omega \to -\omega$, but it is not clear whether we may set $\ln(-\omega) = \ln\omega + i\pi$ or we need some other phase instead of $i\pi$. To resolve this question, we need to analyze the required analytic continuation of the Γ function; a detailed calculation is given in Appendix A1.3.

[1]Because of the carelessly interchanged order of integration while deriving equation (8.40), the integral (A4.28) diverges at $u \to +\infty$ and the definition of $F(\omega, \Omega)$ must be understood in the distributional sense.

Exercise 8.4 (p. 107)

A more direct approach (without using the Γ function) is to deform the contour of integration in equation (A4.28). The contour can be shifted downwards by $-i\pi a^{-1}$ into the line $u = -i\pi a^{-1} + t$, where t is real, $-\infty < t < +\infty$ (see Fig. A4.1). Then $e^{-au} = -e^{-at}$ and we obtain

$$F(\omega, \Omega) = \int_{-\infty}^{+\infty} \frac{dt}{2\pi} \exp\left(i\Omega t + \frac{\pi\Omega}{a} - \frac{i\omega}{a} e^{-at}\right)$$
$$= F(-\omega, \Omega) \exp\left(\frac{\pi\Omega}{a}\right).$$

It remains to justify the shift of the contour. The integrand has no singularities and, since the lateral lines have a limited length, it suffices to show that the integrand vanishes at $u \to \pm\infty - i\alpha$ for $0 < \alpha < \pi a^{-1}$. At $u = M - i\alpha$ and $M \to -\infty$ the integrand vanishes since

$$\lim_{u\to-\infty-i\alpha} \operatorname{Re}\left(\frac{i\omega}{a} e^{-au}\right) = -\lim_{t\to-\infty} \frac{\omega}{a} e^{-at} \sin \alpha a = -\infty. \qquad \text{(A4.29)}$$

At $u \to +\infty - i\alpha$ the integral does not actually converge and must be regularized, e.g. by inserting a convergence factor $\exp\left(-bu^2\right)$ with $b > 0$:

$$F(\omega, \Omega) = \lim_{b\to+0} \int_{-\infty}^{+\infty} \frac{du}{2\pi} \exp\left(-bu^2 + i\Omega u + i\frac{\omega}{a} e^{-au}\right). \qquad \text{(A4.30)}$$

With this (or another) regularization, the integrand vanishes at $u \to +\infty - i\alpha$ as well. Therefore the contour may be shifted and our result is justified in the sense of distributions.

Note that we cannot shift the contour to $u = -i(\pi + 2\pi n)a^{-1} + t$ with any $n \neq 0$ because equation (A4.29) will not hold. Also, with $\omega < 0$ we would be unable to move the contour in the negative imaginary direction. The shift of the contour we used is the only one possible.

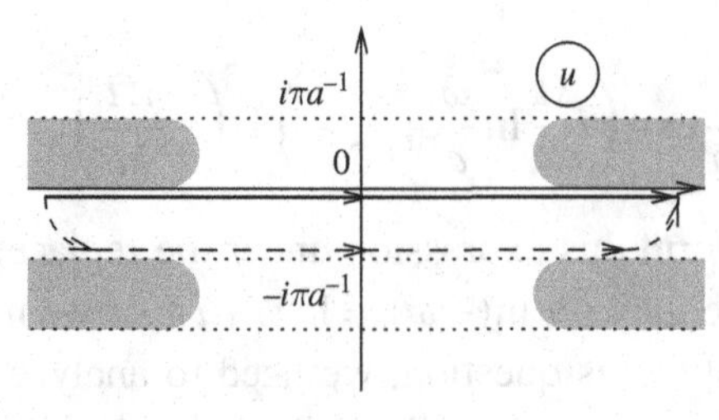

Fig. A4.1 The original and the shifted contours of integration for equation (A4.30) are shown by solid and dashed lines. The shaded regions cannot be crossed when deforming the contour at infinity.

Chapter 9

Exercise 9.1 (p. 114)

Calculations are conveniently done in the radial lightcone coordinates

$$u \equiv t - r, \ v \equiv t + r, \ \eta_{ab} dx^a dx^b = du\, dv,$$

where we omitted the spherical angular coordinates. A suitable coordinate transformation is

$$\tilde{u} = \tanh u, \ \tilde{v} = \tanh v, \ du\, dv = \frac{d\tilde{u}\, d\tilde{v}}{(1 - \tilde{u}^2)(1 - \tilde{v}^2)}.$$

The new coordinates $\tilde{u}, \tilde{v}$ extend from -1 to 1. Multiplying the metric by the conformal factor $\Omega^2(\tilde{u}, \tilde{v}) \equiv (1 - \tilde{u}^2)(1 - \tilde{v}^2)$, we obtain the following conformal transformation of the metric,

$$\Omega^2 du\, dv = d\tilde{u}\, d\tilde{v} = \eta_{ab} d\tilde{x}^a d\tilde{x}^b,$$

$$\tilde{u} \equiv \tilde{x}^0 - \tilde{x}^1, \quad \tilde{v} \equiv \tilde{x}^0 + \tilde{x}^1.$$

The new coordinates $\tilde{x}^a$ have a finite extent, namely $\left|\tilde{x}^0 \pm \tilde{x}^1\right| < 1$, and therefore the resulting diagram has a diamond shape shown in Fig. A4.2. To appreciate the

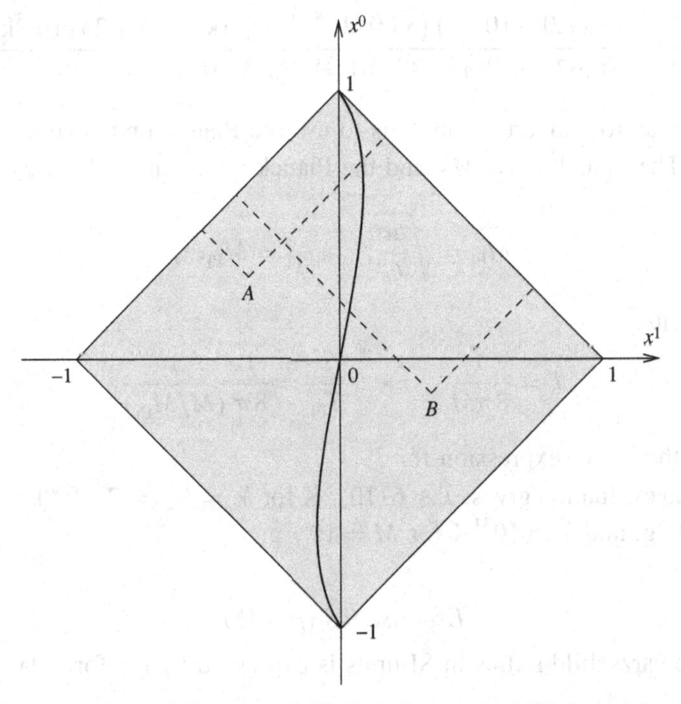

Fig. A4.2 A conformal diagram of the 1+1-dimensional Minkowski spacetime. Dashed lines show lightrays emitted from points A, B. The curved line is the trajectory of an inertial observer moving with a constant velocity, $x^1 = 0.3x^0$.

distortion of the spacetime geometry shown in the conformal diagram, we can draw the worldline of an inertial observer moving with a constant velocity. Note that the angle at which this trajectory enters the endpoints depends on the chosen conformal transformation and thus cannot serve as an indication of the observer's velocity.

Exercise 9.2 (p. 118)

We need to restore the correct combination of the constants c, G, $\hbar$, and k in the equation. The temperature is derived from the relation of the type $\omega = a/(2\pi)$ where $a = (4M)^{-1}$ is the proper acceleration of the observer and ω the frequency of field modes. This relation becomes $\omega = a/(2\pi c)$ in SI units. The relation between a and M contains only the constants c and G since it is a classical and not a quantum-mechanical relation. The Planck constant $\hbar$ enters only as the combination $\hbar\omega$ and the Boltzmann constant k enters only as kT. Therefore we find

$$a = \frac{c^4}{4GM}, \quad kT = \hbar\omega = \frac{\hbar a}{2\pi c} \Rightarrow T = \frac{\hbar c^3}{Gk}\frac{1}{8\pi M}.$$

The relation between temperature in degrees and mass in kilograms is

$$\frac{T}{1°K} \approx \frac{(1.05 \cdot 10^{-34})(3.00 \cdot 10^8)^3}{(6.67 \cdot 10^{-11})(1.38 \cdot 10^{-23})\, 8\pi} \left(\frac{1\text{kg}}{M}\right) = \frac{1.23 \cdot 10^{23}\text{kg}}{M}.$$

Another way to convert the units is to use the Planck units explicitly as in the following. The Planck mass M_{Pl} and the Planck temperature T_{Pl} are defined by

$$M_{\text{Pl}} = \sqrt{\frac{\hbar c}{G}}, \quad kT_{\text{Pl}} = M_{\text{Pl}} c^2.$$

Then we write

$$T = \frac{1}{8\pi M} \quad \Rightarrow \quad \frac{T}{T_{\text{Pl}}} = \frac{1}{8\pi} \frac{1}{(M/M_{\text{Pl}})}$$

and obtain the above expression for T.

Numerical evaluation gives: $T \approx 6 \cdot 10^{-8}$K for $M = M_\odot = 2 \cdot 10^{30}$kg; $T \approx 10^{11}$K for $M = 10^{15}$g; and $T \approx 10^{31}$K for $M = 10^{-5}$g.

Exercise 9.3 (p. 118)

(a) The Schwarzschild radius in SI units is expressed by the formula

$$R = \frac{2GM}{c^2}.$$

The typical wavelength of a photon is

$$\lambda = \frac{2\pi}{\omega}c = \frac{(2\pi c)^2}{a} = \frac{16\pi^2 GM}{c^2}.$$

Note that the ratio of λ to R is independent of M (this can be seen already in the Planck units):

$$\frac{\lambda}{R} = 8\pi^2.$$

(b) The Compton wavelength of a proton is

$$\lambda = \frac{2\pi\hbar}{m_\mathrm{p}c}.$$

The proton mass is $m_\mathrm{p} \approx 1.67 \cdot 10^{-27}\mathrm{kg}$. Protons are produced efficiently if the typical energy of an emitted particle,

$$kT = \hbar\omega = \frac{\hbar c^3}{8\pi GM},$$

is larger than the rest energy of the proton, $m_\mathrm{p}c^2 = \hbar\omega$. The required mass of the BH is

$$M = \frac{\hbar c}{8\pi G m_\mathrm{p}} \approx 1.1 \cdot 10^{10}\mathrm{kg}.$$

The ratio of the Compton wavelength λ to R is (we now use the Planck units, but the dimensionless ratio is independent of units)

$$\frac{\lambda}{R} = \frac{2\pi}{m_\mathrm{p}} 4\pi T = 8\pi^2.$$

Note that this ratio is the same for the massless particles. So the required size of the black hole is about $1/8\pi^2 \approx 0.013$ times the size of a proton.

Exercise 9.4 (p. 120)

The loss of energy due to Hawking radiation can be written as

$$\frac{dM}{dt} = -\frac{1}{BM^2},$$

where B is a constant. Then the lifetime of a black hole of initial mass M_0 is

$$t_\mathrm{L} = \frac{BM_0^3}{3}.$$

In SI units, this formula becomes

$$t_{\mathrm{L}} = \frac{G^2}{\hbar c^4} \frac{BM_0^3}{3}.$$

The dimensionless coefficient B depends on γ, the number of available degrees of freedom in quantum fields. The order of magnitude of B is estimated as

$$B = \frac{15360\pi}{\gamma} \sim 10^4.$$

We find $t_{\mathrm{L}} \sim 10^{74}\mathrm{s}$ for $M = M_{\odot}$; $t_{\mathrm{L}} \sim 10^{19}\mathrm{s}$ for $M = 10^{15}\mathrm{g}$; $t_{\mathrm{L}} \sim 10^{-41}\mathrm{s}$ for $M = 10^{-5}\mathrm{g}$. For comparison, the age of the Universe is of order $\sim 10^{10}$ years or $\sim 3 \cdot 10^{17}\mathrm{s}$; the Planck time is $t_{\mathrm{Pl}} \approx 5.4 \cdot 10^{-44}\mathrm{s}$.

Exercise 9.5 (p. 122)

(a) Here we consider the black hole as a thermodynamical system with a peculiar equation of state. The results are essentially independent of the details of the Hawking radiation, of the kinds of particles emitted by the black hole, and of the nature of the reservoir.

Solution 1: elementary consideration of equilibrium The equilibrium of a black hole with a reservoir is stable if any small heat exchange causes a reverse exchange. It is intuitively clear that in the equilibrium state the temperatures of the black hole T_{BH} and of the reservoir T_{r} must be equal. Suppose that initially $T_{\mathrm{r}} = T_{\mathrm{BH}}$ and the black hole absorbs an infinitesimal quantity of heat, $\delta Q > 0$, from the reservoir. Then the mass M of the black hole will increase by $\delta M = \delta Q$ and the temperatures will change according to

$$\delta T_{\mathrm{r}} = -\frac{1}{C_{\mathrm{r}}}\delta Q, \quad \delta T_{\mathrm{BH}} = \delta \frac{1}{8\pi M} = -\frac{\delta Q}{8\pi M^2} + O\left(\delta Q^2\right).$$

This creates a temperature difference

$$T_{\mathrm{BH}} - T_{\mathrm{r}} = \left(\frac{1}{C_{\mathrm{r}}} - \frac{1}{8\pi M^2}\right)\delta Q + O\left(\delta Q^2\right).$$

If $0 < C_{\mathrm{r}} < 8\pi M^2$, then $T_{\mathrm{BH}} > T_{\mathrm{r}}$ and the black hole will subsequently tend to give heat to the reservoir, restoring the balance. However, for $C_{\mathrm{r}} > 8\pi M^2$ the created temperature difference is negative, $T_{\mathrm{BH}} - T_{\mathrm{r}} < 0$, and the situation is further destabilized since the BH will tend to absorb even more heat.

Similarly, if $\delta Q < 0$ (heat initially lost by the BH), the resulting temperature difference will stabilize the system when $C_r < 8\pi M^2$. Therefore a BH of mass M can be in a stable equilibrium with the reservoir at $T_{\mathrm{BH}} = T_{\mathrm{r}}$ only if the heat capacity C_{r} of the reservoir is positive and not too large, $0 < C_{\mathrm{r}} < 8\pi M^2$.

Solution 2: maximizing the entropy This is a more formal thermodynamical consideration. If a black hole is placed inside a closed reservoir, the total energy of the system is constant and the stable equilibrium is the state of maximum entropy. Let $C_r(T_r)$ be the heat capacity of the reservoir as a function of the reservoir temperature T_r. We shall determine the energy E_r and the entropy S_r of the reservoir which maximize the entropy.

If the reservoir absorbs an infinitesimal quantity of heat δQ, the first law of thermodynamics yields

$$\delta Q = dE_r = C_r(T_r)\, dT_r = T_r dS_r.$$

Therefore

$$E_r(T_r) = \int_0^{T_r} C_r(T) dT, \quad S_r(T_r) = \int_0^{T_r} \frac{C_r(T)}{T} dT.$$

The entropy of a black hole with mass M is

$$S_{BH} = 4\pi M^2 = \frac{1}{16\pi T_{BH}^2}$$

and the energy of the BH is equal to its mass,

$$E_{BH} = M = \frac{1}{8\pi T_{BH}}.$$

This indicates a negative heat capacity,

$$C_{BH}(T) = \frac{dE_{BH}}{dT} = -\frac{1}{8\pi T^2}.$$

Now we have the following thermodynamical situation: two systems with temperatures T_1 and T_2 and heat capacities $C_1(T_1)$ and $C_2(T_2)$ are in thermal contact and the combined energy is constant, $E_1(T_1) + E_2(T_2) = \text{const}$. We need to find the state which maximizes the combined entropy $S = S_1(T_1) + S_2(T_2)$. This problem is solved by standard variational methods. The energy constraint gives T_2 as a function of T_1 such that

$$\frac{dT_2(T_1)}{dT_1} = -\frac{C_1(T_1)}{C_2(T_2)}.$$

The extremum condition $dS/dT_1 = 0$ gives

$$\frac{dS}{dT_1} = \frac{C_1(T_1)}{T_1} + \frac{C_2(T_2)}{T_2}\frac{dT_2}{dT_1} = \left(\frac{1}{T_1} - \frac{1}{T_2}\right) C_1(T_1) = 0.$$

Therefore $T_1 = T_2$ is a necessary condition for the equilibrium. The equilibrium is stable if $d^2S/dT_1^2 < 0$, which yields the condition

$$\left.\frac{d^2S}{dT_1^2}\right|_{T_1=T_2} = \frac{d}{dT_1}\left(\frac{1}{T_1} - \frac{1}{T_2}\right) C_1(T_1) = -\frac{C_1}{C_2}\frac{C_1 + C_2}{T_1^2} < 0.$$

Hence, the stability condition is $C_1 C_2^{-1}(C_1 + C_2) > 0$. Usually heat capacities are positive and the thermal equilibrium is stable. However, in our case $C_1 = C_{BH} < 0$. Therefore the equilibrium is stable if and only if $0 < C_2 = C_r < |C_{BH}|$, in other words

$$0 < C_r < \frac{1}{8\pi T_{BH}^2} = 8\pi M^2.$$

We find that the equilibrium is stable only if the reservoir has a certain finite heat capacity. A combination of a BH and a sufficiently large reservoir is unstable.

(b) The heat capacity of a radiation-filled cavity of volume V is

$$C_r(T_r) = 4\sigma V T_r^3.$$

In equilibrium, we have $T_r = T_{BH} = T$. The stability condition yields

$$C_r = 4\sigma V T^3 < \frac{1}{8\pi T^2} \Rightarrow V < V_{max} = \frac{1}{32\pi\sigma T^5}.$$

A black hole cannot be in a stable equilibrium with a reservoir of volume V larger than V_{max}.

Chapter 10

Exercise 10.1 (p. 125)

(a) We start by assuming that the normalization factor in the mode expansion is $\sqrt{2/L}$ and derive the commutation relation.

We integrate the mode expansion over x and use the identity (10.4),

$$\int_0^L dx\, \hat{\phi}(x,t) \sin\omega_n x = \frac{1}{2}\sqrt{\frac{L}{\omega_n}} \left[\hat{a}_n^- e^{-i\omega_n t} + \hat{a}_n^+ e^{i\omega_n t}\right].$$

Then we differentiate this with respect to t and obtain

$$\int_0^L dx'\, \hat{\pi}(y,t) \sin\omega_n x' = \frac{i}{2}\sqrt{L\omega_n} \left[-\hat{a}_n^- e^{-i\omega_n t} + \hat{a}_n^+ e^{i\omega_n t}\right].$$

Now we can evaluate the commutator

$$\left[\int_0^L dx\, \hat{\phi}(x,t) \sin\omega_n x, \int_0^L dy \frac{d}{dt}\hat{\phi}(x',t) \sin\omega_{n'} x'\right] = i\frac{L}{2}\left[\hat{a}_n^-, \hat{a}_{n'}^+\right]$$

$$= \int_0^L dx \int_0^L dx' \sin\frac{n\pi x}{L} \sin\frac{n'\pi x'}{L} i\delta(x-x') = i\frac{L}{2}\delta_{nn'}.$$

In the second line we used $\left[\hat{\phi}(x,t), \hat{\pi}(x',t)\right] = i\delta(x-x')$. Therefore the factor $\sqrt{2/L}$ indeed cancels and we obtain the standard commutation relations for $\hat{a}_n^{\pm}$.

(b) The Hamiltonian for the field between the plates is

$$\hat{H} = \frac{1}{2}\int_0^L dx \left[\left(\frac{\partial \hat{\phi}(x,t)}{\partial t} \right)^2 + \left(\frac{\partial \hat{\phi}(x,t)}{\partial x} \right)^2 \right].$$

The expression $\langle 0|\hat{H}|0\rangle$ is evaluated using the mode expansion above and the relations

$$\langle 0|\hat{a}_m^- \hat{a}_n^+|0\rangle = \delta_{mn}, \quad \langle 0|\hat{a}_m^+ \hat{a}_n^+|0\rangle = \langle 0|\hat{a}_m^- \hat{a}_n^-|0\rangle = \langle 0|\hat{a}_m^+ \hat{a}_n^-|0\rangle = 0.$$

The first term in the Hamiltonian yields

$$\langle 0|\frac{1}{2}\int_0^L dx \left(\frac{\partial \hat{\phi}(x,t)}{\partial t} \right)^2 |0\rangle$$

$$= \langle 0|\frac{1}{2}\int_0^L dx \left[\sqrt{\frac{2}{L}} \sum_{n=1}^{\infty} \frac{\sin \omega_n x}{\sqrt{2\omega_n}} i\omega_n \left(-\hat{a}_n^- e^{-i\omega_n t} + \hat{a}_n^+ e^{i\omega_n t} \right) \right]^2 |0\rangle$$

$$= \frac{1}{L}\int_0^L dx \sum_{n=1}^{\infty} \frac{(\sin \omega_n x)^2}{2\omega_n} \omega_n^2 = \frac{1}{4}\sum_n \omega_n.$$

The second term yields the same result, and we find

$$\langle 0|\hat{H}|0\rangle = \frac{1}{2}\sum_{n=1}^{\infty} \omega_n.$$

Chapter 11

Exercise 11.1 (p. 137)

(a) Consider a system described by a Hamiltonian $H(\vec{p}, \vec{q})$, where the vector notation stands for a finite number n of degrees of freedom. We follow the derivation of equation (11.15) in the text, replacing p and q by vectors $\vec{p}$, $\vec{q}$ in the n-dimensional space. The evolution operator is approximated as in equation (11.7) as before, and the propagator for a small time interval is

$$K\left(\vec{q}_{k+1}, \vec{q}_k; t_{k+1}, t_k\right) = \langle \vec{q}_{k+1}| \left(1 - \frac{i}{\hbar}\Delta t_k \hat{H} \right) |\vec{q}_k\rangle + O\left(\Delta t_k^2\right).$$

To calculate the matrix element of the Hamiltonian, we now use the decomposition of unity,

$$\hat{1} = \int d^n \vec{p}_k \, |\vec{p}_k\rangle\langle \vec{p}_k|,$$

where we now need to integrate over $d^n \vec{p}_k$ in the n-dimensional momentum space rather than over dp_k in a one-dimensional space. Except for this difference, the calculation proceeds essentially unchanged. We assume that the operators $\hat{\vec{p}}_k$ are ordered to the left of $\hat{\vec{q}}_k$ in the Hamiltonian. Then, similar to equation (11.11), we find

$$\langle\vec{p}_k|\hat{H}|\vec{q}_k\rangle=\frac{1}{(2\pi\hbar)^{n/2}}H(\vec{p}_k,\vec{q}_k)\exp\left(-\frac{i\vec{p}_k\cdot\vec{q}_k}{\hbar}\right),$$

where $\vec{p}_k\cdot\vec{q}_k$ stands for the scalar product of vectors in the n-dimensional space, and the factor $(2\pi\hbar)^{n/2}$ replaces $\sqrt{2\pi\hbar}$. Instead of equation (11.12), we now obtain

$$\langle\vec{q}_{k+1}|\hat{H}|\vec{q}_k\rangle=\int\frac{d^n\vec{p}_k}{(2\pi\hbar)^n}H(\vec{p}_k,\vec{q}_k)\exp\frac{i\vec{p}_k\cdot(\vec{q}_{k+1}-\vec{q}_k)}{\hbar},$$

and then the propagator becomes

$$K\left(\vec{q}_{k+1},\vec{q}_k;t_{k+1},t_k\right)$$
$$=\int\frac{d^n p_k}{(2\pi\hbar)^n}\exp\left[\frac{i\Delta t_k}{\hbar}\left(\vec{p}_k\cdot\frac{\vec{q}_{k+1}-\vec{q}_k}{\Delta t_k}-H(\vec{p}_k,\vec{q}_k)\right)+O\left(\Delta t_k^2\right)\right].$$

Finally, we obtain

$$K\left(\vec{q}_f,\vec{q}_0;t_f,t_0\right)=\int\left[\prod_{k=1}^{N}\frac{d^n\vec{q}_k d^n\vec{p}_k}{(2\pi\hbar)^n}\right]\frac{d^n\vec{p}_0}{(2\pi\hbar)^n}$$
$$\times\exp\left[\sum_{k=0}^{N}\frac{i\Delta t_k}{\hbar}\left(\vec{p}_k\cdot\frac{\vec{q}_{k+1}-\vec{q}_k}{\Delta t_k}-H(\vec{p}_k,\vec{q}_k)\right)+O\left(\Delta t_k^2\right)\right].$$

This expression becomes the path integral in the limit $N\to\infty$,

$$K\left(\vec{q}_f,\vec{q}_0;t_f,t_0\right)=\int\mathcal{D}\vec{p}\mathcal{D}\vec{q}\,\exp\left[\int_{t_0}^{t_f}\frac{i}{\hbar}dt\left(\vec{p}(t)\cdot\dot{\vec{q}}(t)-H(\vec{p}(t),\vec{q}(t))\right)\right],\tag{A4.31}$$

where the integration measure is defined as

$$\mathcal{D}\vec{p}\mathcal{D}\vec{q}\equiv\lim_{N\to\infty}\left[\prod_{k=1}^{N}\frac{d^n\vec{p}_k d^n\vec{q}_k}{(2\pi\hbar)^n}\right]\frac{d^n\vec{p}_0}{(2\pi\hbar)^n}.\tag{A4.32}$$

(b) The transition to an infinite number of degrees of freedom can be performed by treating the field variables $\phi(\mathbf{x},t)\equiv\phi_{\mathbf{x}}(t)$ as an infinite set of time-dependent generalized coordinates $\phi_{\mathbf{x}}$ labeled by the "continuous index" $\mathbf{x}$. The Hamiltonian (4.11) additionally depends on the generalized momenta $\pi_{\mathbf{x}}(t)$, but there is no need to reorder factors of π and ϕ.

We would like to use the results (A4.31)–(A4.32), therefore we need to discretize the space label $\mathbf{x}$, introducing a finite number n of different points $\mathbf{x}_1,\ldots,\mathbf{x}_n$.

The propagator in the discretized approximation is a function of n initial and n final values of ϕ,

$$K(\phi_{\mathbf{x}_1 f}, \dots, \phi_{\mathbf{x}_n f}; \phi_{\mathbf{x}_1 0}, \dots, \phi_{\mathbf{x}_n 0}; t_f, t_0) \equiv K(\phi_f(\mathbf{x}), \phi_0(\mathbf{x}); t_f, t_0).$$

Its path integral representation follows from equations (A4.31)–(A4.32),

$$K(\phi_f(\mathbf{x}), \phi_0(\mathbf{x}); t_f, t_0) = \int \mathcal{D}\phi_{\mathbf{x}} \mathcal{D}\pi_{\mathbf{x}} \exp\left[\int_{t_0}^{t_f} \frac{i}{\hbar} dt \left(\sum_{m=1}^{n} \left(\pi_{\mathbf{x}_m} \dot{\phi}_{\mathbf{x}_m}\right) - H[\pi_{\mathbf{x}}, \phi_{\mathbf{x}}]\right)\right].$$

The integration measure is defined by

$$\mathcal{D}\phi_{\mathbf{x}} \mathcal{D}\pi_{\mathbf{x}} \equiv \lim_{N\to\infty} \frac{\prod_{m=1}^{n} d\pi_{\mathbf{x}_m}(t_0)}{(2\pi\hbar)^n} \prod_{k=1}^{N} \frac{\prod_{m=1}^{n} d\phi_{\mathbf{x}_m}(t_k) d\pi_{\mathbf{x}_m}(t_k)}{(2\pi\hbar)^n}.$$

In the continuous limit ($n \to \infty$) the sum over m becomes an integral,

$$\sum_{m=1}^{n} \pi_{\mathbf{x}_m} \dot{\phi}_{\mathbf{x}_m} \to \int d^3\mathbf{x}\, \pi(\mathbf{x}, t)\, \dot{\phi}(\mathbf{x}, t),$$

while the propagator K becomes a functional of the fields ϕ_f and ϕ_0. Therefore the path integral representation of the propagator is

$$K(\phi_f(\mathbf{x}), \phi_0(\mathbf{x}); t_f, t_0) = \int \mathcal{D}\phi_{\mathbf{x}} \mathcal{D}\pi_{\mathbf{x}} \exp\left[\int_{t_0}^{t_f} \frac{i}{\hbar} dt \left(\int d^3\mathbf{x}\, \pi_{\mathbf{x}}(t) \dot{\phi}_{\mathbf{x}}(t) - H[\pi_{\mathbf{x}}, \phi_{\mathbf{x}}]\right)\right].$$

There will be an additional limit $n \to \infty$ in the integration measure.

Exercise 11.2 (p. 137)

In the case of a time-dependent Hamiltonian, the evolution operator $\hat{U}(t_f, t_0)$ can still be expressed as a product of evolution operators for time intervals,

$$\hat{U}(t_f, t_0) = \hat{U}(t_f, t_n) \hat{U}(t_n, t_{n-1}) \dots \hat{U}(t_1, t_0),$$

where the order is important since these operators do not commute. The propagator can be rewritten as an n-fold integration over q_k as in equation (11.6). The evolution throughout a short time interval Δt_k is approximated as

$$\hat{U}(t_{k+1}, t_k) = 1 - \frac{i\Delta t_k}{\hbar} \hat{H}(\hat{p}, \hat{q}, t_k) + O(\Delta t_k^2),$$

since any corrections due to time dependence of $\hat{H}$ will be of higher order in Δt. Then the derivation of the path integral proceeds as in Chapter 11.

Exercise 11.3 (p. 138)

We follow the derivation of the Lagrangian path integral in the text. The starting point is the discretized form of the Hamiltonian path integral (11.14). We need to reorder the operators in the given Hamiltonian so that $\hat{p}$ is to the left of $\hat{q}$, using the commutation relation

$$[f(\hat{q}), \hat{p}] = i\hbar f'(\hat{q}).$$

We find

$$H(\hat{p},\hat{q}) = \hat{p}^2 f^2 + 2i\hbar \hat{p} f f' + (i\hbar)^2 f f''.$$

Therefore, the propagator is

$$K(q_f, q_0; t_f, t_0) = \lim_{\Delta t \to 0} \int \frac{dp_f}{2\pi\hbar} \prod_{k=1}^{n} \frac{dp_k dq_k}{2\pi\hbar}$$

$$\times \exp\left[\frac{i\Delta t}{\hbar} \sum_{k=0}^{n} \left(p_k \frac{q_{k+1}-q_k}{\Delta t} - p_k^2 f_k^2 - 2i\hbar p_k f_k f_k' + \hbar^2 f_k f_k'' - V_k\right)\right],$$

where we omit the argument q_k and write $V_k \equiv V(q_k)$, $f_k \equiv f(q_k)$, $f_k' \equiv f'(q_k)$, etc. We can now integrate separately over each p_k using the Gaussian formula given in the text, where we set $a \equiv 2if_k^2 \Delta t/\hbar$ and $b = (q_{k+1} - q_k)/\hbar - 2if_k f_k' \Delta t$,

$$\int \frac{dp_k}{2\pi\hbar} \exp\left[\frac{i\Delta t}{\hbar}\left(p_k \frac{q_{k+1}-q_k}{\Delta t} - H(p_k, q_k)\right)\right] = \frac{1}{f_k\sqrt{\pi i \hbar \Delta t}}$$

$$\times \exp\left[-\frac{i\Delta t}{\hbar}\left(V_k - \hbar^2 f_k f_k''\right) - \frac{\hbar}{4if_k^2 \Delta t}\left(\frac{q_{k+1}-q_k}{\hbar} - 2if_k f_k' \Delta t\right)^2\right].$$

This integration is performed $(n+1)$ times over p_k for $k = 0, \ldots, n$, therefore we will obtain the product of $(n+1)$ such terms as shown above. Replacing

$$q_{k+1} - q_k = \dot{q}\Delta t + O(\Delta t^2)$$

and omitting terms of order Δt^2, we get the following expression under the exponential,

$$\frac{i}{\hbar}\left[\frac{1}{4f^2}\left(\dot{q} - 2i\hbar f f'\right)^2 - V + \hbar^2 f f''\right]\Delta t.$$

In the continuous limit $N \to \infty$, the sum over Δt becomes an integral,

$$\frac{i}{\hbar}\int dt \left[\frac{(\dot{q} - 2i\hbar f f')^2}{4f^2} - V + \hbar^2 f f''\right]$$

$$= \frac{i}{\hbar}\int dt \left[\frac{\dot{q}^2}{4f^2} - V + \hbar^2\left(f f'' - f'^2\right)\right],$$

where we integrated by parts using

$$\frac{d}{dt}\ln f(q) = \frac{f'}{f}\dot{q}$$

and discarded the boundary terms. Therefore in the limit $N \to \infty$ we obtain the path integral

$$K\left(q_f, q_0; t_f, t_0\right) = \int \mathcal{D}q \exp\left[\frac{i}{\hbar}\int dt \left(\frac{\dot{q}^2}{4f^2} - V - \hbar^2\left(f'^2 - ff''\right)\right)\right]$$

and the integration measure

$$\mathcal{D}q \equiv \lim_{N\to\infty} \frac{1}{f(q_0)\sqrt{\pi i\hbar\Delta t_0}} \prod_{k=1}^{N} \frac{dq_k}{f(q_k)\sqrt{\pi i\hbar\Delta t_k}}.$$

Exercise 11.4 (p. 138)

We use the results of Exercise 11.1(b). It is clear from the derivation of the Lagrangian path integral that the propagator for a system with n degrees of freedom $\vec{q}(t)$ and a Hamiltonian of the form

$$H(\vec{p}, \vec{q}) = \frac{\vec{p}\cdot\vec{p}}{2m} - V(\vec{q})$$

is expressed as

$$K\left(\vec{q}_f, \vec{q}_0; t_f, t_0\right) = \int \mathcal{D}\vec{q} \exp\left[\frac{i}{\hbar}\int_{t_0}^{t_f} dt \left(m\frac{\dot{\vec{q}}\cdot\dot{\vec{q}}}{2} - V(\vec{q})\right)\right],$$

where the integration measure is

$$\mathcal{D}\vec{q} \equiv \lim_{N\to\infty} \frac{1}{\left(2\pi i\hbar m^{-1}\Delta t_0\right)^{n/2}} \prod_{k=1}^{N} \frac{d^n\vec{q}_k}{\left(2\pi i\hbar m^{-1}\Delta t_k\right)^{n/2}}.$$

This is straightforwardly generalized to the case of a scalar field $\phi(\mathbf{x}, t)$ with the Hamiltonian (4.11). A discretized approximation to the path integral is performed as in Exercise 11.1(b) by introducing n points $\mathbf{x}_1, \ldots, \mathbf{x}_n$ and the n field variables $\phi_{\mathbf{x}_k}(t)$, $k = 1, \ldots, n$, playing the role of the components of the vector $\vec{q}(t)$. Therefore the path integral representation of the propagator is

$$K[\phi_f(\mathbf{x}), \phi_0(\mathbf{x}); t_f, t_0] = \int \mathcal{D}\phi_{\mathbf{x}} \exp\left[\int_{t_0}^{t_f} \frac{i}{2\hbar} dt\, d^3\mathbf{x} \left(\dot{\phi}_{\mathbf{x}}^2 - (\nabla\phi_{\mathbf{x}})^2 - m^2\phi_{\mathbf{x}}^2\right)\right],$$

where the integration measure is

$$\mathcal{D}\phi_{\mathbf{x}} \equiv \lim_{N\to\infty} \lim_{n\to\infty} \frac{1}{(2\pi i\hbar\Delta t_0)^{n/2}} \prod_{k=1}^{N} \frac{\prod_{m=1}^{n} d\phi_{\mathbf{x}_m}(t_k)}{(2\pi i\hbar\Delta t_k)^{n/2}}.$$

Exercise 11.5 (p. 142)

Let us take as the two fundamental independent solutions u and v those that satisfy the boundary conditions

$$u_i = 1, \quad u_f = 0;$$

$$v_i = 0, \quad v_f = 1.$$

They obviously satisfy (11.41); it follows from (11.43) that $W = \dot{v}_i = -\dot{u}_f$. The expression in (11.44) then becomes

$$K = \sqrt{\frac{W}{2\pi i\hbar}} \exp\left[\frac{i}{2\hbar}\left(\dot{v}_f q_f^2 - \dot{u}_i q_i^2 - 2W q_i q_f\right)\right].$$

The solution of the classical equation with the boundary conditions $q(t_i) = q_i$ and $q(t_f) = q_f$ can be written as

$$q(t) = q_i u(t) + q_f v(t),$$

and the action along this classical trajectory is

$$S_{\text{cl}} = \frac{1}{2} q\dot{q}\Big|_{t_i}^{t_f} = \frac{1}{2}\left(\dot{v}_f q_f^2 - \dot{u}_i q_i^2 + (\dot{u}_f - \dot{v}_i) q_i q_f\right).$$

Hence

$$\frac{\partial^2 S_{\text{cl}}}{\partial q_f \partial q_i} = \frac{1}{2}(\dot{u}_f - \dot{v}_i) = -W,$$

and it is clear that the expression for K can be written as (11.45).

Exercise 11.6 (p. 144)

The solution consists of essentially the same calculation as in Section 11.4.2, except for Euclidean quantities replacing Lorentzian ones. See also the paper by L. Moriconi, *Am. J. Phys.* **72** (2004), 1258 (preprint `arxiv:physics/0402069`).

Choose the solution $u(\tau)$ of the Euclidean equation of motion, $\ddot{u} - \omega^2 u = 0$, as

$$u(\tau) = e^{-\omega\tau},$$

and introduce new variables Q and η instead of x and τ by

$$Q(\tau) \equiv u^{-1}x(\tau) = e^{\omega\tau}x(\tau), \quad \eta \equiv \frac{1}{2\omega}e^{2\omega\tau}, \quad e^{2\omega\tau}d\tau = d\eta.$$

This is a well-defined change of variables since $u \neq 0$ and $d\eta/d\tau \neq 0$ for all τ. The function $Q(\tau)$ can be expressed through η and is denoted again $Q(\eta)$; then we have

$$\dot{x} = \frac{d}{d\tau}\left(e^{-\omega\tau}Q\right) = e^{-\omega\tau}\left(\frac{dQ}{d\tau} - \omega Q\right)$$
$$= e^{-\omega\tau}\left(e^{2\omega\tau}\frac{dQ}{d\eta} - \omega Q\right) = e^{\omega\tau}Q' - \omega x,$$

where we denoted the derivative with respect to η by a prime. Now we can rewrite the classical Euclidean action through $Q(\eta)$,

$$S_{\mathrm{E}}[x(\tau)] = \frac{m}{2}\int_{\tau_i}^{\tau_f}\left(\dot{x}^2 + \omega^2 x^2\right) d\tau = \frac{m}{2}\int_{\tau_i}^{\tau_f}\left((\dot{x}+\omega x)^2 - 2\omega x\dot{x}\right) d\tau$$
$$= \frac{m}{2}\int_{\eta_i}^{\eta_f} Q'^2 d\eta - \frac{m\omega}{2}\left(x^2(\tau_f) - x^2(\tau_i)\right)$$
$$\equiv S_{\mathrm{E}}^{(\mathrm{free})}\left[Q(\eta)\right] - \frac{m\omega}{2}\left(x^2(\tau_f) - x^2(\tau_i)\right),$$

where

$$S_{\mathrm{E}}^{(\mathrm{free})} \equiv \frac{m}{2}\int_{\eta_i}^{\eta_f} Q'^2 d\eta$$

is the action of a free particle of mass m. Apart from the terms depending on the boundary conditions $x_i \equiv x(\tau_i)$ and $x_f \equiv x(\tau_f)$, the action $S_{\mathrm{E}}[x]$ is now reduced to the action for a free particle described by the coordinate $Q(\eta)$.

The next step is to relate the measure $\mathcal{D}x$ in the Euclidean path integral (11.48),

$$\mathcal{D}x = \lim_{N\to\infty}\frac{1}{\sqrt{2\pi\hbar\Delta\tau_0}}\prod_{k=1}^{N}\frac{dx_k}{\sqrt{2\pi\hbar\Delta\tau_k}},$$

to the corresponding measure for the variable $Q(\eta)$,

$$\mathcal{D}Q = \lim_{N\to\infty}\frac{1}{\sqrt{2\pi\hbar\Delta\eta_0}}\prod_{k=1}^{N}\frac{dQ_k}{\sqrt{2\pi\hbar\Delta\eta_k}}.$$

The relationship between the intervals $\Delta\eta_k$ and $\Delta\tau_k$ needs to be derived up to second order in $\Delta\tau_k$. This is done by expanding $\eta = \frac{1}{2\omega}e^{2\omega\tau}$ in Taylor series,

$$\Delta\eta_k = e^{2\omega\tau_k}\Delta\tau_k + \omega e^{2\omega\tau_k}\Delta\tau_k^2 + O(\Delta\tau_k^3)$$
$$= e^{2\omega\tau_k}\Delta\tau_k e^{\omega\Delta\tau_k} + O(\Delta\tau_k^3).$$

Note that we replaced $1+\omega\Delta\tau_k$ by $e^{\omega\Delta\tau_k}$, which is admissible up to terms of order $\Delta\tau_k^2$. Hence

$$\prod_{k=0}^{N}\Delta\eta_k=\prod_{k=0}^{N}\Delta\tau_k e^{2\omega\tau_k+\omega\Delta\tau_k}=e^{\omega(\tau_f-\tau_i)}\prod_{k=0}^{N}\Delta\tau_k e^{2\omega\tau_k},$$

where we relabeled $\tau_0\equiv\tau_i$ and used the property $\sum_{k=0}^{N}\Delta\tau_k=\tau_f-\tau_i$. Since $dQ_k=e^{\omega\tau_k}dx_k$, we have

$$\begin{aligned}\mathcal{D}Q&=\lim_{N\to\infty}\prod_{k=0}^{N}\frac{1}{\sqrt{2\pi\hbar\Delta\eta_k}}\prod_{k=1}^{N}dQ_k\\&=\exp\left(-\frac{\omega}{2}\left(\tau_f-\tau_i\right)\right)\lim_{N\to\infty}\frac{\prod_{k=0}^{N}e^{-\omega\tau_k}}{\sqrt{2\pi\hbar\Delta\tau_0}}\prod_{k=1}^{N}\frac{dx_k e^{\omega\tau_k}}{\sqrt{2\pi\hbar\Delta\tau_k}}\\&=\exp\left(-\frac{\omega}{2}\left(\tau_f-\tau_i\right)-\omega\tau_i\right)\mathcal{D}x=\exp\left(-\omega\frac{\tau_f+\tau_i}{2}\right)\mathcal{D}x.\end{aligned}$$

Finally, we can compute the path integral,

$$\begin{aligned}&\int_{x(\tau_i)=x_i}^{x(\tau_f)=x_f}\mathcal{D}x\exp\left(-\frac{S_{\mathrm{E}}[x(\tau)]}{\hbar}\right)\\&=\exp\left(\omega\frac{\tau_f+\tau_i}{2}+\frac{m\omega}{2\hbar}\left(x_f^2-x_i^2\right)\right)\\&\quad\times\int_{Q(\eta_i)=x_i e^{\omega\tau_i}}^{Q(\eta_f)=x_f e^{\omega\tau_f}}\mathcal{D}Q\exp\left(-\frac{S_{\mathrm{E}}^{(\mathrm{free})}[Q(\eta)]}{\hbar}\right).\end{aligned}$$

It remains to use the known formula for the free propagator,

$$\begin{aligned}&\int_{Q(\eta_i)=Q_i}^{Q(\eta_f)=Q_f}\mathcal{D}Q\exp\left(-\frac{S_{\mathrm{E}}^{(\mathrm{free})}[Q(\eta)]}{\hbar}\right)\\&=\frac{1}{\sqrt{2\pi\hbar\left(\eta_f-\eta_i\right)}}\exp\left(-\frac{m\left(Q_f-Q_i\right)^2}{2\hbar\left(\eta_f-\eta_i\right)}\right),\end{aligned}$$

and to substitute the values η_i, η_f, Q_i, Q_f:

$$\eta_f-\eta_i=\frac{1}{2\omega}\left(e^{2\omega\tau_f}-e^{2\omega\tau_i}\right),\quad Q_i=x_i e^{\omega\tau_i},\quad Q_f=x_f e^{\omega\tau_f}.$$

To simplify calculations, we note that

$$\eta_f - \eta_i = \frac{1}{\omega} e^{\omega(\tau_f+\tau_i)} \sinh\left(\omega\left(\tau_f - \tau_i\right)\right) \equiv \frac{1}{\omega} e^{\omega(\tau_f+\tau_i)} \sinh \omega T.$$

Thus, we obtain

$$\int_{x(\tau_i)=x_i}^{x(\tau_f)=x_f} \mathcal{D}x \exp\left(-\frac{S_{\mathrm{E}}[x(\tau)]}{\hbar}\right)$$

$$= \frac{\sqrt{\omega}}{\sqrt{2\pi\hbar \sinh \omega T}} \exp\left[\frac{m\omega}{2\hbar}\left(x_f^2 - x_i^2\right) - \frac{m\omega\left(x_f e^{\omega\tau_f} - x_i e^{\omega\tau_i}\right)^2}{\hbar e^{\omega(\tau_f+\tau_i)} \sinh \omega T}\right],$$

which coincides with the formula (11.50) after some rearrangements and upon setting $m = 1$.

Chapter 12

Exercise 12.1 (p. 147)

The general solution of an inhomogeneous equation such as equation (12.1) is a sum of a particular solution and the general solution of the homogeneous equation. We need to use the boundary conditions to select the correct solution.

Elementary solution For $t \neq t'$, i.e. separately in the two domains $t > t'$ and $t < t'$, the Green's function satisfies the homogeneous equation

$$\left(\frac{\partial^2}{\partial t^2} + \omega^2\right) G_{\mathrm{ret}}(t, t') = 0.$$

The general solution is

$$G_{\mathrm{ret}}(t, t') = A \sin \omega (t - \alpha),$$

where A and α are constants that are different for $t > t'$ and for $t < t'$. So we may write

$$G_{\mathrm{ret}}(t, t') = \begin{cases} A_- \sin \omega\left(t - \alpha_-\right), & t < t' \\ A_+ \sin \omega\left(t - \alpha_+\right), & t > t' \end{cases}$$

$$= A_- \sin \omega\left(t - \alpha_-\right) \theta(t' - t) + A_+ \sin \omega\left(t - \alpha_+\right) \theta(t - t').$$

The boundary condition $G_{\mathrm{ret}}(t, t') = 0$ for $t < t'$ forces $A_- = 0$. Therefore by continuity $G_{\mathrm{ret}}(t', t') = 0$ and $\alpha_+ = t'$. To find A_+, we integrate equation (12.1) over a small interval of t around t' and obtain

$$\int_{t'-\Delta t}^{t'+\Delta t} \left[\frac{\partial^2 G_{\mathrm{ret}}}{\partial t^2} + \omega^2 G_{\mathrm{ret}}\right] dt = \int_{t'-\Delta t}^{t'+\Delta t} \delta(t - t') dt = 1.$$

For small $\Delta t \to 0$, this gives

$$\lim_{t\to t'+0} \frac{\partial G}{\partial t} - \lim_{t\to t'-0} \frac{\partial G}{\partial t} = 1.$$

Therefore $A_+ = \omega^{-1}$, and we find the required solution.

Solution using Fourier transforms A Fourier transform of equation (12.1) defines the Fourier image $g(\Omega)$,

$$g(\Omega) = \int_{-\infty}^{+\infty} dt\, G_{\text{ret}}(t,t') e^{-i\Omega(t-t')}.$$

The function $g(\Omega)$ must satisfy the equation

$$g(\Omega)\left(\omega^2 - \Omega^2\right) = 1. \tag{A4.33}$$

Here $g(\Omega)$ must be treated as a distribution (see Appendix A1.1). The general solution of equation (A4.33) in the space of distributions is

$$g(\Omega) = \mathcal{P}\frac{1}{\omega^2 - \Omega^2} + a_+\delta(\omega - \Omega) + a_-\delta(\omega + \Omega), \tag{A4.34}$$

where $\mathcal{P}$ denotes the Cauchy principal value and $a_\pm$ are unknown constants.

The general form of Green's function with arbitrary constants corresponds to the freedom of choosing a solution of the homogeneous equation. The values $a_\pm$ must be determined from the boundary condition $G_{\text{ret}}(t,t') = 0$ for $t < t'$. The inverse Fourier transform of equation (A4.34) gives

$$\begin{aligned} G_{\text{ret}}(t,t') &= \frac{1}{2\pi}\int_{-\infty}^{+\infty} d\Omega e^{i\Omega(t-t')} g(\Omega) \\ &= \frac{1}{2\pi}\left[\mathcal{P}\int_{-\infty}^{+\infty} \frac{e^{i\Omega(t-t')}}{\omega^2-\Omega^2} d\Omega + a_+ e^{i\omega(t-t')} + a_- e^{-i\omega(t-t')}\right]. \end{aligned}$$

This expression confirms our expectation that the terms with $a_\pm$ represent the as-yet unspecified solution of the homogeneous oscillator equation. Now the principal value integral is computed using contour integration. For $t < t'$ the contour must be deformed into the lower half-plane Im $\Omega < 0$, while for $t > t'$ one must use the upper half-plane. We find

$$\frac{1}{2\pi}\mathcal{P}\int_{-\infty}^{+\infty} \frac{e^{i\Omega(t-t')}}{\omega^2-\Omega^2} d\Omega = \text{sign}(t-t')\frac{\sin\omega(t-t')}{2\omega} = \frac{1}{2\omega}\sin\omega\left|t-t'\right|.$$

To satisfy the boundary conditions, the constants must be chosen as

$$a_\pm = \pm\frac{\pi}{2i\omega}, \tag{A4.35}$$

hence

$$G_{\text{ret}}(t,t') = \theta(t-t')\frac{\sin\omega(t-t')}{\omega}. \tag{A4.36}$$

Exercise 12.2 (p. 149)

An elementary solution (without using Fourier transforms) can be found as in Exercise 12.1.

Solution using Fourier transforms A Fourier transform of equation (12.13) yields

$$g(\Omega)\left(\omega^2+\Omega^2\right)=1$$

for the Fourier image $g(\Omega)$ defined by

$$g(\Omega)=\int_{-\infty}^{+\infty} d\tau\, G_E(\tau,\tau')e^{-i\Omega(\tau-\tau')}. \tag{A4.37}$$

We obtain

$$g(\Omega)=\frac{1}{\Omega^2+\omega^2}.$$

Now, $g(\Omega)$ does not need to be treated as a distribution since there are no poles on the real Ω line. The inverse Fourier transform presents no problems,

$$G_E(\tau,\tau')=\frac{1}{2\pi}\int_{-\infty}^{+\infty} d\Omega e^{i\Omega(\tau-\tau')}g(\Omega)=\frac{1}{2\pi}\int_{-\infty}^{+\infty} d\Omega\frac{e^{i\Omega(\tau-\tau')}}{\Omega^2+\omega^2}.$$

The integral is evaluated using contour integration and yields the answer (12.12). The boundary condition $G_E(\tau\to\pm\infty,\tau')=0$ is satisfied automatically. In fact, by treating $g(\Omega)$ as a usual function rather than a distribution we *implicitly assumed* that the Green's function $G_E(\tau,\tau')$ tends to zero at large $|\tau|$. If $G_E(\tau,\tau')$ did not tend to zero at large $|\tau|$, the Fourier transform (A4.37) would not exist in the usual sense (or $g(\Omega)$ would have to be treated as a distribution).

Exercise 12.3 (p. 158)

According to the approach explained in the main text, we expect to compute the in-out matrix element by considering the ratio of the path integrals

$$\frac{\int a^+(t)a^-(t)e^{iS[q,J]}\mathcal{D}q}{\int e^{iS[q,J]}\mathcal{D}q} \tag{A4.38}$$

and by replacing the Feynman Green's function $G_F(t,t')$ by the retarded Green's function $G_{\text{ret}}(t,t')$ in the effective action. To compute the path integrals in equation (A4.38), we consider the action

$$S\left[q,J^+,J^-\right]=\int\left(\frac{1}{2}\dot{q}^2-\frac{\omega^2}{2}q^2+J^+a^++J^-a^-\right)dt,$$

where $J^+(t)$ and $J^-(t)$ are two external forces. This action is real-valued since $J^- = (J^+)^*$. The Lorentzian effective action is

$$\exp\left(i\Gamma_L\left[J^+, J^-\right]\right) = \int \mathcal{D}q\, e^{iS[q,J^+,J^-]},$$

where the integration is over all paths $q(t)$ satisfying $q(t \to \pm\infty) = 0$, as in the main text. If we compute this effective action, the path integral ratio (A4.38) will be

$$\exp\left[-i\Gamma_L\right] \frac{\delta}{i\delta J^+(t)} \frac{\delta}{i\delta J^-(t)} \exp\left[i\Gamma_L\right]. \tag{A4.39}$$

We now express $a^\pm$ through q and $\dot{q}$ as

$$a^+(t) = \sqrt{\frac{\omega}{2}}\left(q(t) - \frac{i}{\omega}\dot{q}(t)\right),$$

$$a^-(t) = \sqrt{\frac{\omega}{2}}\left(q(t) + \frac{i}{\omega}\dot{q}(t)\right).$$

Then after integrations by parts we obtain

$$\int \left(J^+a^+ - J^-a^-\right) dt = \int \sqrt{\frac{\omega}{2}}\left(J^+ + \frac{i}{\omega}\frac{dJ^+}{dt} + J^- - \frac{i}{\omega}\frac{dJ^-}{dt}\right) q(t)dt,$$

and therefore the Lorentzian effective action can be copied from the text,

$$\Gamma_L[J] = \frac{1}{2}\iint J(t_1)\, J(t_2)\, G_F(t_1, t_2)\, dt_1 dt_2, \tag{A4.40}$$

if we use for $J(t)$ the expression

$$J(t) \equiv \sqrt{\frac{\omega}{2}}\left(J^+ + \frac{i}{\omega}\frac{dJ^+}{dt} + J^- - \frac{i}{\omega}\frac{dJ^-}{dt}\right). \tag{A4.41}$$

Now we need to substitute equation (A4.40) into equation (A4.39), using $J(t)$ given by equation (A4.41), and then replace G_F by G_{ret}. Then we will find the required matrix element according to the recipe presented in the text.

First, the expression (A4.39) is simplified to

$$\frac{\delta\Gamma_L}{\delta J^+(t)} \frac{\delta\Gamma_L}{\delta J^-(t')} - i\frac{\delta^2\Gamma_L}{\delta J^+(t)\delta J^-(t')}.$$

The functional derivatives are evaluated like this,

$$\frac{\delta\Gamma_L}{\delta J^+(t)} = \int d\tilde{t}\frac{\delta\Gamma_L[J]}{\delta J(\tilde{t})}\frac{\delta J(\tilde{t})}{\delta J^+(t)}.$$

Using equation (A4.40) and the fact that $G_F(t,t')$ is a symmetric function of t and t', we find

$$\frac{\delta\Gamma_L}{\delta J(t)} = \int J(t_1)\, G_F(t,t_1)\, dt_1.$$

To compute the functional derivative $\delta J/\delta J^+$, we write J as a functional of J^+ in an integral form:

$$J(\tilde{t}) = \int J^+(t)\sqrt{\frac{\omega}{2}}\left[\delta(t-\tilde{t}) - \frac{i}{\omega}\delta'(t-\tilde{t})\right] dt + c.c.,$$

where "$c.c.$" denotes the complex conjugate terms with $J^- = (J^+)^*$. Then

$$\frac{\delta J(\tilde{t})}{\delta J^+(t)} = \sqrt{\frac{\omega}{2}}\left[\delta(t-\tilde{t}) - \frac{i}{\omega}\delta'(t-\tilde{t})\right],$$

and we obtain

$$\frac{\delta\Gamma_L}{\delta J^+(t)} = \sqrt{\frac{\omega}{2}}\int dt_1 J(t_1)\left[G_F(t,t_1) - \frac{i}{\omega}\frac{\partial}{\partial t}G_F(t,t_1)\right].$$

Now replacing G_F by G_{ret} and simplifying

$$G_{ret}(t,t_1) - \frac{i}{\omega}\frac{\partial}{\partial t}G_{ret}(t,t_1) = \theta(t-t_1)\frac{e^{i\omega(t-t_1)}}{i\omega},$$

we get

$$\frac{\delta\Gamma_L}{\delta J^+(t)} = -\frac{i}{\sqrt{2\omega}}\int_0^T dt_1 J(t_1)\, e^{i\omega t_1} \equiv -J_0.$$

The functional derivative $\delta\Gamma_L/\delta J^-(t)$ is the complex conjugate of this expression, so

$$\frac{\delta\Gamma_L}{\delta J^+(t)}\frac{\delta\Gamma_L}{\delta J^-(t)} = |J_0|^2.$$

The second functional derivative

$$\frac{\delta^2\Gamma_L}{\delta J^+(t_1)\,\delta J^-(t_2)}$$

yields terms independent of J because $\Gamma_L[J]$ is a quadratic functional of J. But the expectation value we are computing cannot have any terms independent of J since it should be equal to 0 when $J \equiv 0$. Therefore any terms we get from this functional derivative are spurious and we ignore them. Note that one of the ignored terms is proportional to $\delta(t_1 - t_2)$ and would diverge for $t_1 = t_2$.

Finally, we obtain

$$\langle 0_{in}|\, \hat{a}^+(t)\hat{a}^-(t)\, |0_{in}\rangle = |J_0|^2.$$

This agrees with the answer obtained in equation (3.20).

Chapter 13

Exercise 13.1 (p. 171)

We omit the trivial term $-V(x)\delta(x-x')$ from $\hat{M}$, compute the generalized function $g^{1/4}\Box_{g(x)}g^{-1/4}\delta(x-x')$ and substitute $g_{\alpha\beta}=\delta_{\alpha\beta}+h_{\alpha\beta}$ into the result. Denoting $\partial_\mu\equiv\partial/\partial x^\mu$ and suppressing the arguments of $\delta(x-x')$ for brevity, we find

$$\partial_\mu\left(g^{-1/4}\delta\right)=g^{-1/4}\left[\delta_{,\mu}-\frac{1}{4}(\ln g)_{,\mu}\delta\right],$$

$$g^{-1/4}\partial_\nu\left[g^{\mu\nu}\sqrt{g}\partial_\mu\left(g^{-1/4}\delta\right)\right]=g^{\mu\nu}\left[\delta_{,\mu}-\frac{1}{4}(\ln g)_{,\mu}\delta\right]_{,\nu}$$
$$+\left[g^{\mu\nu}_{,\nu}+\frac{1}{4}g^{\mu\nu}(\ln g)_{,\nu}\right]\left[\delta_{,\mu}-\frac{1}{4}(\ln g)_{,\mu}\delta\right].$$

This expression splits into terms with different orders of derivatives of $\delta(x-x')$. The derivatives of $g_{\alpha\beta}$ are replaced with the derivatives of $h_{\alpha\beta}$, but otherwise we keep $g_{\alpha\beta}$. (Therefore we do not actually need to make any approximations in this calculation and in particular do not need to assume that $h_{\alpha\beta}$ is small.) The term with the second derivative is

$$g^{\mu\nu}\delta_{,\mu\nu}=\delta^{\mu\nu}\partial_\mu\partial_\nu\delta(x-x')+h^{\mu\nu}\partial_\mu\partial_\nu\delta(x-x').$$

This corresponds to the operator expression $\Box+\hat{h}$. The term with the first derivative of δ is

$$-\frac{1}{4}g^{\mu\nu}(\ln g)_{,\mu}\delta_{,\nu}+g^{\mu\nu}_{,\nu}\delta_{,\mu}+\frac{1}{4}g^{\mu\nu}(\ln g)_{,\nu}\delta_{,\mu}=h^{\mu\nu}_{,\mu}\delta_{,\nu}.$$

This corresponds to the operator $\hat{\Gamma}$. Finally, the term without derivatives of $\delta(x-x')$ is $P(x)\delta(x-x')$, where

$$P(x)=-\frac{1}{4}g^{\mu\nu}(\ln g)_{,\mu\nu}-\frac{1}{4}\left[g^{\mu\nu}_{,\nu}+\frac{1}{4}g^{\mu\nu}(\ln g)_{,\nu}\right](\ln g)_{,\mu}$$
$$=-\frac{1}{4}g^{\mu\nu}g^{\alpha\beta}h_{\alpha\beta,\mu\nu}-\frac{1}{4}g^{\mu\nu}h^{\alpha\beta}_{,\mu}h_{\alpha\beta,\nu}$$
$$-\frac{1}{4}h^{\mu\nu}_{,\nu}g^{\alpha\beta}h_{\alpha\beta,\mu}-\frac{1}{16}g^{\mu\nu}g^{\alpha\beta}g^{\kappa\lambda}h_{\alpha\beta,\mu}h_{\kappa\lambda,\nu}.$$

Here we substituted

$$(\ln g)_{,\mu}=g^{\alpha\beta}g_{\alpha\beta,\mu}=g^{\alpha\beta}h_{\alpha\beta,\mu}.$$

It remains to add the omitted term $-V(x)\delta(x-x')$ to $\hat{P}$ to obtain the required result.

Exercise 13.2 (p. 175)

We rewrite the argument of the exponential as a complete square,

$$\begin{aligned}
&-A\,|\mathbf{x}-\mathbf{a}|^2 - B\,|\mathbf{x}-\mathbf{b}|^2 + 2\mathbf{c}\cdot\mathbf{x} \\
&= -(A+B)\,|\mathbf{x}|^2 + 2\mathbf{x}\cdot(A\mathbf{a}+B\mathbf{b}+\mathbf{c}) - \left(Aa^2+Bb^2\right) \\
&\equiv -(A+B)\,|\mathbf{x}-\mathbf{p}|^2 + Q.
\end{aligned}$$

Here we introduced the auxiliary vector $\mathbf{p}$ and the constant Q:

$$\mathbf{p} \equiv \frac{A\mathbf{a}+B\mathbf{b}+\mathbf{c}}{A+B}, \quad Q \equiv (A+B)p^2 - \left(Aa^2+Bb^2\right).$$

The Gaussian integration yields the required expression:

$$\int d^{2\omega}\mathbf{x}\,\exp\left[-(A+B)\,|\mathbf{x}-\mathbf{p}|^2 + Q\right] = \frac{\pi^\omega \exp[Q]}{(A+B)^\omega},$$

$$Q = -\frac{AB}{A+B}\,|\mathbf{a}-\mathbf{b}|^2 + 2\mathbf{c}\cdot\frac{A\mathbf{a}+B\mathbf{b}}{A+B} + \frac{|\mathbf{c}|^2}{A+B}.$$

Exercise 13.3 (p. 175)

Following the method used in the text for the calculation of $\langle x|\,\hat{K}_1^\Gamma\,|y\rangle$, we find

$$\langle x|\,\hat{K}_1^h(\tau)\,|y\rangle = \frac{\partial^2}{\partial y^\mu \partial y^\nu}\,\langle x|\,\hat{K}_1^P(\tau)\,|y\rangle\bigg|_{P(z)\to h^{\mu\nu}(z)}.$$

Now we use equation (13.19) to evaluate the second derivative and then substitute $y=x$. We find

$$\frac{\partial^2}{\partial y^\mu \partial y^\nu}\bigg|_{y=x} \exp\left[-\frac{(x-y)^2}{4\tau}\right] = -\frac{\delta_{\mu\nu}}{2\tau},$$

$$\frac{\partial^2}{\partial y^\mu \partial y^\nu}\bigg|_{y=x} h^{\mu\nu}\left(x+\frac{\tau-\tau'}{\tau}(y-x)\right) = \left(\frac{\tau-\tau'}{\tau}\right)^2 h^{\mu\nu}_{\ ,\mu\nu}.$$

The terms with first derivatives are proportional to $(x^\mu - y^\mu)$ and vanish in the limit $y\to x$. Therefore we obtain the required expression.

Exercise 13.4 (p. 177)

Note that $h^{\mu\nu} = -h_{\alpha\beta}g^{\mu\alpha}g^{\nu\beta} + O(h^2)$ and since we may omit terms of order $O(h^2)$, we can convert covariant components $h_{\mu\nu}$ to contravariant $h^{\mu\nu}$ by a change of sign.

The first required identity is derived by

$$\frac{\partial\sqrt{g}}{\partial g^{\mu\nu}} = -\frac{1}{2}\sqrt{g}g_{\mu\nu} \Rightarrow \sqrt{\det\left(\delta_{\mu\nu}+h_{\mu\nu}\right)} = 1-\frac{1}{2}\delta_{\mu\nu}h^{\mu\nu}+O\left(h^2\right).$$

Expanding the metric according to equation (13.1), we get

$$\Gamma^{\nu}_{\alpha\beta} = \frac{1}{2}g^{\mu\nu}\left(h_{\alpha\mu,\beta}+h_{\beta\mu,\alpha}-h_{\alpha\beta,\mu}\right),$$

$$g^{\alpha\beta}\partial_{\nu}\Gamma^{\nu}_{\alpha\beta} = \delta^{\alpha\beta}\delta^{\mu\nu}\left(h_{\alpha\mu,\beta\nu}-\frac{1}{2}h_{\alpha\beta,\mu\nu}\right)+O(h^2),$$

$$g^{\alpha\beta}\partial_{\beta}\Gamma^{\nu}_{\alpha\nu} = \frac{1}{2}\delta^{\alpha\beta}\delta^{\mu\nu}h_{\mu\nu,\alpha\beta}+O(h^2).$$

Using equation (5.17), we compute the scalar curvature as

$$\begin{aligned} R &= g^{\alpha\beta}\partial_{\nu}\Gamma^{\nu}_{\alpha\beta}-g^{\alpha\beta}\partial_{\beta}\Gamma^{\nu}_{\alpha\nu}+O(h^2) \\ &= \delta^{\alpha\beta}\delta^{\mu\nu}\left(h_{\alpha\mu,\beta\nu}-h_{\alpha\beta,\mu\nu}\right)+O(h^2) \\ &= -\delta_{\alpha\beta}\delta_{\mu\nu}\left(h^{\alpha\mu,\beta\nu}-h^{\alpha\beta,\mu\nu}\right)+O(h^2) \\ &= \delta_{\alpha\beta}\Box h^{\alpha\beta}-h^{\mu\nu}_{\ \ ,\mu\nu}+O(h^2). \end{aligned}$$

Here we have used the relation $g^{\alpha\beta}=\delta^{\alpha\beta}+O(h)$.

Exercise 13.5 (p. 178)

The required values are found by computing the following limits:

$$f_1(0) = \lim_{\xi\to 0}\int_0^1 e^{-\xi u(1-u)}du = 1;$$

$$\lim_{\xi\to 0}\frac{f_1(\xi)-1}{\xi} = \left.\frac{df_1}{d\xi}\right|_{\xi=0} = -\int_0^1 u(1-u)du = -\frac{1}{6};$$

$$\lim_{\xi\to 0}\frac{f_1(\xi)-1+\frac{1}{6}\xi}{\xi^2} = \frac{1}{2}\left.\frac{d^2 f_1}{d\xi^2}\right|_{\xi=0} = \frac{1}{2}\int_0^1 u^2(1-u)^2du = \frac{1}{60}.$$

Chapter 14

Exercise 14.1 (p. 185)

The task is to verify the integral

$$I_0 \equiv \frac{1}{32}\int_0^{\infty}dx\left(f_1(x)+4\frac{f_1(x)-1}{x}+12\frac{f_1(x)-1+\frac{1}{6}x}{x^2}\right) = \frac{1}{12},$$

where the function $f_1(x)$ is defined by

$$f_1(x) \equiv \int_0^1 dt\, e^{-xt(1-t)}.$$

Since

$$\int_0^1 t(1-t)dt = \frac{1}{6},$$

we may rewrite I_0 as

$$I_0 = \frac{1}{32}\int_0^\infty dx \int_0^1 dt \left(e^{-xt(1-t)} + 4\frac{e^{-xt(1-t)} - 1}{x} + 12\frac{e^{-xt(1-t)} - 1 + xt(1-t)}{x^2} \right).$$

It is impossible to exchange the order of integration because of the nonuniform convergence of the double integral at large x. Therefore we add a regularization factor e^{-ax} with $a > 0$ and evaluate the limit $a \to 0$ at the end of the calculation,

$$I_0 = \lim_{a\to 0} \frac{1}{32}\int_0^1 dt \int_0^\infty dx\, e^{-ax} \left(e^{-xt(1-t)} + 4\frac{e^{-xt(1-t)} - 1}{x} + 12\frac{e^{-xt(1-t)} - 1 + xt(1-t)}{x^2} \right).$$

Let us denote $q \equiv t(1-t)$. Then the integration over x can be performed using the auxiliary integrals

$$I_1(a, q) \equiv \int_0^\infty dx\, e^{-ax} \frac{e^{-qx} - 1}{x} = \ln \frac{a}{a+q},$$

$$I_2(a, q) \equiv \int_0^\infty dx\, e^{-ax} \frac{e^{-qx} - 1 + qx}{x^2} = -q - (a+q)\ln \frac{a}{a+q}.$$

The functions $I_{1,2}(a, q)$ are easily found by integrating the equations

$$\frac{\partial I_1}{\partial q} = -\frac{1}{a+q}, \quad I_1(a, q=0) = 0;$$

$$\frac{\partial I_2}{\partial q} = -I_1(a, q), \quad I_2(a, q=0) = 0.$$

Then we express the integral I_0 as

$$I_0 = \frac{1}{32} \lim_{a \to 0} \int_0^1 dt \left(\frac{1}{a+t(1-t)} + 4 \ln \frac{a}{a+t(1-t)} \right.$$
$$\left. +12\,(a+t(1-t)) \ln \frac{a}{a+t(1-t)} - 12t(1-t) \right).$$

The last integral is elementary although rather cumbersome to compute. While performing this last calculation, it helps to decompose

$$a+t(1-t) = (a_1 - t)\,(t - a_2)\,, \quad a_{1,2} \equiv \frac{1}{2} \pm \sqrt{\frac{1}{4} + a}.$$

The limit $a \to 0$ should be performed *after* evaluating the integral. The result is

$$I_0 = \frac{1}{32} \lim_{a \to 0} \left[\frac{8}{3} - 16a - 32a^2 \ln a + o\left(a^2\right) \right] = \frac{1}{12}.$$

Exercise 14.2 (p. 187)

In this and the following exercise, the symbol $\delta g^{\mu\nu}$ stands for the variation of the *contravariant* metric $g^{\mu\nu}$. Since $g_{\alpha\nu} g^{\alpha\mu} = \delta^\mu_\nu$, we have

$$0 = \delta\left(g_{\alpha\nu} g^{\alpha\mu}\right) = g_{\alpha\nu} \delta g^{\alpha\mu} + g^{\alpha\mu} \delta g_{\alpha\nu}.$$

Thus the variation $\delta g_{\mu\nu}$ of the covariant tensor $g_{\mu\nu}$ is

$$\delta g_{\mu\nu} = -g_{\alpha\mu} g_{\beta\nu} \delta g^{\alpha\beta}.$$

(a) First we prove that the variation $\delta\Gamma^\alpha_{\mu\nu}$ of the Christoffel symbol is a tensor quantity even though $\Gamma^\alpha_{\mu\nu}$ itself is not a tensor. Indeed, the components

$$\Gamma^\alpha_{\mu\nu} = \frac{1}{2} g^{\alpha\beta} \left(\partial_\mu g_{\beta\nu} + \partial_\nu g_{\beta\mu} - \partial_\beta g_{\mu\nu} \right)$$

change under a coordinate transformation according to the (non-tensorial) law

$$\Gamma'^\alpha_{\mu\nu} = \frac{\partial x'^\alpha}{\partial x^\beta} \frac{\partial x^\rho}{\partial x'^\mu} \frac{\partial x^\sigma}{\partial x'^\nu} \Gamma^\beta_{\rho\sigma} + \frac{\partial x'^\alpha}{\partial x^\beta} \frac{\partial^2 x^\beta}{\partial x'^\mu \partial x'^\nu}. \tag{A4.42}$$

However, it follows from equation (A4.42) that the variation $\delta\Gamma^\alpha_{\mu\nu}$ transforms as

$$\delta\Gamma'^\alpha_{\mu\nu} = \frac{\partial x'^\alpha}{\partial x^\beta} \frac{\partial x^\rho}{\partial x'^\mu} \frac{\partial x^\sigma}{\partial x'^\nu} \delta\Gamma^\beta_{\rho\sigma}$$

and is therefore a tensor.

We can always choose a locally inertial frame such that $\tilde{\Gamma}^\alpha_{\mu\nu}(x) = 0$ at a given spacetime point x. In that frame, the covariant derivative coincides with the ordinary derivative, i.e. $\tilde{\nabla}_\mu = \tilde{\partial}_\mu$, where the tilde means that the quantities are computed in the locally inertial frame at point x. Then the variation of the Christoffel symbol $\tilde{\Gamma}^{\alpha}_{\mu\nu}$ is

$$\begin{aligned}\delta\tilde{\Gamma}^{\alpha}_{\mu\nu} &= \frac{1}{2}\delta\tilde{g}^{\alpha\beta}\left(\tilde{\partial}_{\mu}\tilde{g}_{\beta\nu}+\tilde{\partial}_{\nu}\tilde{g}_{\beta\mu}-\tilde{\partial}_{\beta}\tilde{g}_{\mu\nu}\right)\\ &\quad+\frac{1}{2}g^{\alpha\beta}\left(\tilde{\partial}_{\mu}\delta\tilde{g}_{\beta\nu}+\tilde{\partial}_{\nu}\delta\tilde{g}_{\beta\mu}-\tilde{\partial}_{\beta}\delta\tilde{g}_{\mu\nu}\right)\\ &=\frac{1}{2}g^{\alpha\beta}\left(\tilde{\partial}_{\mu}\delta\tilde{g}_{\beta\nu}+\tilde{\partial}_{\nu}\delta\tilde{g}_{\beta\mu}-\tilde{\partial}_{\beta}\delta\tilde{g}_{\mu\nu}\right)\\ &=\frac{1}{2}\left(\tilde{\nabla}_{\mu}\delta\tilde{g}_{\beta\nu}+\tilde{\nabla}_{\nu}\delta\tilde{g}_{\beta\mu}-\tilde{\nabla}_{\beta}\delta\tilde{g}_{\mu\nu}\right),\end{aligned} \tag{A4.43}$$

because in the locally inertial frame we have

$$\tilde{\partial}_{\mu}\tilde{g}_{\beta\nu}+\tilde{\partial}_{\nu}\tilde{g}_{\beta\mu}-\tilde{\partial}_{\beta}\tilde{g}_{\mu\nu}=\tilde{\Gamma}^{\alpha}_{\mu\nu}\tilde{g}_{\alpha\beta}=0.$$

Since the last expression in equation (A4.43) involves explicitly tensorial quantities, the tensor $\delta\Gamma^{\alpha}_{\mu\nu}$ is equal to

$$\begin{aligned}\delta\Gamma^{\alpha}_{\mu\nu} &= \frac{g^{\alpha\beta}}{2}\left(\nabla_{\mu}\delta g_{\beta\nu}+\nabla_{\nu}\delta g_{\beta\mu}-\nabla_{\beta}\delta g_{\mu\nu}\right)\\ &\equiv\frac{g^{\alpha\beta}}{2}\left(\delta g_{\beta\mu;\nu}+\delta g_{\beta\nu;\mu}-\delta g_{\mu\nu;\beta}\right)\end{aligned} \tag{A4.44}$$

in *every* coordinate system. (Here is a more rigorous argument: We first consider the tensor (A4.44) which happens to coincide with equation (A4.43) in a particular coordinate system and only at the point x. However, two tensors cannot coincide in one coordinate frame but differ in another frame. Therefore the tensor $\delta\Gamma^{\alpha}_{\mu\nu}(x)$ is given by equation (A4.44) in all coordinate systems. Since the construction is independent of the chosen point x, it follows that the formula (A4.44) is valid for all x.)

Note that an explicit formula for the covariant derivative $\nabla_{\beta}\delta g_{\mu\nu}$ is

$$\nabla_{\beta}\delta g_{\mu\nu}=\partial_{\beta}\delta g_{\mu\nu}-\Gamma^{\alpha}_{\beta\nu}\delta g_{\alpha\mu}-\Gamma^{\alpha}_{\beta\mu}\delta g_{\alpha\nu}.$$

The trick of choosing a locally inertial frame helps us avoid cumbersome computations with such expressions.

(b) Since

$$\Box_g\Box_g^{-1}=\hat{1},$$

we have

$$0=\delta\left(\Box_g\Box_g^{-1}\right)=\Box_g\left(\delta\Box_g^{-1}\right)+\left(\delta\Box_g\right)\Box_g^{-1}$$

and hence

$$\delta\left(\Box_g^{-1}\right)=-\Box_g^{-1}\left(\delta\Box_g\right)\Box_g^{-1}. \tag{A4.45}$$

The covariant Laplace operator $\Box_g$ acting on a scalar function $\phi(x)$ is

$$\Box_g \phi \equiv g^{\mu\nu} \nabla_\mu \nabla_\nu \phi = g^{\mu\nu} \phi_{;\mu\nu} = g^{\mu\nu} \left(\phi_{,\mu\nu} - \Gamma^\alpha_{\mu\nu} \phi_{,\alpha} \right).$$

The variation of this expression with respect to $\delta g^{\mu\nu}$ is

$$\begin{aligned} \delta \Box_g \phi &= \left(\delta g^{\mu\nu}\right) \phi_{;\mu\nu} + g^{\mu\nu} \delta \left[\phi_{,\mu\nu} - \Gamma^\alpha_{\mu\nu} \phi_{,\alpha} \right] \\ &= \left(\delta g^{\mu\nu}\right) \phi_{;\mu\nu} - g^{\mu\nu} \left(\delta \Gamma^\alpha_{\mu\nu} \right) \phi_{,\alpha}. \end{aligned}$$

This can be rewritten as an operator identity

$$\delta \Box_g = \left(\delta g^{\mu\nu}\right) \nabla_\mu \nabla_\nu - g^{\mu\nu} \left(\delta \Gamma^\alpha_{\mu\nu} \right) \nabla_\alpha. \tag{A4.46}$$

We emphasize that this identity holds only when the operators act on a *scalar* function $\phi(x)$. For vector- or tensor-valued functions, the formula would have to be modified.

(c) To derive

$$\delta R^\alpha{}_{\beta\mu\nu} = \nabla_\mu \delta \Gamma^\alpha_{\beta\nu} - \nabla_\nu \delta \Gamma^\alpha_{\mu\beta}, \tag{A4.47}$$

we again pass to a locally inertial frame in which $\tilde{\Gamma}^\alpha_{\mu\nu} = 0$ and $\tilde{\nabla}_\mu = \tilde{\partial}_\mu$. Then the Riemann tensor (in the Landau–Lifshitz sign convention) is

$$\tilde{R}^\alpha{}_{\beta\mu\nu} = \tilde{\partial}_\mu \tilde{\Gamma}^\alpha_{\beta\nu} - \tilde{\partial}_\nu \tilde{\Gamma}^\alpha_{\mu\beta}. \tag{A4.48}$$

Note that the RHS of this expression is not a tensor. Varying both sides of the relation (A4.48), we obtain

$$\delta \tilde{R}^\alpha{}_{\beta\mu\nu} = \tilde{\partial}_\mu \delta \tilde{\Gamma}^\alpha_{\beta\nu} - \tilde{\partial}_\nu \delta \tilde{\Gamma}^\alpha_{\mu\beta} = \tilde{\nabla}_\mu \delta \tilde{\Gamma}^\alpha_{\beta\nu} - \tilde{\nabla}_\nu \delta \tilde{\Gamma}^\alpha_{\mu\beta}. \tag{A4.49}$$

Note that both sides of equation (A4.49) are written in an explicitly covariant form and are tensors. Therefore equation (A4.49) holds in every coordinate system and equation (A4.47) follows.

Exercise 14.3 (p. 187)

Here we derive the expectation value of the energy-momentum tensor

$$\langle T_{\mu\nu} \rangle = \frac{2}{\sqrt{-g}} \frac{\delta \Gamma_L}{\delta g^{\mu\nu}}$$

from the (Lorentzian) Polyakov effective action

$$\begin{aligned} \Gamma_L[g_{\mu\nu}] &= \frac{1}{96\pi} \int d^2x \sqrt{-g} R \Box_g^{-1} R \\ &\equiv \frac{1}{96\pi} \int d^2x \sqrt{-g(x)} \int d^2y \sqrt{-g(y)} R(x) G_g(x,y) R(y), \end{aligned} \tag{A4.50}$$

where $G_g(x, y)$ is the (retarded) Green's function of the Laplace operator $\Box_g$. Recall that we have

$$\left(\Box_g^{-1}\Phi\right)(x) \equiv \int d^2y\sqrt{-g(y)}G_g(x, y)\Phi(y),$$

where Φ is a scalar field, and the Green's function satisfies

$${}^x\Box_g G_g(x, y) = \frac{1}{\sqrt{-g(x)}}\delta(x-y).$$

The variation of $\sqrt{-g}R\,\Box_g^{-1}R$ can be written as

$$\frac{1}{\sqrt{-g}}\delta\left(\sqrt{-g}R\,\Box_g^{-1}R\right) = \frac{\delta\sqrt{-g}}{\sqrt{-g}}R\Box_g^{-1}R \\ +(\delta R)\,\Box_g^{-1}R + R\Box_g^{-1}(\delta R) + R\left(\delta\Box_g^{-1}\right)R,$$

where we introduced the prefactor $1/\sqrt{-g}$ for convenience. This expression is integrated over d^2x, while the operator $\Box_g^{-1}$ is self-adjoint, hence

$$\int d^2x\sqrt{-g}\,(\delta R)\,\Box_g^{-1}R = \int d^2x\sqrt{-g}R\,\Box_g^{-1}(\delta R)\,.$$

Thus, for our purposes it is sufficient to compute the auxiliary quantity

$$\delta I \equiv \frac{\delta\sqrt{-g}}{\sqrt{-g}}R\,\Box_g^{-1}R + 2\,(\delta R)\,\Box_g^{-1}R + R\left(\delta\Box_g^{-1}\right)R. \tag{A4.51}$$

The EMT will be expressed through δI using the formula

$$\int d^2x\sqrt{-g}\langle T_{\mu\nu}\rangle\,\delta g^{\mu\nu} = \frac{1}{48\pi\sqrt{-g}}\int d^2x\sqrt{-g}\,\delta I\,(x)\,. \tag{A4.52}$$

Now we shall evaluate the expression (A4.51) term by term. The variation of $\sqrt{-g}$ is (see equation (A4.14) on p. 229)

$$\frac{\delta\sqrt{-g}}{\sqrt{-g}} = -\frac{1}{2}g_{\mu\nu}\delta g^{\mu\nu}. \tag{A4.53}$$

In two dimensions, the Ricci tensor $R_{\mu\nu}$ is related to the Ricci scalar as

$$R_{\mu\nu} \equiv R^{\alpha}{}_{\mu\alpha\nu} = \frac{1}{2}g_{\mu\nu}R,$$

therefore with help of equation (A4.47) we get

$$\delta R = R_{\mu\nu}\delta g^{\mu\nu} + g^{\mu\nu}\delta R_{\mu\nu} = \frac{1}{2} R g_{\mu\nu}\delta g^{\mu\nu} + \nabla_\alpha \left(g^{\mu\nu}\delta\Gamma^\alpha_{\mu\nu} - g^{\mu\alpha}\delta\Gamma^\nu_{\mu\nu} \right). \quad \text{(A4.54)}$$

Using the formula (A4.44) and the relation $g^{\alpha\beta}\delta g_{\beta\gamma} = -g_{\beta\gamma}\delta g^{\alpha\beta}$, we derive the necessary expressions

$$g^{\mu\nu}\delta\Gamma^\alpha_{\mu\nu} = -\nabla_\mu \delta g^{\alpha\mu} + \frac{1}{2} g_{\mu\nu}\nabla^\alpha \delta g^{\mu\nu},$$

$$g^{\mu\alpha}\delta\Gamma^\nu_{\mu\nu} = -\frac{1}{2} g_{\mu\nu}\nabla^\alpha \delta g^{\mu\nu}.$$

Thus the variation δR is reduced to

$$\begin{aligned}
\delta R &= \frac{1}{2} R g_{\mu\nu}\delta g^{\mu\nu} + \nabla_\alpha \left(-\nabla_\mu \delta g^{\alpha\mu} + \frac{1}{2} g_{\mu\nu}\nabla^\alpha \delta g^{\mu\nu} + \frac{1}{2} g_{\mu\nu}\nabla^\alpha \delta g^{\mu\nu} \right) \\
&= \left(\frac{1}{2} R g_{\mu\nu} - \nabla_\mu \nabla_\nu + g_{\mu\nu}\Box_g \right) \delta g^{\mu\nu},
\end{aligned}$$

while the variation of the Laplace operator becomes

$$\delta \Box_g = \left(\delta g^{\mu\nu} \right) \nabla_\mu \nabla_\nu + \left(\delta g^{\alpha\mu}{}_{;\mu} - \frac{1}{2} g_{\mu\nu}\delta g^{\mu\nu;\alpha} \right) \nabla_\alpha.$$

Now we can put all terms in δI together,

$$\begin{aligned}
\delta I =& -\frac{1}{2} g_{\mu\nu}\delta g^{\mu\nu} R \Box_g^{-1} R \\
& + 2 \left(\frac{1}{2} R g_{\mu\nu}\delta g^{\mu\nu} - \delta g^{\mu\nu}{}_{;\mu\nu} + g_{\mu\nu}\Box_g \delta g^{\mu\nu} \right) \Box_g^{-1} R \\
& - R \Box_g^{-1} \left[\delta g^{\mu\nu}\nabla_\mu \nabla_\nu \Box_g^{-1} R + \left(\delta g^{\alpha\mu}{}_{;\mu} - \frac{1}{2} g_{\mu\nu}\delta g^{\mu\nu;\alpha} \right) \nabla_\alpha \Box_g^{-1} R \right].
\end{aligned}$$

It is now straightforward to compute the functional derivative (A4.52). For arbitrary scalar functions $A(x)$ and $B(x)$, we have the identity

$$\int d^2x \sqrt{-g} A \Box_g^{-1} B = \int d^2x \sqrt{-g} B \Box_g^{-1} A,$$

and that integration by parts yields for arbitrary tensors X and Y the formula

$$\int d^2x \sqrt{-g} X \nabla_\alpha Y = -\int d^2x \sqrt{-g} Y \nabla_\alpha X.$$

Thus we compute (up to a total divergence)

$$\frac{\delta I}{\delta g^{\mu\nu}} = \frac{1}{2} g_{\mu\nu} R \Box_g^{-1} R - 2\left(\Box_g^{-1} R\right)_{;\mu\nu} + 2 g_{\mu\nu} R - \left(\Box_g^{-1} R\right)\left(\Box_g^{-1} R\right)_{;\mu\nu}$$

$$+ \left[\left(\Box_g^{-1} R\right)_{;\nu} \Box_g^{-1} R\right]_{;\mu} - \frac{1}{2} g_{\mu\nu} \left[\left(\Box_g^{-1} R\right)_{;\alpha} \Box_g^{-1} R\right]^{;\alpha}$$

$$= -2\left(\Box_g^{-1} R\right)_{;\mu\nu} + 2 g_{\mu\nu} R + \left(\Box_g^{-1} R\right)_{;\mu}\left(\Box_g^{-1} R\right)_{;\nu}$$

$$- \frac{1}{2} g_{\mu\nu} \left(\Box_g^{-1} R\right)_{;\alpha} \left(\Box_g^{-1} R\right)^{;\alpha}.$$

Thus the final result is

$$\langle T_{\mu\nu} \rangle = \frac{1}{48\pi} \Big\{ -2 \nabla_\mu \nabla_\nu \Box_g^{-1} R + \left[\nabla_\nu \Box_g^{-1} R\right]\left[\nabla_\mu \Box_g^{-1} R\right]$$

$$+ 2 g_{\mu\nu} R - \frac{1}{2} g_{\mu\nu} \left(\nabla_\alpha \Box_g^{-1} R\right)\left(\nabla^\alpha \Box_g^{-1} R\right) \Big\},$$

which coincides with equation (14.10).

Index